Praise for

STARBORN

'A stunning and unforgettable voyage through the stars. Almost every page will make you gulp in astonishment. To be so authoritative and yet so readable and companionable is a rare and priceless achievement'

Stephen Fry, actor

'We are in danger of losing our night skies, our first and most primal connection to the greater universe around us. Roberto Trotta's rich and poetic book is a powerful call to preservation. By tracing the intimate connections between human history and the stars above, he reminds us the skies are not only filled with beauty, but also with meaning and promise'

Sean Carroll, author of *The Biggest Ideas in the Universe*

'A sweeping tour of humanity's relationship with the night sky, *Starborn* soars from the historical to the personal. Trotta reveals how our lives are intertwined with the stars, from the exploration of our own planet and the birth of the sciences to how the human gaze turned inward. He also offers fantastical vignettes of what might have been—imagining a world without the heavens—and a clear-eyed view of humankind's current and future connection with Earth and the cosmos'

Professor Emily Levesque, author of *The Last Stargazers*

'Who would've thought the stars were so decisive for humanity? Fascinating and wondrous, the untold starry tale of how we came to be and a stark warning of the starless desolation ahead, should we be unwise enough to neglect our cosmic heritage'

Professor Carissa Véliz, author of *Privacy Is Power*

'A fascinating insight into how and why the study of the stars has been central to the human story, and a book for anyone who cares about human culture and where it will be heading next'

Andy Lawrence, author of *Losing the Sky*

'The most universal feature of our environment, the starry sky has been wondered at by all human societies since prehistoric times. Trotta draws on a trove of historical, scientific, and literary sources to reveal the often-surprising influences of a cosmic perspective on human lives. A fascinating book, admirable for erudition and style, that will leave readers viewing the stars with fresh eyes'

Professor Martin Rees, Astronomer Royal

'Beautifully written and intensely personal, Trotta has produced the perfect guide to how the star-speckled sky has shaped the human story. A lyrical hymn to the cosmos and our intimate connection to the heavens'

Professor Lewis Dartnell, author of *Being Human*

'Stimulating and sobering, a very readable inquiry into the night sky with a broad sweep'

Dame Jocelyn Bell Burnell, University of Oxford

STARBORN

ALSO BY ROBERTO TROTTA

The Edge of the Sky: All You Need to Know About the All-There-Is

STARBORN

HOW THE STARS MADE US –
AND WHO WE WOULD BE WITHOUT THEM

ROBERTO TROTTA

BASIC
BOOKS

LONDON

First published in Great Britain in 2023 by Basic Books UK
An imprint of John Murray Press

2

Copyright © Roberto Trotta 2023

Text design by Linda Mark

A CIP catalogue record for this title is available from the British Library

Hardback ISBN 9781529346084
Trade Paperback ISBN 9781529346091
ebook ISBN 9781529346114

Typeset in Janson Text LT Std

Printed and bound in Great Britain by Clays Ltd, Elcograf S.p.A.

John Murray Press policy is to use papers that are natural, renewable and
recyclable products and made from wood grown in sustainable forests.
The logging and manufacturing processes are expected to conform to the
environmental regulations of the country of origin.

Basic Books UK
Carmelite House
50 Victoria Embankment
London EC4Y 0DZ

www.basicbooks.uk

John Murray Press, part of Hodder & Stoughton Limited
An Hachette UK company

To Elisa, Benjamin, and Emma

As above, so below.

Teach me your mood, O patient stars!
Who climb each night the ancient sky,
Leaving on space no shade, no scars,
No trace of age, no fear to die.

—Ralph Waldo Emerson,
"Fragments on Nature and Life, Nature"

CONTENTS

PROLOGUE   1

    The Night That Changed My Life *1*

    Shaped by the Stars *3*

    The Midwife of Science *6*

CHAPTER 1: A PALE BLUE DOT   11

    A Postcard from Outer Space *11*

    Down the Pit of History *12*

    The Secret Ingredient *15*

    The Caligo Tales: Remembering *20*

CHAPTER 2: THE LOST SKY   23

    Looking for Totality *23*

    The Harbinger of Doom *26*

    The Principle of Life *29*

    Broom Stars and Newton's Kites *31*

    A Jolt of Awe *34*

    The Forgotten Sky *35*

    A Hank of Light *39*

    The Caligo Tales: Cloud-Watcher's Tale *42*

Contents

CHAPTER 3: LIFE UNDER A CLOUD   45

   Among the Standing Stones *45*

   The Sun-Starved City *49*

   Love and Sulfuric Acid *53*

   When a Cloud Won't Do *55*

   Where Clouds Never Part *57*

   The Caligo Tales: Freshwater's Tale *62*

CHAPTER 4: THE WEIGHT OF STARLIGHT   65

   A Palaeolithic Face-Off *65*

   Meat from the Moon *70*

   Lunar Tallies *73*

   Songlines and Peepholes *77*

   The Lost Sister *81*

   The Sky-Savvy Sapiens *86*

   The Caligo Tales: Once-Upon-A-Glow's Tale *91*

CHAPTER 5: CELESTIAL CLOCKS   93

   On a Clock Face *93*

   Twelve Diamonds in the Sky *95*

   The Heavenly Clock *98*

   Fallen Angels *102*

   The Heavens in a Shoebox *104*

   Ping! A Star Goes By *108*

   The Caligo Tales: Shepherd's Tale *112*

CHAPTER 6: TRIPLE BRONZE AND OAK   115

   Navigating by the Stars *115*

   Fifteen Hundred Years of Fame *119*

   The Polynesian Masters *121*

   A Moonshot Prize *123*

Contents

For the Pride of the Empire   *127*

The Lord of the Lunars   *129*

A Clash of Cultures   *132*

Our Trusty Friend, the Watch   *137*

Time for a Change of Time   *140*

The Caligo Tales: Way-Finder's Tale   *144*

CHAPTER 7: FROM BEAUTY, ORDER   147

A Golden Flower in Space   *147*

From Humble Beginnings   *149*

A Four-Pronged Assault   *151*

The Castle of Urania   *155*

The Measure of All Things   *157*

The Shapes of Cynthia and Other Wonders   *159*

Isaac's Dials   *164*

A Fiery Return   *169*

The Trackless Abysses of Space   *171*

The Clockwork Universe   *176*

The Caligo Tales: Bison-Seeker's Tale   *180*

CHAPTER 8: THE DEMON UNLEASHED   183

The Mathematics of Uncertainty   *183*

The Birth of Average Man   *187*

Laplace's Demon Goes Places   *189*

Feeding the Demon   *191*

Unexpected Connections   *197*

The Human Machine   *199*

The Immensity of Time   *202*

The Demon Reborn   *205*

The Caligo Tales: Fire-Keeper's Tale   *209*

Contents

CHAPTER 9: A MIRROR TO OURSELVES   211

The Sun Worshippers   *211*

Sol Invictus   *213*

Swear Not by the Moon   *217*

The Great Memory   *220*

Led by One's Stars   *222*

The Estranged Mother   *226*

The Last Ripple   *228*

To Every Man His Star   *229*

The Caligo Tales: Mist-Catcher's Tale   *233*

CHAPTER 10: TO REBEHOLD THE STARS   235

The Broom Star Encounter   *235*

A Black Canvas   *237*

The Last Global Commons   *240*

Beam Me Up, Scotty!   *244*

A Loss of Happiness   *247*

The Price We'll Pay   *249*

Becoming Good Ancestors   *252*

The Caligo Tales: The Skeleton's Dance   *257*

EPILOGUE: SO SPOKE THE SILENT STARS   261

Acknowledgments   265

Notes   267

Bibliography   291

Index   321

PROLOGUE

Do you remember still the falling stars
that like swift horses through the heavens raced
and suddenly leaped across the hurdles
of our wishes—do you recall?

—Rainer Maria Rilke, "Falling Star"

THE NIGHT THAT CHANGED MY LIFE

The night that changed my life began in underwhelming circumstances.

The play had turned out to be a *commedia degli errori*, a slapstick piece by Eduardo de Filippo in which three men vie to cover up their affairs, get married, and avoid taking responsibility for getting each other's fiancées pregnant. Not exactly conducive to romance, but I had two cheap student tickets, and the local theater, with its tiered seats covered in plush red velvet, felt like an upscale destination for an evening with my date.

This wasn't our first time out together. We had taken long walks on the lakeshore, listened to the poplars murmuring, and watched their leaves cover the waters as the ducks took flight. June afternoons at the swimming pool seemed to stop the Sun at its highest point, its rays reflected by the water into her smiling green eyes. Strawberry ice creams, our bare feet dangling from a wooden jetty, had given way to hot teas and *vermicelle*

1

cakes. We had been to the local English film club's screenings and worked through the subtitles of *The English Patient*, *Howard's End*, and *Thelma and Louise*. We had walked and talked, then talked some more into the autumn nights, delighting in each other's company, our mysterious resonance like a planetary nebula awaiting the spark that will ignite a star. We were falling in love.

That November evening, wrapped in a cocoon of sideways glances, we paid scant attention to the stage, the play unfolding as if in a distant galaxy. Our hands touched over the velvet armrest, her elegant pianist's fingers fluttering like small animals exploring new territory. We clapped dutifully when the actors bowed, a sardonic smile on her lips saying that she, like me, had no recollection of the past two hours beyond the electrical current through our fingertips. Her simple black dress contrasted with the blonde hair pinned back to highlight her delicate cheekbones.

We left with the crowd, spilling out into the deserted main square. We zigzagged through the parked cars dotting the Piazza Grande, the sound of our steps ricocheting off nineteenth-century archways, heading for the pedestrian bridge over the river.

I don't know what compelled us to stop shortly before the bridge. Perhaps we felt that the evening was slipping away and wanted, despite the bitter cold, to hold on to our cocoon. We stood in the darkness, the murmur of the river washing away our thoughts. The night sky was moonless, the shapes of mountains like paper cutouts against a dense field of stars. We were alone, the only characters on a stage billions of years in the making.

We turned to face each other, holding hands through our thick gloves. Above her beret, the constellation Orion hovered between the inky mountains. I gently turned her by the shoulders and pointed out the outline of the giant: on his right shoulder, the orange sparkle of Betelgeuse; on his left foot, the blue glare of Rigel; and for his belt, the perfect slash of Mintaka, Alnilam, and Alnitak, from which hung a sword made of stars. Orion's nebula adorned the sword, and from thirteen hundred light-years away, it appeared the same size and consistency as the white breath condensing before our lips.

Five thousand years ago, the Sumerians looked at these same stars and saw in them a portrait of Gilgamesh, the mythical hero who sought immortality and slayed the Great Bull of Heaven, the nearby constellation Taurus. Four thousand years ago, the Chinese called Orion's belt Shen (the three stars). Three thousand years ago he represented Osiris for the Egyptians, who may have designed their pyramids to guide the pharaoh toward him after death. And two thousand years ago, the Greeks saw the nephew of Zeus: Orion, a peerless hunter who, according to different legends, either offended the goddess Artemis or forced his unrequited love onto the daughters of Atlas and Pleione, the Pleiades, and as a punishment died from the sting of a scorpion. Since then, Orion and Scorpio have inhabited opposite hemispheres in the sky, the giant setting as his nemesis rises.

I told my date that I saw a man holding a shield in his left, outstretched hand—actually a lion's pelt in the Greek myth. In place of the unbreakable bronze club described by Homer—with which Orion pursues "wild beasts over the field of asphodel" in Hades—she imagined the giant in the gesture of drawing an arrow with his right hand.[1] Our heads almost touched as we spoke.

All of a sudden a streak of light opened a gash in the darkness, draping the shoulders of Orion in a blazing ribbon—as if a giant brush, dipped in white-blue light, had embellished the constellation. It lasted only a second or two, but it felt as if the meteor were traveling in slow motion, so much longer was the appearance than anything else I have ever witnessed. There was no need to draw attention to it, since we were both looking in the same direction at the same time. We wouldn't have been able to raise the alert anyway, each too surprised to speak until it was over. We just marveled as the falling star burned up and vanished, leaving no trace of its passage but wonder.

When we turned toward each other, the afterimage of the meteor was still shimmering, it seemed to me, at the bottom of her eyes. We had expressed the same silent wish at the moment it was granted.

SHAPED BY THE STARS

Decades later, I had the opportunity to look back on the meandering path on which the stars have led me.

It is customary for professors at Imperial College London, the university where I have spent most of my academic career, to give a public lecture on their subject within a year or so of their appointment. These inaugural lectures are often the occasion for the newly minted professor to reflect on the academic and personal trajectory that got them there—for me, the fulfillment of a goal twenty years in the making.

In my January 2020 inaugural lecture, I recounted how, early on in my undergraduate physics studies, I came to realize that I wasn't cut out for observational astronomy. I had been looking forward to measuring the properties of a stellar binary system, but bad weather had scuppered our weeklong observing trip in the Swiss mountains. When the snowstorms let up a little, we looked for sunspots through the clouds instead, as nighttime observing was out of the question. "Don't ever look at the Sun directly through the eyepiece of the telescope!" our professor warned sternly. "This is a mistake you will only make twice in your life—once for each eye!"

After shoveling a path through waist-deep snow and setting up the scope, I found myself unable to project the Sun onto a piece of cardboard for safe observation. I couldn't correctly aim the scope at the largest astronomical object in the sky! Embarrassed and frustrated, without thinking, I did what I was told to avoid at all costs: I bent down and looked through the eyepiece to try to understand what was wrong with my pointing. A fraction of a second too late I realized my mistake. I knew that blinding could be almost instantaneous and painless. I straightened up in panic and blinked. My eyesight had been saved by my ineptitude: the lens cap had been on the whole time.

Leaving behind observational astronomy, I pursued a PhD in theoretical cosmology. If physics was the science of the fundamental workings of nature, nothing was grander than the whole universe. It wasn't the stars, or the galaxies, or the galaxy clusters that hooked me. It was a simpler, earlier, smoother universe: the cosmos as it was before stars were born, before galaxies were formed, before the building blocks of life came into existence. It was the universe as it emerged from the Big Bang itself, the primordial explosion that marks the beginning of time and of the expanding cosmos as we know it.

I was never a pure-bred theorist. I lacked the mathematical virtuosity of fellow graduate students who were at home in eleven dimensions of space-time. But I came of age, scientifically, when cosmology was becoming a precision science. New, detailed measurements of the cosmic microwave background (the afterglow of the Big Bang), the distribution of galaxies, and the explosion of stars were coming in thick and fast, carrying information on the age of the universe, the recent acceleration in its expansion, the properties of dark matter, and many more aspects of the fundamental nature of reality. Bigger, faster telescopes equipped with then new digital technology were seeing ever farther. Almost overnight, cosmology went from data starved to data driven, and the urgent question was how to extract scientific knowledge from this Niagara of numbers. My career became one of looking at the cosmos through a shower of digits on a computer screen rather than one of freezing nights in a dark dome.

As I prepared my lecture, I had hardly looked through a telescope since the dismal failure of my solar observations. With the benefit of hindsight, however, I began to realize just how much the stars had shaped my life. I was part of a community of scientists intent on revealing the fundamental nature of the cosmos. We were using instruments whose lineage stems from a wooden tube with hand-ground lenses made by Galileo Galilei in 1610. Our thinking had been molded by Isaac Newton, who pried his law of universal gravitation from the motion of comets, then reshaped by Albert Einstein, who, aged sixteen, had his first inklings of relativity by imagining riding a beam of starlight through space.[2] Astronomical problems had even directly inspired many of our mathematical tools.

My field of academic endeavor was perched on an edifice built out of thousands of years of human curiosity. "Since men walked the Earth, the sky had admirers,"[3] wrote Italian poet Giacomo Leopardi. In time, those same admirers would become the natural philosophers of the Scientific Revolution first and then the space engineers driving NASA's rovers on Mars today.

The stars had also done more for me than trace my path to academia. As my lecture ended, my children ran from the audience and hugged me. Misty-eyed, I took the bunch of white lilacs they offered me, and squeezing them tight, I lifted my gaze onto the public.

A cloud of white hair, carefully combed back, adorned an old man sitting in the front row. My dad clapped slowly, deliberately, his hands still powerful after a lifetime of manual work. Next to him, his daughter-in-law whispered something in his ear that made him smile. When she turned to face me again, her green eyes flashed with the memory of the meteor trail that had brought us together so long ago. I nodded to my wife, the mother of my children, and knew that nothing would have been the same in my life in a world without stars.

THE MIDWIFE OF SCIENCE

What do we owe the stars? They had silently guided my own life. Just how much, I wondered, had they steered the course of humanity?

Science was my natural starting point. The questions driving my own research—What is the universe made of? How did it emerge from the Big Bang? What will its ultimate destiny be?—are the tip of a branch that can be traced back all the way down the tree of modern science to its mighty base, when Newton affirmed that all that circled above obeys the same intelligible laws as what lies below. "The stars could not be cajoled or perverted," writes historian of science Lewis Mumford. "Their courses were visible to the naked eye and could be followed by any patient observer."[4] Astronomy was, so to speak, the midwife to all the sciences, with physics, geology, chemistry, and biology successively illuminating a universe subject to comprehensible laws.

With the notion of nature as law-abiding came the ambition to bend it to our will: no longer supplicants, cowering from capricious sky gods, we became masters, reshaping the Earth, defeating diseases, and riding rockets through the nonexistent crystalline spheres imagined by Aristotle. But the tree of science has borne fruit both sweet and poisonous: our voracity is today threatening the stability of nature on the planet, and, having brought the atomic fire that powers the stars down to Earth, we are prepared to unleash it with the flick of a finger. We are mightier than Zeus, fiercer than Ares, and fickle as any of the old gods. The heavens have bestowed on us the means to fulfill our wildest dreams, but also to bring forth our own destruction.

Clearly, the stars have a great deal to answer for.

The influence of the stars on human history did not begin with the Scientific Revolution. Its roots extend far deeper than that era's springboard into technological modernity, into the dark, fertile loam of myth and religion, and eventually into prehistory. The question of the origin of the world, and of our place within it, has preoccupied humankind since well before we had spaceborne microwave detectors capable of revealing the faint whisper of the Big Bang.

The procession of the stars, the reappearance of the morning Sun, the cyclical phases of the Moon, the bright light of the "wanderers"—the ancient Greek meaning of the word *planets*—became in every culture self-evident expressions of higher powers that ruled the cosmos from above the clouds, unreachable, eternal, and all-mighty. "The regions above man's reach, the starry places, are invested with the divine majesty of the transcendent, of absolute reality, of everlastingness," writes historian of religion Mircea Eliade.[5] Tellingly, the ancient Greek word *kosmos* means both "order" and "ornament"—hence our words *cosmology* (the study of the orderly beauty of the universe) and *cosmetology* (the orderly ornamenting and beautifying of face, hair, and skin)

Ancient gods could hide in the thunder, haunt the forest, surge with the foam of the sea, ride with the wind, or take the shape of a bull. But the highest gods ruled over the sky, reigning over the cosmos from their celestial seat. Zeus, the king of the Olympians, took his name from an older, Sanskrit root word meaning "to shine," "day," and "sky"; the Romans called him Jupiter, from the Sanskrit *Dyaus pitar*, "Father sky." "Heavens" in both Hebrew (*shamyim*) and Greek (*ouranos*) also means "sky"—the place where, according to Christian theology, God resides: "Our Father who art in heaven." From one end of the world to the other, across millennia, impressed in clay tablets as cuneiform script, gilded inside pharaohs' tombs, or painted on leather in Aztec manuscripts, *sky* and *god* became almost synonyms. The Finns called their supreme sky being Jumala; the Sumerians, An; the Iroquois, Oke. The most powerful of the gods was called Wakan by the Sioux, Tengri by the Mongols, and Mawy by the Ewe peoples of Togo. The names differ, but the sentiment is the same, as expressed by a member of the Ewe tribe: "I have always looked up to the visible sky as to God. When I spoke of God, I spoke of the sky, and when I spoke of the sky, I thought of God."[6]

From on high, the sky gods could see everything and therefore know everything—like the Great Spirit Father of the aborigines of New South Wales or the Egyptian god falcon Horus, who surveyed the world with his two eyes, the Sun and the Moon. Thousands of years later, we still raise our eyes to the sky in despair and lift our arms in a sign of victory. In the final scene of *The Truman Show*, the all-powerful director who for decades controlled the life of the protagonist, a man unknowingly trapped since birth on the gigantic set of a television show, booms from the fake sky like a contemporary demigod. To this day, Navajo shamans use rituals and altered states of consciousness to ascend to the sky, where they are put in direct contact with the divine. Above the nine spheres of Heaven, Dante Alighieri informs us in *The Divine Comedy*, is the Empyrean, where the poet at the end of his journey comes face-to-face with God, "the Love which moves the sun and the other stars."[7]

Is it then a wonder that the ancients scrutinized the stars when nothing less than the will of the gods was at stake? That they put so much effort into predicting their movements, on which the destinies of kings rested? That they were terrified when the order of the universe was subverted as day turned into night during a total solar eclipse—that word derived from the Greek for "abandonment"?

The practice of astrology stemmed from the belief that what happened in the sky had a direct influence on the lives of human beings on Earth, as we shall see in Chapter 9. Astrology pushed Sumerian priests, Egyptian hour-watchers, and Chinese mandarins to keep detailed records of notable heavenly events. And the interest in astrology of pharaohs, kings, and emperors created a caste of men dedicated to studying the stars—the precursors of today's scientists. Thanks to astrology, Tycho Brahe built and ran the first professional research observatory in history, whose data fed Johannes Kepler's calculations of the motion of Mars.

What had started as a personal reflection on my path in life became an exploration of questions I had never considered, revealing connections I never suspected. As I followed bifurcating roots deeper into the past,

I found only more connections to modern cosmology. Stars and planets spurred the invention of mathematics; the Moon, that of the calendar. And could it be that paying attention to the heavens was the secret weapon that gave *Homo sapiens* supremacy over the Neanderthals fifty thousand years ago?

Colleagues looking for life on planets around other stars described the wildly different conditions that might reign there: planets with multiple suns, planets in permanent twilight, planets wrapped in never-parting clouds. Planets, in other words, where stars would not be visible to any lifeforms that might evolve there. As I followed seminar after lecture on these fantastical alt-Earths, I also wondered, What if our destiny had been to never see the stars?

This idea of a world without stars began to haunt me. Insofar as my gaze could travel, the heavens appeared to have touched much more than just physics or just science. How diminished would we be, I wondered, without the poetry, music, and artwork the heavens have inspired? What would spirituality be like without our sky gods? How different would our legends, our great novels, our conceptions of the universe be—how different our very selves—in a world without stars?

Almost thirty centuries ago, the blind bard Homer imagined what the lack of stars would inflict upon humanity. His hero Odysseus, upon leaving the island of the sorceress Circe after a year of lust and luxury, sails northwest beyond the confines of the known world, where he reaches the "deep-flowing Oceanus, that bounds the Earth." On these far-flung shores, Circe has told him, the diaphragm separating us from the underworld runs thin: with libations and sacrifices, Odysseus can lure the spirits of the dead out of Hades and enlist their help to return to his home island of Ithaca.

Homer describes this liminal place poised between worlds as "the land and city of the Cimmerians," a mythical people probably inspired by the inhabitants of the shores of the Black Sea, in present-day Crimea, a region infamous among the Greeks for its foggy, cold, and inhospitable weather.[8] For this reason Homer, used to the blue sea and clear skies of the Mediterranean, calls the Cimmerians "wretched mortals," for their land is forever "wrapped in mist and cloud. Never does the bright sun look down on them with his rays either when he mounts the starry heaven or when he turns again to earth from heaven, but baneful night is spread over [them]."[9]

"Wrapped in mist and clouds," without ever seeing the Sun or the stars, the Cimmerians are only semihuman, a step removed from becoming dead themselves, like the spirits who visit Odysseus, attracted by his offerings.

I began to suspect that the same would be true of humanity as a whole, had we been fated never to see the stars.

1

A PALE BLUE DOT

Think how diminished humanity would be if, under
heavens constantly overclouded, as Jupiter's must be,
it had forever remained ignorant of the stars. Do you
think that in such a world we should be what we are?

—HENRI POINCARÉ, *The Value of Science*

A POSTCARD FROM OUTER SPACE

On Valentine's Day 1990, the Voyager 1 spacecraft received an unusual command from NASA's Deep Space Network. Twenty-three years earlier the space probe had been launched on a trajectory past Jupiter and Saturn, and in three years it had revolutionized our understanding of the gas giants and their moons with detailed images and other data. Its mission accomplished, Voyager 1 was fated to continue its solitary journey outward, sniffing around interplanetary space for another five years or so.

More than two decades later, Voyager 1 had surpassed everybody's expectations. From the far reaches of the solar system, it was still calling home.

The command was the brainchild of astronomer and author Carl Sagan, and it took six requests and eight years before NASA was convinced to use Voyager 1's delicate cameras for something other than science. By the time it received NASA's peculiar instructions, Voyager 1 had passed the orbit of Neptune and was receding into deep space at the rate of five hundred million kilometers per year. The probe was to silently turn around, power its cameras back up ten years after they had been last used, and point them back toward home. Panning and pointing, our distant sentinel searched the darkness to tease out six barely visible dots, creating a unique portrait of six of the eight planets that make up our cosmic family around the Sun. (Mercury is missing as it was too close to the Sun to be visible, while Mars turned out to be too dim.)

It is one of the most iconic images in space exploration. Seen from the cosmic vantage point of Voyager 1, our beautiful planet is nothing but a speck of dust, a moth floating in a beam of sunshine. "Look again at that dot. That's here. That's home. That's us. On it everyone you love, everyone you know, everyone you ever heard of, every human being who ever was, lived out their lives," wrote Sagan, underscoring the importance of taking care of that pale blue dot, the only place in the universe we can call home.[1]

With its brief career as travel photographer over, Voyager 1 switched off its cameras to conserve its dwindling energy for the journey ahead. It is today the farthest man-made object in the universe, having left the fuzzy confines of our solar system in 2012. Over four decades after it left Earth, the spacecraft is still sending back data. It will not encounter another star for forty thousand more years, by which time its creators might well have been long forgotten.

DOWN THE PIT OF HISTORY

How did a biped life form, which only the blink of an eye ago scuttered in caves, manage to send a spacecraft the size of a compact car to take a picture of their home planet from six billion kilometers away? How did lifeless molecules click together just so to jump from stolid matter to the quivering thing we call life in the oceans of a primitive Earth? How did gravity assemble and ignite our life-giving star, the Sun, out of a swirling cloud of gas five billion years ago? How did the universe come into being, creating space, time, particles, and light out of nothing, just shy of fourteen billion years ago? To understand humankind's cosmic condition, we must retrace the story of the world all the way to the beginning.

Insights from theoretical physics, astronomy, evolutionary biology, chemistry, paleontology, anthropology, and neuroscience combine to reveal how the atoms in your body became capable of understanding these words. As in a pointillist painting by Georges Seurat, only from a distance does the picture emerge as a whole—if one looks closely, it explodes in a myriad of discrete disciplines. But to gain perspective, we must first shrink the cosmic timeline down to a scale we can comprehend.

On average, a human life lasts today in Western countries around four thousand weeks. A week is a useful chunk of time, as we all have a good grasp of such a time span. We can each picture a year as a collection of fifty-two weeks, and even four thousand of them is a relatively small number, within reach of our intuition. Let us now squeeze a whole week into a millisecond: seven nights of sleep, seven breakfasts, seven lunches and seven dinners, five working days, maybe an evening out on Saturday night, a restful Sunday, all compressed into just a thousandth of a second. A millisecond is too short for us to have a direct grasp of its duration: even the proverbial "blink of an eye" takes a ponderous three hundred of them. But we can just about imagine a millisecond as the time it takes for our car to cover three centimeters as we speed down the highway.

If a whole week is packed into a millisecond, then an average life flickers out in just over four seconds. On this compressed timeline, human civilization has blossomed for about nine minutes. *Homo sapiens* strutted onstage an hour and a half ago. Life on Earth sprouted six and a half years ago, and the universe came into being twenty-two and a half years ago.

Let us take one last glance around us, admiring the bright light of twenty-first-century Earth, before embarking on a journey of discovery down the pit of our past. We step over the edge, and our grasp is secure at first: the written record of civilization easily takes us back to classical antiquity (one hundred seconds ago in our timescale) and, with some more difficulty, to Babylonian times (another one hundred seconds deeper into our past). As we descend further, the light of writing disappears and is replaced by the fainter trace of artifacts, stone tools, fossilized bones, and cave paintings. We are now in the realm of the paleontologist, whose chisel and brushes are replaced by DNA sequencing and microscopes, rock hammers and core drills as, one and a half hours into our descent, we enter the province of the evolutionary biologist and then that of the geologist.

On our compressed timeline, we are vertiginously sliding toward the beginning of life on Earth. We pick up speed, as three days down the human shape blurs into that of a great ape; sloths the size of kangaroos and armored armadillos larger than hippos blink into view, before birds lose their feathers as they morph back into dinosaurs while sharks swim on unchanged. At this depth into our past, entire days on our timeline—corresponding to millions of years of real time—go by as life on our planet soldiers on, eon after eon: glaciers grow and shrink, and countless beings have a try at the lottery of life, competing in the ruthless race of evolution. At times a catastrophic change tilts the course of life, like the asteroid that wipes out the dinosaurs forty days down our timeline and opens the way for warm-blooded mammals. Occasionally a variation blossoms with long-lasting ramifications. Flowering plants first appear some eighty days down and, with them, a new ecosystem of insects, birds, fruits, and grassland that will, in time, feed humankind.

As we slide yet deeper into Earth's past, we notice the climate changing: ice ages come and go, sea levels rise, continents move and merge into a single landmass called Pangaea, and the atmosphere becomes unbreathable, as we dip before the evolution of oxygen-producing blue-green algae. Six and a half years down, a profound change: the slimy scum we call "life," simple cells capable of controlling their own chemical environment and having chanced onto the magic trick of reproduction, vanish. We abandon the biologist's stomping ground and enter the domain of the planetary scientist, in which the Earth is a molten ball of lava and the Moon might have been lopped off by a giant impact with another protoplanet. And down we go.

Some seven and a half years down into the pit, the chief instruments that illuminate our way are physics and computer simulation. We sense around us the solar system forming from a swirling cloud of gas, the Sun ignited by gravity as it collapses into itself. As we widen our gaze, the astrophysicist now leads the way, as billions of years of real time, each compressed to one and a half years on our timeline, tumble away. We ignore dark energy and its mysterious repulsive force that speeds up cosmic expansion some ten years down the timeline—cosmologists are still puzzling this one out. Eons peel away, and we cannot stop: we are almost twenty years in, and the Milky Way, our galaxy, is beginning to form, helped along by dark matter, as young blue stars blaze with heavy elements recycled after the death of earlier, more massive ancestors.

Toward the very bottom of the well of history, the Big Bang lurks—whether a soft or a hard boundary we don't know. Here, the theoretical physicist and the cosmologist join forces to push further back the veil of darkness. The leftover light from the Big Bang glows faintly a mere five and a half hours from the bottom, the last solid rung in our descent. Farther down, further back in time, our view becomes fuzzier, our grasp of reality uncertain. Particle physicists discern clearly the formation of helium and hydrogen in the primordial furnace; they glimpse three of the four fundamental forces merging earlier on; they speculate about the first infinitesimal fraction of a second; they fumble in blackness when it comes to the very beginning, unable to bring gravity together with the other three forces. The physics that has guided our path for the greatest part of the way fails us when confronted with time zero, the moment of our ultimate inception. Beyond the stars, at the beginning of time, at the origin of the universe, religion and science meet and collide in mystery.

THE SECRET INGREDIENT

To concoct the conditions for our existence, the universe needed vast quantities of time and space. After the Big Bang left one in a billion matter particles standing while all others disappeared in pure energy, the universe needed time to cool off as it expanded; time to grow minuscule wrinkles in the primordial cosmic soup of particles into galaxies, long-lived stars, and habitable planets; time to throw the dice of random variation over and over again, so that evolution could play its improbable hand. Above us, every night, is the literal record of all that time and space: a visible cosmos that is today one hundred billion light-years across, containing fifty billion galaxies, each with some three hundred billion stars, half of which may sport planets, one of which is ours. A universe that can create

15

the complexity necessary to become conscious of itself must be billions of years old and therefore—since space is expanding—vast. Grandiosely so.

Our pale blue dot has been lucky indeed: lucky to form at the right distance from the Sun to sport liquid water; lucky to have sufficient gravity to hang on to its atmosphere; lucky to have a large Moon that helped in keeping our climate stable over time and whose strong tides encouraged life to leap from its oceanic beginnings onto the shore; lucky to have a big brother of a planet like Jupiter that acted as shepherd of asteroids, keeping the threat of a collision largely at bay. Dark matter, elusive as it is, may have played an important role in hurling comets at us at regular intervals, thus periodically resetting conditions for evolution to tinker with until-then inferior specimens, such as mammals—and ourselves.[2]

And oh, how lucky we have been to take our first upright steps on a planet with a clear view of the stars. Ever since *Homo sapiens* first craned their necks upward while squatting around a fire, the night sky has been our constant companion, a trusted guide, an awesome sight; a source of wonder, a place of mystery, and often a feared ruler. The same grandiosity necessary for our physical existence also primed us for reverence: when we look up to the night sky, it "directly reveals a transcendence, a power and a holiness," which, according to Mircea Eliade, may have produced the first religious experience in the human mind.[3] Not only our biological body but also our soul, it seems, required a grand cosmological stage to come forth.

It is often said that we are made of stardust, the atoms of our bodies having been created in the atomic furnaces of long-dead stars. But more than that, the simple fact that we could see the stars, adore them, and study them is the secret ingredient that made us who we are today. To those who knew how to read it, the sky was once a watch, a calendar, an almanac, and a map. Astrologers used it to foretell the future, kings and pharaohs to justify their power. The mysteriously regular movements of the heavens may have molded our ancestors' brains for rational thinking; they led to major breakthroughs in mathematics and set off the Scientific Revolution; they have inspired some of the highest expressions of human art.

A highly specialized caste, today as in Babylonian times, is charged with reading the stars. Astronomers diagnose from tiny wobbles and blemishes in the starlight the presence of pale blue dots around other suns, while astrophysicists tune into the space-time tremor of two black holes embracing; doctors of cosmology auscultate the heartbeat of the Big Bang, as data science alchemists conjure up in their computers the slow-motion blooming of a planetary nebula. But we no longer remember which star heralds the arrival of the harvest, no longer marvel at the meandering shape of the Milky Way, no longer await the young crescent of the Moon as a symbol of hoped for rebirth, no longer ask the constellations for guidance when navigating the high seas.

It is all too easy to forget about the stars, hidden from our view by light pollution, peppered with artificial satellites, or confined to a desktop background. Yet taut threads still tie us to them, faint like a spiderweb glittering in the moonlight. Every time you glance at the twelve subdivisions of the hours on a clock's face, you are staring at the star groups that the Egyptians used four thousand years ago to mark the passage of time at night. The GPS inside our phones and cars would be helplessly inaccurate without Albert Einstein's theory of general relativity, which explained the mysterious petal-shaped trail of the orbit of Mercury and was tested with the reshuffling of stars during a total solar eclipse and the gentle slowing down of cosmic lighthouses in the form of spinning neutron stars. None of this would have been possible in a world without stars.

By simply being there in the sky for us to admire, puzzle over, and wonder about, the stars led us from the Stone Age to artificial intelligence, shaping in a remarkable but often unrecognized manner the course of human history. This book is the story of how.

It is also the story of how things might have been otherwise. In his 1905 book *The Value of Science*, French physicist, mathematician, and polymath Henri Poincaré celebrates astronomy as the most fundamental of all the sciences. He argues that astronomy's usefulness goes far beyond its practical applications, as important as they are: "Astronomy is useful because it raises us above ourselves; it is useful because it is grand," he declares.[4] Poincaré then introduces a fictional world covered in clouds, perhaps inspired by Homer's Cimmerians:

Think how diminished humanity would be if, under heavens constantly overclouded, as Jupiter's must be, it had forever remained ignorant of the stars. Do you think that in such a world we should be what we are? . . .

Does anyone believe that, without the lessons of the stars, under the heavens perpetually overclouded that I have just supposed, [our souls] would have changed so quickly? Would the metamorphosis have been possible, or at least would it not have been much slower?[5]

This book takes up Poincaré's provocation: "Do you think that in such a world we should be what we are?" We will imagine the rippling consequences on humankind's trajectory of removing our view of the sky by traveling to an alternative version of our pale blue dot, one that, seen from a distance, looks like a milky marble. On this counterfactual Earth, covered in clouds that obliterate the view of the stars, at all times, everywhere, the absence of visible heavens changes everything. By removing something taken for granted, we will, I hope, see it anew—a lesson many of us learned during the coronavirus pandemic, when freedoms we thought unassailable were curtailed overnight. In a world without stars, which we shall call "Caligo," humanity is sent down untried forks in the road of history. "To see things as they really are, you must imagine them for what they might be."[6] What American civil rights activist Derrick Bell intended as a device to foreground racial injustice, we will use to better understand how singular our star-strewn path has been.

We will begin our journey in the disenchantment of the present, in Chapter 2, and call upon our ancestral links to the sky to reawaken the awe the stars elicited in earlier generations. In Chapter 3 we will draw a veil over the sky, setting the premise of a world without stars alongside our investigation of humanity's actual cosmic history.

We will slide down the pit of history once again in Chapter 4, all the way back to the dawn of *Homo sapiens*, teasing out the importance of the "disciplined army of luminous points" in the sky, in the words of Poincaré, for our prehistoric ancestors.[7] The cyclical changes in the firmament were humanity's first timepiece, a crucial precursor of "the key-machine of the modern industrial age" (according to historian Lewis Mumford): the clock.[8] The many consequences of sky-based timekeeping, shaping nearly every aspect of our lives today, are explored in Chapter 5.

The view of the sky gave our ancestors the means to orient themselves in space as well as in time and to navigate the vast oceans that make our pale dot blue. In Chapter 6 we will encounter the horizons-expanding possibilities afforded by a clear view of the sky, crucial for navigating beyond coastal waters and spreading *Homo sapiens* to the four corners of the planet. We will witness the unique near-encounter of the Polynesian wayfinding tradition with the Western approach to navigation, illuminating how the stars were crucial for both, albeit in very different ways.

Astronomy will become the midwife to all Earth's sciences in Chapter 7—not so for the less fortunate Caligoans. Nicolaus Copernicus, Galileo Galilei, Isaac Newton, Pierre-Simon Laplace, Carl Friedrich Gauss: they invented and perfected the scientific method to understand and describe the motion of planets and the laws that moved the sky. In doing so, they came up with new kinds of mathematics but, just as importantly, with a new way of looking at nature, one that focused on regularities, measurement, and prediction. As we will discuss in Chapter 8, this mind-set not only spread like wildfire to all sorts of scientific and technical endeavors but engulfed the human sphere, too, influencing the way we think of ourselves and the kind of societies we built.

In Chapter 9, we will turn our attention inward and explore how the symbolism of the heavenly bodies has influenced our very selves and still permeates our lives today. Belief in astrology changed the course of history, and its study provided the means of support and often the spark of inspiration for those who set the Scientific Revolution in motion.

As we reemerge from our journey through history, we will look at how we are beginning to remake the sky: as our sphere of influence encroaches into space, our view of the sky is rapidly changing, and not only because of light pollution. As our star-inspired technology risks wrecking the whole ecosystem, some are thinking of turning to the stars for salvation. Chapter 10 is concerned with the future of our cosmic heritage, and our own: Is aiming for the stars the answer, or should we be looking at the stars for a different kind of inspiration?

With the very survival of life on Earth today at stake in the wake of climate change, loss of biodiversity, and the rumbling of war, once again, I hope, the stars can show us the way.

THE CALIGO TALES

Remembering

The Glow was gone—it was time for Remembering.

I crushed the last morsel of charcoal between my fingers and daubed it on the cloud, darkening it to mask the glittering of crystals dotting the vault. I stepped back and took it all in: the carpet of cloud ran up the walls of the cave, slithered across the ceiling and back down the other side, as far as the eye could see. In the flickering light, it appeared to be flowing and rolling. I felt proud.

One after the other, they were all filing in: Bison-Seeker was first, followed by Shoe-Sewing, Way-Finder, Spear-Thrower, Beehive, Once-Upon-A-Glow, and, last of all, Cloud-Watcher. Beehive stopped for a beat to look at my work. As always, she gave me her opinion on some detail: too fat here, a bit more gray there, a touch less heavy over there.

As we joined the others around the fire and I took my usual place, I couldn't help but stare at the empty sitting stone. Freshwater was still out from yesterglow, and who knew whether we'd see her again. I had told her not to go; when she left, there was Blackness in her eyes. Fire-Keeper was tending the logs, and the flames jumped higher, licking the blackened vault. Shepherd's flute filled the cave with music and our hearts with mist. As always at the beginning of Remembering, Shepherd took the bowl full of charcoals from Fire-Keeper and, calling in the power of Lightning, poured water on it. A dense billow of white smoke puffed up from the bowl, and the Cloud was among us.

Shepherd's figure had nearly disappeared in the whiteness, but his voice boomed as if the Cloud itself were speaking.

"Daughters and sons," he intoned. "We are gathered at the end of the Glow so that we may Remember. But first, let's send help to our sister Freshwater, who is out in the Darkness, and may be lost: may Lightning pierce the Darkness and show her the way!"

To this, we all shouted our best invocation of Rumble. When silence returned, Shepherd resumed.

"Daughters and sons, let us Remember!"

He walked around the fire, emerging from the white screen as he approached each of us. He stooped to look me in the eyes, so close that our brows almost touched. He held my gaze for a long while, then moved on. He passed Beehive without pausing, then abruptly pointed at Cloud-Watcher, and commanded, "Cloud-Watcher, you are closer to the Cloud than any of us. Remember the way of the Cloud!"

Cloud-Watcher stood, and thus began her story.

2

THE LOST SKY

But if a man would be alone, let him look at the stars. . . . One might think the atmosphere was made transparent with this design, to give man, in the heavenly bodies, the perpetual presence of the sublime.

—RALPH WALDO EMERSON, *Nature*[1]

LOOKING FOR TOTALITY

We have traveled over five thousand miles to get here, crossing oceans and mountains, but the success of the whole expedition hinges on these last five hundred yards.

The dirt track switches back and forth as it ascends, flanked by rustling high grass, which, we have been told, can hide venomous snakes. The scent of dry earth and pine fills the mid-morning air, as our steps kick up small puffs of dirt that cling to our sweaty bodies. Our son, Benjamin, not quite two years old, is enjoying the ride in the baby carrier on my back, a lookout from a crow's nest on a miniature ship onto the dusty inlands. Emma, our five-year-old, proudly hangs on to a walking stick fashioned out of a dead branch, a pink bucket hat shielding her from the blazing Sun. In a few minutes' time, she won't need it any longer.

We have come to this remote part of Oregon in search of the perfect total solar eclipse. Dubbed "the Great American Eclipse," the August 21, 2017, event is the first solar eclipse to traverse the United States since 1918. The path of totality, the narrow strip on Earth where the Moon lines up exactly with the Sun, will cross the country from the Pacific coast in Newport, Oregon, to the Atlantic coast in Charleston, South Carolina, touching thirteen states and giving millions of people the opportunity to witness one of the most dramatic spectacles the sky can offer.

I had tried to see a total eclipse once before. When the Earth, Moon, and Sun obligingly swung into alignment on my doorstep in August 1999, when I had been a physics undergraduate student, I dutifully traveled from Zurich to southern Germany, a comfortable train trip followed by buses to the countryside and a hike to the top of a gentle hill, tripods, cameras, solar filters, and rolls of film at the ready. Just like the Swiss train we rode on, the eclipse was on time that morning, but my fellow physics students and I could only divine it through the thick clouds thanks to the telephoto lenses on our cameras. Later, we would brag that we had been there, with trophy pictures of the eclipsed Sun to prove it, but our smiles were laced with disappointment, the gray sky a mocking backdrop to our trio—it didn't feel as if we had experienced the real thing.

This time, I have done my research, and I know that the best conditions are to be found in eastern Oregon, where the Cascade Range keeps the humid air drifting inland from the Pacific at bay and the summer sky is highly likely to be clear at 10:21 a.m. on August 21, 2017. Unfortunately for us, in the age of the internet, millions of Americans and eclipse chasers have come to the same conclusion, and I have quickly discovered that hotels, vacation rooms, camping sites, RV parks, and even open fields from Bend to Portland are full—everybody wants to bathe in the shadow of the Moon as it glides across the continent.

But I have an ace up my sleeve: astronomer Tyler Nordgren, a colleague of mine, is a member of that rarefied group of people who call themselves "coronaphiles"—eclipse fiends. With dozens of eclipses under his belt, he is a man with a plan for the Great American Eclipse, and he doesn't disappoint: when I reached out to him, he invited us to join him on a private ranch near Kimberley, where he has organized an eclipse camp.

There will be barbeques, there will be mules and horses, there will be the great American wilderness, and, most importantly, there will be the highest probability of one minute, fifty-eight seconds of totality under perfectly blue skies. Phone calls were made, savings rustled up, and transport booked. We have flown from London into San Francisco, adjusted our body clocks to Pacific time in Marin County, picked up during rush hour on a Friday afternoon in Newark an RV more spacious than my first student flat, shot due north up Highway 5, turned west at Reading, slept among fragrant pine trees in Lassen County, barreled through deserts with cobalt blue lakes, filled up in Wagontire (pop. 2), and cooked *pasta al sugo* under a canopy of stars. This is no time to dally—we have a cosmic appointment to keep. Highway 19 took us to Kimberley, past the John Day Fossil Beds National Monument, and then it was a question of looking out for a certain milepost and a red mailbox with a mule on top, gently leading the RV down a dirt track, and pulling into what in the headlights looked like an open field.

When I opened my eyes this morning, the rush of adrenaline was immediate. I parted the RV curtains and looked out: the sky was an auspicious shade of peach, heralding a cloudless day. Success was within reach. I gazed out on ochre buttes dotted with dark conifers, silver-leafed sagebrush, yellow-flowered Scotch broom bushes, and spring-green thicket that I didn't recognize. This landscape is so iconic of the American West that a version of it is painted on the sides of our RV.

We have decided to climb one of the buttes to get a view of the landscape as the lunar shadow approaches and to get away from the six-hundred-strong crowd of fellow eclipse chasers who have transformed the remote ranch into a good-humored party site. If we are to connect with the sky, we need isolation and silence as much as a clear view. But the plateau we have elected as our observation site is farther away than we had imagined, the slopes steeper than we expected, and prepping our two mini-explorers for the last leg of our adventure has taken longer than planned this morning. As a result, it's ten past nine, and the start of the eclipse finds us about halfway up. We've missed first contact, when the Moon takes its first, tentative bite of the Sun's disc, and we need to speed up if we are to make it to the top in good time for the main show.

THE HARBINGER OF DOOM

My friend and colleague Mark McCaughrean, a man of science not inclined to lyricism, told me once that totality had transformed him into "a gaping ape." Isaac Asimov went further, using a solar eclipse to drive an entire civilization to collapse in his 1941 novelette *Nightfall*. The story takes place on a perpetually sunbathed planet, Lagash, whose inhabitants are unaware of the existence of stars. Every 2,050 years, however, a moon brings about a total solar eclipse that plunges the planet into darkness and thus reveals the star-studded night sky. A handful of scientists predict the spectacle and raise the alarm but are ridiculed by the public and assaulted by members of a sect who believe the eclipse will be the end of the world. When darkness engulfs the day and thousands of stars appear, the minds of the Lagashians bend and crack under the weight of an unchecked fear, and the horizon turns crimson with fires destroying their cities.

The extreme reaction of Lagashians is perhaps only a little exaggerated. Total solar eclipses have always been seen as portents and often as harbingers of doom. When the Sun, provider of heat, light, and life, disappeared without warning from the sky, the natural order of the world was subverted. It was often believed that eclipses were an attack by some evil force: a dragon for the Chinese, a serpent for the Maya, a pair of wolves for the Vikings, an evil spirit taking the form of a toad for the Shan people of Annam (in today's Vietnam), a vampire for the Tatars of Siberia. The prophet Joel warns, "The sun shall be turned to darkness, and the moon to blood, before the great and the terrible day of the Lord come."[2]

Our forebears saw events in the sky as connected with human affairs, and a dramatic, unannounced disturbance like a solar eclipse was perceived as a threat to be dispelled with whatever magical means were available. Remedies included shouting to scare away the Sun's attacker, throwing stones and shooting coal-tipped arrows into the sky, and singing and chanting. Only slightly less frightening was the spectacle of the full Moon dimming and turning blood-red during a lunar eclipse, which called for corresponding magical countermeasures. Only rarely was the joining of Sun and Moon in the sky interpreted as the joyful mating of husband and wife, as Australian First People and Tlingit Native Americans believed.

One terrifying aspect of a solar eclipse was that no one in antiquity could predict it reliably—not until Edmond Halley in the eighteenth century. A solar eclipse occurs when the Moon passes in front of the Sun, which happens on average every eighteen months.[3] But the type of eclipse depends on the orbital vagaries of the Moon. Total (the Sun disappears entirely) eclipses are startling, while partial (the Sun's face is not fully covered) and annular (a ring of light surrounds the dark lunar disc) eclipses are not nearly as dramatic: a partial eclipse often goes unnoticed unless one is specifically looking for it. Less than a third of solar eclipses are total, and what's more, the path of totality, the pencil line on the surface of the Earth along which one can observe Moon and Sun in perfect alignment, is quite narrow, typically only one or two hundred kilometers wide. Outside that region, a total eclipse is demoted to partial. So the chance to experience the hair-rising moment of a total solar eclipse from any one location on the surface of our planet is narrow—on average, only two or three times per millennium. By contrast, lunar eclipses, which occur when the Moon is obscured by the shadow of the Earth, are observable from everywhere on the night side of the Earth, and thus a total lunar eclipse can be seen about once a year on average from any location.

Many ancient civilizations anxiously recorded the patterns of solar and lunar eclipses in the hope of divining their future occurrence—an effort that commanded sustained attention, precise recordkeeping, and astronomical knowledge passed down through many generations. Since the second millennium BC, the Chinese have attempted to predict lunar eclipses as part of their systematic calendrical method, called *lifa*, while solar eclipses were considered part of *tianwen*, the unpredictable phenomena of the heavens. Some managed to crack the code: Babylonian astrologers noticed in the second millennium BC that lunar eclipses repeat exactly every 6,585 days (slightly more than eighteen years). This cycle is today called a *saros*, and it encompasses 223 new Moons, a period after which the relative locations of the Sun and the Moon in the sky are almost exactly reproduced. The Babylonians, like the Chinese, could not foretell the ominous solar eclipses, but they might have noticed that a total solar eclipse is always preceded or followed by a lunar eclipse, with the two events fourteen days apart. So knowing when a lunar eclipse would next happen would have helped them prepare for the potential occurrence of a total solar eclipse.

This might have been the strategy adopted by Greek astronomer and mathematician Thales of Miletus, who, according to Herodotus, foretold the total solar eclipse of 585 BC, which occurred in modern-day northern Turkey during the sixth year of a bloody battle between the Lydians and the Medes. But when the fighting armies "saw the day turned to night, they stopped fighting, and both were the more eager to make peace."[4] The Stonehenge megalithic circle might have been used as an eclipse predictor, indicating when the Moon was approaching the orbital danger zone in which it could obscure the Sun.

The premodern terror of solar eclipses is brought to life in an account by Fray Bernardino de Sahagún, a Franciscan missionary who spent over sixty years among the Aztecs of Mexico, recording their customs and beliefs before they got wiped out by the Spanish conquest. The Aztecs believed that the underworld was located in the sky and could not be seen because of the shining Sun. Evil skeletal creatures of darkness, called *Tzitzimime*, were said to be unleashed during totality and would bring destruction onto the world. The Aztecs' reaction to the obliteration of the Sun's light during the eclipse in the sixteenth century is reminiscent of that of Asimov's Lagashians: "There was shouting everywhere. People of light complexion were slain; captives were killed. All offered their blood; they drew straws through the lobes of their ears, which had been pierced. And in all the temples there was the singing of fitting chants; there was an uproar; there were war cries. It was thus said: 'If the eclipse of the sun is complete, it will be dark forever! The demons of darkness will come down; they will eat men!'"[5]

Total solar eclipses came to be seen as connected with or auguring the demise of powerful men: kings, emperors, popes, and prophets. The Gospel of Luke describes the Sun as "darkened" for three hours after Jesus's crucifixion, but this could not have been an astronomical eclipse, as we know that Jesus died on the day of Passover, which coincides with a full Moon, not with the new Moon that would be required for a solar eclipse.[6] Perhaps Luke was speaking symbolically or used the power of an eclipse to underline rhetorically the dramatic moment of Christ's death, thus fulfilling the prophecy in the Bible about the moment of reckoning for Israel:

"And it shall come to pass in that day, saith the Lord God, that I will cause the sun to go down at noon, and I will darken the earth in the clear day."[7] The son of Charlemagne, Louis the Pious, was said to have died of the terror inspired in him by the solar eclipse of AD 840. And according to popular wisdom recorded in the Anglo-Saxon Chronicle, the total eclipse of August 2, 1133, did it for Henry I, king of England and Normandy: "In this year King Henry went over sea at Lammas, and the second day as he lay and slept on the ship the day darkened over all lands; and the Sun became as it were a three-night-old Moon, and the stars about it at mid-day. Men were greatly wonder-stricken and were affrighted, and said that a great thing should come thereafter. So it did, for the same year the king died on the following day after St Andrew's Mass-day, Dec 2 in Normandy."[8]

That King Henry I died more than two years after the eclipse (on December 1, 1335) and of a more mundane reason (indigestion of lamprey, on which he gorged against his doctors' advice) shows how deep-seated the connection between heavenly portents and earthly events was in observers' minds. The Assyrians of Mesopotamia sought to deflect the ill consequences of an eclipse by replacing their king with a surrogate, whose role was to face the danger of the bad omen in place of the sovereign.[9] Usually a prisoner, a criminal, or a commoner, the puppet king was appointed on the recommendation of the court astrologers; in the wake of an eclipse, such as that of June 15, 763 BC (the first unambiguously recorded solar eclipse in history), he would be enthroned, dressed in regal robes, and given a consort queen.[10] He was made to swear a special oath in which he "took all the celestial and terrestrial portents on himself," and for up to one hundred days he would be dined, entertained, and surrounded by bodyguards (lest he escape), while the real king remained in hiding. When the celestial danger was deemed to have passed, the substitute was executed and the king reinstated. The eclipse did invariably bring death for the substitute king.

THE PRINCIPLE OF LIFE

As I sweat under the blazing Sun, climbing the butte as fast as the short legs of my daughter and the weight of my sleeping son in the baby carrier will allow, I reflect on how much our relationship with our star has changed. Today, the term *sun worshipper* evokes nothing more than ice-cold cocktails on a white, sandy beach at sunset, the whiff of SPF 50 sunscreen hovering in the air. But in ancient times, when humanity's survival was at the mercy of scorching drought, torrential rains, or devastating floods, rituals to propitiate the supreme god were no laughing matter.

Many ancient civilizations worshipped the Sun as a major deity, recognizing in him (and it was often—although not always—a masculine god) the principle of life itself. This was all the more reason to be terrified by his sudden disappearance from the sky. No civilization was more obsessed with, or more fervently devoted to, the Sun than the ancient Egyptians. In ancient Egypt, the falcon-headed Sun god Ra triumphantly traverses the sky in his boat, flanked by minor deities keeping watch and fighting off any foes that might stand in the way. At sunset, the Sun's bark plunges into the western horizon, celebrated by the chattering of monkeys, to face a danger-strewn return journey through the underworld. There, Ra must overcome great serpents with monstruous bodies, navigate twelve pitch-dark compartments (one every hour, something we'll return to) across the depths of the Earth, and cross lakes of fire, until he finally confronts his great rival, Apophis, the serpent of death, harboring a knife in each of its coils. Helped along by allies of the light, cheered on by monkeys and the spirits of the dead, Ra finally prevails after a gruesome battle, but to be reborn, he must enter with his boat the tail of a snake and be regurgitated from its mouth, transformed into a sacred scarab beetle, who rolls the rejuvenated sphere of the Sun upward from the eastern horizon. This daily cycle of death and resurrection embodied, for the Egyptians and many others, the struggle of existence and the hope for a new life after death. Remarkably the Egyptians didn't leave any records of eclipses; perhaps such deep disturbances in the normal course of Ra were considered too shameful for the pharaoh, the god's incarnation on Earth, to be shared with posterity.[11]

The second protagonist of the cosmic union unfolding over our heads played just as important a role in shaping our ancestors' beliefs. Insofar as we can sound the oceanic depths of our oldest myths, the Moon was the original focus of religious life, not the Sun. Its cycle of waning and waxing extended time reckoning past the simple alternation of day and night. As we will see in Chapter 4, the earliest calendars were Moon-based. But just as importantly, the lunar phases came to be associated with the ever-recurring cycle of life, the universal law of becoming, the inescapable wheel of human birth, growth, decline, and death—something we will also explore further.

Brother and sister, mother and son, left and right eye of the heavens, bride and groom: whatever their imagined relation, the ancient symbolism connected with the Sun and the Moon relies on their almost identical apparent size in the sky—the Sun being approximately four hundred times larger than the Moon but as much farther away. This happy astronomical coincidence treats us to total solar eclipses and also forms the basis for a myriad of myths where the Moon and the Sun are, if not equals, at least of similar status in their respective roles. But this delicate match won't last forever: as the Moon recedes from the Earth, at a steadily increasing rate of four centimeters per year, at some point it will be too far away for its disc to fully cover the Sun. In about six hundred million years, the day will come when the last total solar eclipse will grace the Earth.[12]

BROOM STARS AND NEWTON'S KITES

Eclipses were the heavens' most dramatic unannounced visitations, but there were others that our ancestors connected with human drama. Matthew the Evangelist tells us that the Magi visited King Herod in Jerusalem, looking for the newborn king of the Jews, who would fulfill the prophecy of Balaam: "There shall come a Star out of Jacob, and a Sceptre shall rise out of Israel."[13] "We have seen his star in the east," they said, "and are come to worship him." In Matthew's account, the star then reappears in the south to guide the Magi to Bethlehem and the Nativity: "And, lo, the star, which they saw in the east, went before them, till it came and stood over where the young child was."[14]

Efforts to identify the Star of Bethlehem with an actual heavenly phenomenon started with German astronomer Johannes Kepler, who, in 1604, described in detail a new star he saw appear in the foot of the constellation Ophiuchus. This was, we today know, a supernova explosion—marking the death of a compact, dense star torn apart by a runaway thermonuclear reaction. Happening a mere twenty thousand light-years away, the appearance of Kepler's star was the last such event to occur in our own galaxy.[15] The new star had burst forth in the region of the sky that Kepler was monitoring, looking for a rare conjunction of the planets Saturn, Jupiter, and Mars. When he calculated that a similar planetary configuration would have occurred around the birth of Jesus, Kepler suggested that "the star which led the Magi to the manger of Christ . . . could be compared with our star." To Kepler, his nova was likely the Star of Bethlehem returned, preannounced to him, as to the Magi, by the conjunction of Jupiter and Saturn: "What else could the Chaldeans [Magi] conclude from their, and still existing, rules of their art [astrology], but that some event of the greatest moment was imminent?"[16]

Many others would go on to pore over astronomy books, ancient records, and commentaries to lost manuscripts; translations were queried, orbital trajectories revised, and theological arguments debated in an effort to identify the Star of Bethlehem: Venus was too mundane, Halley's Comet years too early, a meteor too ephemeral. Could it be that the portent that had convinced the Magi to saddle their camels was a "broom star," a comet reported on a Chinese silk scroll dated to around the birth of Jesus? That this interpretation of the Star of Bethlehem survives to this day on Christmas cards may be due to Giotto's being inspired by the sight of Halley's Comet in 1301 to represent it as such in the Nativity of the Scrovegni Chapel in Padua. Whether astronomical event, supernatural sign, or allegorical fabrication, the Star of Bethlehem tells us a bigger story: that of marking a miraculous birth with a miraculous apparition in the sky, an augur from the heavens for humanity on Earth, a sign of good fortune that opposes the "darkened Sun" accompanying the crucifixion.

Comets were considered versatile astronomical omens: like a cosmic exclamation mark, they could herald the birth of the son of God or drag an emperor to death. The elite and the populace alike respected comets as prodigies, monstrous apparitions that heralded some unusual event of cosmic importance. Danish astronomer Tycho Brahe said of them, "Such unnatural births in the heavens . . . always have had something great to deliver to this lower world."[17] The difficult question was, what exactly did the comet mean? The great comet of May 44 BC, one of the brightest in history, was believed by the Romans to be the soul of Julius Caesar ascending to heaven after his assassination two months earlier. Mindful of Caesar's fate, the terminally ill Emperor Vespasian sought to divert onto others the augury of death seemingly portended by the comet of AD 79. Referring to the comet as a "hairy star," he said, "This is an omen, not for me, but for the Persian king; for he has long hair, whereas I am bald."[18] The ruse didn't work, and a few weeks later Vespasian was dead. According to Saint Thomas Aquinas, Saint Jerome saw comets as one of the signs leading up to the apocalypse: "on the 'seventh' day all the stars, both planets and fixed stars, will throw out fiery tails like comets."[19] Luke tells us that Jesus himself warned that, before the coming of the Lord, "there shall be signs in the sun, and in the moon, and in the stars."[20]

More than underlining historical events, comets also shaped them. The year is 1066, the comet is Halley's, insouciantly appearing almost six hundred years before the birth of its namesake. Visible even in full daylight, the display is magnificent at night: "a comet with three long extended rays [illuminating] a very great part of the South for the space of fifteen nights"; others described its triple tail as "streaming like smoke." Four months into his precarious reign, Harold Godwinson, the last Anglo-Saxon king of England, marvels at the "new star" while his people offer prayers to prevent disaster. A monk at Malmesbury Abbey, having perhaps witnessed the passage of Halley's Comet in 989, which had heralded the Viking invasion of Britain, exclaims, "Here art thou again, cause of tears to many mothers. It is a long time since I saw thee last, but I see thou now, more terrible than ever, thou threatenest my country with utter ruin."[21]

Two hundred miles to the south, William of Normandy marvels at the new star too, but in its hairy tail pointing toward Harold's castle, he sees a divine augury of victory for his military campaign. His army rallies to the cry "A new star, a new king!" Five months later, as Halley's Comet wearily heads home to the Kuiper belt, dim and cold, a Norman arrow extinguishes Harold's life in the Battle of Hastings and makes William king.

Later, the triple tail of Halley at the height of its splendor will adorn an exquisite linen commemorating the campaign of William the Conqueror.

The Bayeux tapestry, a scroll seventy meters long and finely embroidered throughout, was commissioned by William's half brother to commemorate the Norman conquest of England on the occasion of the inauguration of Bayeux Cathedral in 1077.[22] The apparition of Halley's Comet opens a sequence of scenes that unfold like a comic strip until the death of Harold.

The power of comets was not lost on the boy who, six centuries later, would pin them on parabolic orbits and thus tame their willful trajectories. As a schoolboy, Isaac Newton used to fly paper lanterns tied to the tails of homemade kites, which many a villager mistook for a real comet on dark winter nights. Later on, as a scholar in Cambridge, Newton "sate up so often long in the year 1664 to observe a comet that appeared then, that he found himself much disordered." Another refulgent comet at the end of 1680 played a central role in his masterwork, the theory of universal gravitation, as we will see in Chapter 7.[23]

A JOLT OF AWE

I pick up my pace up the side of the butte, sweat pouring down my back, my wife and I each holding one of our daughter's hands. We must make it to the top soon or we will be surprised by totality halfway up. I know that we can never experience a total solar eclipse like our ancestors did, but if we are to reclaim at least some of the awe that gripped them, we must find an isolated spot, away from the crowds. We pause frequently to catch our breath, turn around, don our eclipse glasses, and check the progress of the dragon's bite (or is it a werewolf's?) in its relentless march across the solar disc. Suddenly, the steep slope falls away: we are on the top ridge, a stark landscape of striped buttes dotted with sagebrush and pine all around us. The Sun has another half hour to climb before it locks in perfect alignment with the encroaching black circle. I put my son down, grateful to be relieved of the baby carrier and grateful to be here.

A few minutes before totality, the colors are sucked out of the landscape, and shadows become eerie, sharpened on one side and crooked on the other, for we are standing in a giant spotlight in the shape of a crescent. I had heard it described as "sunset during the day," but it's nothing like that: at sunset, colors shift toward red, and the world is bathed in a warm, soft, glowing light. Just before totality, the light turns ominous, gray, accusing; the Sun seems helpless as darkness encroaches. The Moon shadow is gliding toward us at over two thousand miles per hour, faster than the speed of sound, molding itself onto indifferent mountain ranges, pine forests, the glittering cities of man, and millions of expectant eyes behind mylar glasses. It's cold now, as if the life energy of the universe is being withdrawn from our planet. The birds sense it too: their chirping dies down to silence. "The birds dropped to the ground," reported the chronicle of an eclipse over Scotland in 1652.[24] Are they sleeping or quietly terrified? Flowers close, tricked into their nocturnal aspect by a phenomenon they won't witness again.

Then it happens: the Sun slides fully behind the dark disc. The stillness of a moment beyond time is broken by our exclamations, soon joined by the cheers and shouts of the crowds at the ranch down below us. It's impossible to keep it in—the wonder has to be voiced. We yell, we scream, we bawl, we howl: we regress to a more primitive state. Through cracks in the sky, the stars appear as if eager to witness the impending doom. The corona—the outer part of the Sun's atmosphere, hundreds of times hotter than its surface yet normally invisible—radiates around the unnatural black hole in the sky. It's a ring of fire where the Sun used to be, fierce, ominous. I know that the appearance of the corona doesn't change in a matter of minutes; yet I find it hard to shake off the impression that it is vibrating with cold energy. I know that totality will last one minute and fifty-eight seconds; yet time feels expanded. What if the Sun never reappears? I am standing in the Moon's shadow, and the beam of darkness connects me to the stars. I touch my cheek, and my hand comes away damp.

THE FORGOTTEN SKY

In recent times, with long-distance air travel affordable to many, the drama of solar eclipses and the dance of the Northern Lights have been drawing millions of worshippers. Some have gone to extreme lengths: Concorde flights were chartered to chase at supersonic speed the path of a solar eclipse and stretch totality to over an hour; oceanic cruises deftly maneuvered into a parting of the clouds for forty seconds of totality; packs of snowmobiles unleashed into the Arctic night, full of sleep-deprived aurora chasers glued to unreliable magnetic storm forecasts. In the age of social media, being there is also not enough—even more important is broadcasting the fleeting moment to the world with real-time updates. The lived experience of the sky, which took so much effort to achieve, is downgraded to the pixelated view of a smartphone screen, not dissimilar to (if far more expensive than) what is only a couple of clicks away for anyone with an internet connection.

For the rest of us, "the heavenly bodies, the perpetual presence of the sublime," in the words of Ralph Waldo Emerson, retreat behind thicker and thicker curtains of artificial light. In 1869, astronomer Edwin Dunkin published his now classic book *The Midnight Sky*, showing in beautifully hand-drawn figures the Milky Way shining majestically over St. Paul's Cathedral in central London, then the largest metropolis in the world. Electric street lighting was still twenty years away, and on a moonless, clear night, Londoners could recognize the description given by John Milton in the seventeenth century:

> *A broad and ample rode, whose dust is Gold*
> *And pavement Starrs, as Starrs to thee appeer,*
> *Seen in the Galaxie, that Milkie way*
> *Which nightly as a circling Zone thou seest*
> *Pouderd with Starrs.*[25]

Milton wrote this description from memory, after having gone completely blind in his mid-forties. In this part of his masterpiece epic poem, *Paradise Lost*, he conjures up a view of the heavens of which he had been wholly deprived.

Dunkin remarked that up to two thousand stars could be seen from London at any given time, a heavenly spectacle whose "picturesque beauty [was] amply sufficient to leave upon the mind of the most indifferent observer such deep impressions of the majesty of creation as ought not be easily effaced."[26] On a moonless summer night, I stood in the largest common in London—twelve hundred acres of ancient heathland and bogs near Wimbledon that have been inhabited for millennia. I counted no more than 150 stars, trembling and almost lost in the orange haze exuding from the nine-million-strong city all around me.

With effort, we can still recapture some of what our ancestors saw. Camp in the mountains, hike through a desert, sail until the last flicker of coastline has disappeared below the horizon. Switch off all lights and wait until your eyes have found their bearings. Then you will see it: an intricate embroidery with appliqués made of glowing hydrogen, imaginary stitches connecting fusion reactors floating in the void, coal-black patches that a pair of binoculars reveals to be full of glitter powder, refulgent little discs ablaze with the promise of other worlds.

To our forebears, stars weren't a mere decorative tapestry with a dazzling array of brightness and colors; they delineated in their mind's eye shapes of bears, baboons, eagles, scorpions, and snakes, princesses and dragons, young maidens and heroes, old kings and humble shepherds. Heroic characters, mythical beings, sacred animals, and magical objects filled the sky in the form of constellations.

Some of these products of our imagination are remarkably common across cultures, like the Bull or Orion. The night sky has no depth: stars that appear next to each other in a constellation might be thousands of light-years apart in three-dimensional space. Castor and Pollux, the two brightest stars in the constellation Gemini (the Twins), are actually twenty light-years apart, their apparent proximity an illusion. Constellations also morph over the ages, as the movement of stars in space deforms their shape beyond recognition. We would not identify the constellations the Neanderthals saw in their sky a million years ago had they preserved them in stone (they didn't).

I once met a physics graduate student from New York, whom I'll call Max, on a Mediterranean island where I was teaching an advanced course. I had dined with a group of students on a terrace overlooking the sea at the bottom of dramatic cliffs, and we were walking back together to the isolated site where the school was being held. It was a dark night, and as we walked, we craned our necks to the stars, the Milky Way as distinct a road as the pavement we trod on. I casually commented on its beauty, to which Max responded that he had never seen the stars before in his life. I thought at first he meant that he had never experienced them like this, but he explained that, growing up near New York City, he had believed that stars could only be seen with a telescope until he started graduate school.

I was astounded: a young man in his mid-twenties, with a degree in physics and specializing in cosmology, was hardly uncultivated. Yet Max had had no inkling that stars and nebulae shone for the naked eye, patiently waiting, away from the big-city lights. Stars are but the latest casualty in our relentless conquest of darkness, steadily retiring under the advance of lampposts, searchlights, digital billboards, solar-powered fairy lights in gardens, and high-intensity discharge lamps in stadiums.

The first to retreat was the Milky Way of Milton, which long ago disappeared from urban areas to fall back to the countryside, where it was encircled and pursued by the skyglow of cities. It has now fled to remote deserts, national parks, and small islands. Today's city dwellers are unlikely to even recognize its rare appearances. On the night of November 9, 1965, a statewide blackout brought it back to tower over the usually glittering Manhattan skyscrapers. Some characters in Don De-Lillo's novel *Underworld* can't help but take notice, even though they can't quite name what "the streaked sky" actually is: "People talked to each other and looked up periodically. They looked toward the midtown sky or tried to look toward the tip of the island, blocked off of course by clustered buildings, but always up, skywatching, and they pointed and talked. . . . I could see the silhouetted towers of midtown, exact and flat against the streaked sky."[27]

When the starry sky blitzed over Los Angeles at 4:31 a.m. on January 17, 1994, in the aftermath of the Northridge earthquake that cut power to the city, bewildered inhabitants called the Griffith Observatory, puzzled by the "strange sky" they were experiencing—strange because full of stars.[28]

The paradox of our age is that while the real sky recedes, a few clicks can transport us to the farthest corners of the visible universe, where the pixelized ghosts of photons that chanced to fall onto the mirrors of our giant telescopes after a few billion years' journey through space are served up from the cloud in an instant. If we do not recognize the frayed constellations of our urban skies, our smartphones' augmented reality fills in our diminished experience.

But our telescopes, space observatories, and digital cameras are not the mere extensions of our senses we often imagine them to be. Powerful and reliable, they have unlocked vistas and produced insights that would have seemed fantastical to Galileo, who first used a telescope to study the cosmos over four hundred years ago. They are also designed to capture a certain phenomenon and disregard all others; the data they produce are scrubbed of imperfections, cleaned of irrelevant signals, boosted above the noise. If the aim is scientific investigation, the digital stream is fed into complex computational machinery that squeezes out the answer we are after. If the pictures are for public consumption, they are framed, colorized, and hung in virtual galleries we can visit all over the web.

The personal encounter with the deep sky is another matter altogether.

A HANK OF LIGHT

A few years ago, I organized a conference at a center in the Swiss town of Ascona. Having grown up less than a mile away from the beautiful site, I was aware of its history stretching back to the early twentieth century. It's easy to understand why Monte Verità (Hill of Truth), nestled in peaceful woods, sitting on a hilltop overlooking a subalpine lake, was once chosen by a group of idealists as the place to start a new life based on freedom, vegetarianism, and proximity to nature (which included nudism). The stimulating environment of Monte Verità charmed an eclectic mix of artists and intellectuals, including Carl Gustav Jung, Paul Klee, Rudolf Steiner, Walter Gropius, Marcel Breuer, and Hermann Hesse. Bauhaus artist Xanti Schawinsky captured its bohemian character as "the place where brows reached the sky and bottoms travelled third class."[29]

By 2009, Monte Verità had become a first-class conference center run by ETH Zürich, my alma mater, and I invited fifty colleagues from around the world, astrophysicists and statisticians, to discuss the data-analysis challenges posed by dark matter and dark energy to our understanding of the universe. I felt that the esoteric topic would have pleased the original inhabitants of the site. The invitation to reach the sky was powerfully present, thanks to the relatively dark nights, so I organized a public stargazing session in collaboration with the local astronomical society. With the telescopes set up in the garden almost hanging over the lake, the starry canopy sparkling above us in the cloudless sky, something remarkable happened.

Once the local astronomers had exhausted the list of sights they wanted to show, Chris Genovese, a statistician at Carnegie Mellon, took hold of one of the reflectors and smoothly maneuvered it to point in what appeared a random direction. People bent to peer through the telescope, and a chorus of exclamations rose. In two deft movements, Chris re-aimed the telescope, and more cries of surprise pierced the silence. I pressed my eye to the eyepiece, the first time I had done so since I had risked losing my sight as a student. A delicate nebula jumped out of the trembling darkness, a floating, spidery hank of light that seemed simultaneously within reach and impossibly far away. I knew that the glowing gases that made up the bluish haze I was seeing stretched over unimaginably vast swaths of space, perhaps ten thousand billion kilometers across. Yet it also seemed as if I were peering at some microscopic lifeform. Distance and size were warped out of proportion by the conjuring trick of the telescope. I understood then why many a prelate had simply refused to believe what they saw through Galileo's telescope. It was nothing like the sharp, high-definition, full-color images from the Hubble Space Telescope but rather a diminutive shadow of light fluttering on the edge of perception. I looked at the thing until it crawled out of the field of view, as Monte Verità, the telescope, and I were spun away by Earth in the opposite direction.

A 2015 study of American university students found that on average nonscience majors recognize just two constellations (Orion and Ursa Major are the constellations most frequently identified correctly) and a single star (the North Star and Betelgeuse in Orion are the only two stars recognized at a rate better than random). Surprisingly, having grown up under dark skies makes no statistical difference to students' ability to name stars and constellations: perhaps we have just lost interest.[30] By 2050, nine out of ten Americans will live in urban areas, where a dome of artificial light will cap their nights and banish stars from their sight, and their minds, altogether.

America's most famous evangelist of direct experience of nature, Henry David Thoreau, explored his connection with the sky in his retreat at Walden Pond in the mid-nineteenth century. In his famous meditations Thoreau writes of fishing by moonlight in the "clear and deep green well" of the pond and of how, on dark nights, his thoughts wandered to other spheres as he felt he might cast his line "upward into the air, as well as downwards into this element."[31] Ralph Waldo Emerson, Thoreau's friend and mentor, similarly felt that "the rays that come from these heavenly worlds will separate [man] and vulgar things."[32] If I strain my imagination, I can just about grasp a faint reflection of Thoreau's experience, faded as it reverberates down the five or six generations between us. It's still harder to picture walking through the dark streets of London, as Edwin Dunkin must have done around the same time as Thoreau was fishing at Walden, and looking up to see the Milky Way.

What do we lose by no longer connecting with the stars?

To answer this, we must learn what we've gained from all that neck-craning awe. We must cast our minds back thousands of generations, return to that vanished world when mythical monsters, great heroes, and beautiful princesses populated the night, take a shallow dive into deep time.

But first, we'll visit a world without stars.

THE CALIGO TALES

Cloud-Watcher's Tale

Cloud-Watcher stood, and her singing filled the cave. We all remembered the words, but still our hearts began trembling as if full of Rumble.

> *Daughters and sons!*
> *Who we are, we owe it to the Cloud.*
> *That sends the rain that makes leaves unfurl.*
> *That blows the wind that dries the land.*
> *That gives life to the Disc with the power of Lightning.*
> *How to watch the Cloud is what I Remember.*
> *The Cloud that brings the birds back.*
> *That makes the apples grow.*
> *That makes the nuts fall.*
> *Its song carried onto its breath*
> *from the edge of the Disc*
> *lulls the Bear into his Long Sleep;*
> *its Rumble gives us strength in war;*
> *its Glow keeps the Darkness at bay.*
> *To watch the Cloud is to understand the Cloud.*
> *To watch the Cloud is to feel the Cloud.*
> *To watch the Cloud is to be alive.*
> *When our sleep gets shorter, and the Glow gets stronger,*

that's when the baby deer come into the Disc,
that's when the blue-speckled eggs are sweet onto the tongue
and we crunch tender bones full of marrow, happy.
But when the grass turns brown, and the Bison is gone
when the Cloud's Glow is short and weak
that's when we dig out the bitter root,
and chew it slowly, so it lasts longer.
When Darkness grows, and the Glow abandons us,
we sit around the fire, the stone working the hides,
so we may fight off the cold.
All the Glow long, I watch the Cloud,
I watch it flowing, its shape shifting, its color changing.
I watch the Cloud so it may tell us
when the baby deer will come, and Bear goes to sleep.
I watch the Cloud so I may tell you
"Rain in the mountains" and when the Rumble is heard,
   faintly,
I know then that boulders are rolling on the Roof of the
   Disc,
above the Cloud,
far away, where the mountaintops hold up the Roof.
I watch the Cloud, looking for the nest of Lightning,
the white fire that sometimes shimmers in the Darkness.
I watched the Cloud all Glow long, and this is what I saw:
Rumble is on its way, and it will bring Lightning along.

3

LIFE UNDER A CLOUD

I bind the Sun's throne with a burning zone
And the Moon's with a girdle of pearl;
The volcanos are dim and the stars reel and swim
When the whirlwinds my banner unfurl.
From cape to cape, with a bridge-like shape,
Over a torrent sea,
Sunbeam-proof, I hang like a roof—
The mountains its columns be!

—Percy B. Shelley, "The Cloud"

AMONG THE STANDING STONES

I opened the door and walked out into the inky night.

The gale-force wind almost threw me to the ground. I pulled up the wool-lined hoodie of my storm jacket and took a few more steps down the gravel driveway. The darkness engulfed me. Just a few yards away, the soft, warm light spilling out of the cottage windows seemed to recede much faster than the laws of physics would allow.

I could only just make out the edge of the driveway, beyond which the machair began. The wind screeched in concert with the roar of the unseen North Atlantic Ocean. I scanned the horizon for directions: the undulating moorland was a deeper purple darkness where the night sky ended. As I walked on, hunched against the wind, a disquiet gripped my stomach: the sense of moving shadows, the night bearing down on me. Searching for a presence, I craned my neck backward and scoured the sky—not a light in sight. I felt lost.

I had traveled to the Outer Hebrides to present at the Hebridean Dark Skies Festival, a two-week celebration of the night sky mixing astronomy, music, comedy, food, and art. Located at fifty-eight degrees north, the sparsely populated Isle of Lewis enjoys one of the darkest skies in the United Kingdom, offering the opportunity to admire the Milky Way and even the Andromeda Galaxy with the naked eye. The Northern Lights are a regular occurrence too—when the skies are clear. This is not guaranteed in February in the north of Scotland, and that year Storm Dennis, one of the strongest extratropical cyclones on record, was threatening to spoil the stargazing parties.

Dennis arrived, and it erased the night sky.

Days were only a little better. The cloud cover was so thick that not even the disk of the Sun was visible. The diffuse light cast no shadows, and colors were sucked out of the moorland and lochs. I remembered then the endlessly gray winter days in Geneva, Switzerland, where I had lived as a graduate student. Clouds remained trapped for weeks on end between the Alps to the south and the Jura mountains to the north—a veil that muted my inner spirits too. *Gloomy* can, tellingly, describe equally a cloudy and dull sky, a dismal-looking person, or a somber, poorly lit room. It was as if the lack of sunshine outside insinuated itself into my inner landscape. When sadness and depression regularly clutch one's spirit during the gloomy part of the year, the condition is called seasonal affective disorder (or, fittingly, SAD). We don't know for sure what causes SAD, but it is more prevalent the farther away from the equator people live: SAD is nine times more common in Alaska than in Florida, the "Sunshine State." I wondered what permanent clouds would do to my mental health. Nothing good, was the answer.

One day, as the storm raged on, I visited the prehistoric stone circle of Callanish, or Calanais, as it is called in Gaelic, a monumental arrangement of forty-eight standing stones, or menhirs (Breton for "long stone"), plus a central megalith towering sixteen feet over a burial chamber. An inhabitant of Lewis, John Morisone, wrote in 1680 that, according to legend, the menhirs are "a sort of men converted into stones by an Inchanter."[1] Some are pointed at their tops; some have a crooked appearance, arching up like petrified tentacles of a forgotten prehistoric creature; some resemble squat tombstones, some shark teeth veined with white quartz, some giant phalluses defying the eons. All were erected at great cost to withstand the elements and the injuries of time some five thousand years ago, in a pattern resembling a Celtic cross, with the shorter arm closely aligned in the east-west direction. The menhirs will probably still stand in five thousand more years, long after our steel-and-glass skyscrapers, the glory of the Anthropocene, have been reduced to dust.

Sometimes called "the Stonehenge of the Hebrides," Calanais is, in many ways, more impressive than its world-famous, bigger cousin in the south of England—perhaps more striking at first sight because less familiar. As you approach Callanish from the west, the menhirs are silhouetted against the sky on a ridge, giant gneiss fingers protruding from the Earth and reaching for the sky. The Greek polymath Eratosthenes, the first to measure the radius of the Earth by using the shadow of the Sun, wrote of a "winged temple" of the Hyberboreans, the inhabitants of the Far North of Europe, perhaps in reference to Calanais's east-west rows of stones. An intriguing mention by the Greek historian Diodorus of Sicily in 55 BC hints at Calanais's raison d'être: astronomical observation.

Diodorus writes, "It is also said that in this island [that of the Hyperboreans] the Moon appears very near to the Earth; that the god (Apollo) visits the island once in the course of 19 years, in which period the stars complete their revolutions. . . . During the season of his appearance, the god plays upon the harp, and dances every night from the Vernal Equinox [around March 21] to the rising of the Pleiades [on April 10 in 1750 BC], pleased with his own successes."[2] Diodorus seems to refer to a remarkable phenomenon that an observer standing in the middle of the main row of menhirs can witness every 18.6 years, when the Moon rises and sets at its southernmost points, called "a major standstill." The architects of Calanais, according to astronomers starting with Norman Lockyer in 1909, could have designed and oriented its layout as a cross between an astronomical observatory and a place of worship with this singular spectacle in mind. The full Moon occurring around the summer solstice of a major standstill year rises in the southeast over a range of hills in the distance resembling the outline of a lying woman (in Gaelic called Cailleach na Mointeach, meaning "the old woman of the moors"), then moves westward, just skimming the horizon. The Moon then disappears behind an undulation of the hills, before resurfacing in perspective, strikingly framed by the menhirs. To the assembled astronomer-priests of neolithic times, it would have seemed as if the Moon (Apollo's sister in Greek mythology) had materialized among the stones, perhaps the signal for Apollo to join. As anyone who has witnessed the rise of the full Moon knows well, the disc of our satellite appears huge near the horizon. This supermoon illusion, maintained along all the trajectory as the goddess appears to play hide-and-seek between the menhirs, must have been as striking for our neolithic ancestors as it is for us. And the period of nineteen years mentioned by Diodorus matches well with the actual astronomical cycle of 18.6 years between major standstills, making this highly charged moment a rare sight, experienced maybe once or twice in a lifetime.[3]

I advanced toward the central circle along the main alleyway, a solitary visitor flanked by the menhirs on a stormy day. A lone sheep munched on as I approached, oblivious to me and to my cloud-filled questions. As I reached the central ring of thirteen stones (a reference to the maximum number of full Moons in a solar year? I wondered), I walked around it, not quite ready to enter the innermost part of what felt like a sanctuary. I continued along to the end of the alleyway and looked out over the moors, trying to reconstruct the shape of the lying woman in the distant hills, only to be pushed back by the howling wind at the exposed tip of the outcrop. As I retraced my steps, I paused next to a menhir and put my hand on its veined gneiss. The surface was rough and cold. The image formed in my head of another hand, in another time and another world, poised in a similar gesture: a neolithic man or woman, biologically identical to me in all essential respects, but so unreachably different. What did those stones mean to her? What motives lay behind the monument?

Why did you build this? I screamed the question into the wind, my words snatched away as soon as they left my lips. The depths of time offered no answer.

I stepped inside the circle, my back to the central megalith, elegantly reaching for the billowing clouds, unmoved by the squalls. As I took out my slick smartphone to take a picture, I felt like a mirror image of the opening scene in *2001: A Space Odyssey*. There, a band of prehuman primates is surprised to find a tall, black, smooth megalith, which inspires them to use bones as rudimentary tools, the first step toward technology and supremacy over nature. In February 2020 on the Isle of Lewis, the space megalith had morphed into a handheld device packed with technology indistinguishable from magic. And just like the apes in Stanley Kubrick's movie, I was clueless as to the purpose and meaning of the menhirs of Calanais, as inscrutable to me as the black monolith to the band of monkeys.

I studied the sky, as it picked new hues in a palette of grays. *Sky* is the Old Norse word for "cloud cover," and in many modern Nordic languages the word for "sky" comes from an ancient root meaning "to conceal." In southern European languages, by contrast, a sunnier climate has led to an affinity between the word for "sky" (*cielo* in Italian, for example) and that for light blue, which also means "heavenly" (*celeste* in Italian). What would life be like, I asked myself, in a world where *sky* is a synonym for *cloud*?

To answer this, I had to find the cloudiest place on Earth.

THE SUN-STARVED CITY

I grew up in the almost subtropical climate of southern Switzerland, accustomed to over two thousand hours of sunshine a year, so my move to Oxford and then London came with a touch of SAD: a downbeat mood that got hold of me during summers that felt nothing like the blue-sky affairs my psyche had been conditioned to expect. My initial bemusement about the English obsession about the weather had given way, over the years, to a weary acceptance of the customary small talk. I joined in many an exchange about gray skies, playing along to the tune of "at least it's not raining." But I had never stopped to consider the finer points of cloud classification.

Perhaps unsurprisingly, the systematic classification of clouds is the brainchild of an Englishman, the pharmacist Luke Howard, who became fascinated by the spectacular sunsets caused by the eruption of the Krakatoa volcano in August 1883. The explosion filled the atmosphere with volcanic ash and dust, producing for over a year afterward sunsets with otherworldly bloody-red hues and "sanguinary flushes" that people initially mistook for great fires burning over the horizon. It has been suggested that such displays inspired the dramatic background in Edvard Munch's painting *The Scream* a decade later.[4]

Howard devised a similar classification system for clouds as had been in use for animals and plants, achieving immortality as the "namer of clouds." Today clouds are divided into ten main groups, or genera. Within most genera, the structure of the cloud determines its species, within which variations in shape and transparency mark subspecies. The resulting names are some of the most sensually evocative descriptors of nature: a cumulonimbus (a "heavy and dense cloud, with a considerable vertical extent, in the form of a mountain or huge towers," looking like a towering heap of dense whipped cream floating in the sky) can be *calvus* (Latin for bald) or *capillatus* (hairy—with protuberances with fibrous or striated structure, "a vast, more or less disorderly mass of hair"); it might have mamma dangling from it (udder-like protuberances hanging under the cloud). A cirrus might be *fibratus*, *uncinus*, or *spissatus*. A stratus might be *fractus*, *undulatus*, and *homogenitus* (created by humans, in this case as consequence of airplane contrails). Descriptors are endless, in a Quixotic attempt to systematize what is infinitely variable.

Certain clouds are particularly effective at blotting out the sky. Amid the dreamy shapes of *The International Cloud Atlas*, I learned that the culprit for my gloomy mood as a student in Geneva was stratus nebulosus, a uniform, gray layer capable of drizzle or snow; its equally depressing sidekick was nimbostratus, similarly diffuse but more likely to be spotted in dark gray. Both will erase the blue from the sky and, if sufficiently thick, might hide the disk of the Sun. At night, even thinner clouds can draw an impassable curtain over the feeble light of the stars.

The gray world of nimbostratus is often associated with London, in its alleged permanent fog and drizzle—perhaps owing to the collective memory of the bygone years when the sky was filled with smog, a term blending the words *smoke* and *fog* and coined in 1905. Surprisingly, it turns out that the inhabitants of Miami, Sydney, and Rio de Janeiro (among others) endure more rainy days per year than Londoners. But to find the city that comes closest to losing the sky to clouds, we need to travel to the middle of the North Atlantic, roughly halfway between the tip of Scotland and Iceland.

In the middle of the North Atlantic Ocean, buffeted by the winds, the tiny Faroe Islands sit at sixty-two degrees north, an archipelago of eighteen islands created by volcanic eruptions around fifty-five million years ago. Their dramatic birth is still visible in rocky cliffs plunging vertically into the sea, cragged mountains, basaltic rock formations, and fjords filling ancient volcanos. Today, the Faroe Islands are a self-governing nation of fifty thousand inhabitants, eighty thousand sheep, and one Nobel Prize winner—Niels Ryberg Finsen, whose Nobel Prize in Physiology in 1903 gives the Faroes the distinction of being the country with the most winners per inhabitant, almost twenty times more than the United States.

The Faroese capital, Torshavn, sports the unenviable medal for the city with the least sunshine in the world. Edward Gryspeerdt, a climatologist and cloud expert at Imperial College London to whom I turned for advice in locating the cloudiest place on the face of the planet, cautions that "the least sunshine" does not necessarily mean "the cloudiest": thin white clouds might let sufficient light through to show up as "sunny" on ground-based weather stations, while localized cloud cover would wrongly be recorded as "cloudy." Combing the planet from geostationary orbit, Gryspeerdt identified the cloudiest place on Earth as El Danubio, a region sandwiched between the Pacific Ocean and the Andes in western Colombia, where the moist air coming in from the ocean gets trapped and produces an almost constant cloud cover. With a stunning 98.6 percent cloud cover, El Danubio beats Torshavn easily: satellite images of El Danubio for 2020 show, day after day, an almost uninterrupted blanket of whiteness.[5]

Still, the Faroes certainly have plenty of cloud cover, and a longer history of sky records than El Danubio, thanks to a crystal ball–resembling instrument known as a Campbell-Stokes recorder, which measured the Faroes' sunshine from the late 1800s until 2007.[6] Looking through the sunshine recorder measurements from the Faroese capital, I discovered that December 1942 and January 1967 had no sunshine at all. The clouds might have parted during the long nights, revealing a glimpse of the stars and perhaps of the Northern Lights—the recorder cannot tell. But for weeks, the connection of Faroese people to the sky was stretched to the point of rupture. While I could find no documentation of the psychological impact of this separation, a hint is perhaps offered by the other local claim to fame: Finsen, who won his Nobel Prize for discovering the beneficial effect of "chemical rays from the sun" (i.e., ultraviolet light) in treating skin diseases. Finsen suffered from a disabling condition that pushed him to study what his homeland lacked the most: sunshine.[7]

The weather in Torshavn during those two sunless months is a good blueprint for a world without stars, but there are other, man-made ones. Atmospheric pollution in England's cities during the second half of the nineteenth and first half of the twentieth centuries reached scarcely imaginable levels. Coal-fired manufacturing plants spewed their fumes next door to terraced houses erected to accommodate workers, adding to the noxious smoke from open-flame heating with coal fires banked with potato peelings and coal dust. London was regularly choked by the deadly "pea soup fog," so called because of its yellowish or greenish color and soup-like consistency—a toxic mix of soot and sulfur dioxide that caused widespread respiratory problems and thousands of deaths. American writer Jack London, visiting the English capital in 1903, remarked on the lack of strength, unhealthy pallor, unusual slang, and habit of talking loudly of Londoners—the latter due to their permanently congested sinuses and need to be heard in the smog.[8] The smog covered everything with black soot: not just laundry but animals and trees too. "I thought sheep were gray until we traveled away on holiday and I saw white ones!" confessed a Yorkshire woman.[9] A rare mutation of the white peppered moth that turned its body and wings completely black became a major evolutionary advantage in post–Industrial Revolution England, as black insects were camouflaged perfectly against soot-covered tree trunks. In a few decades, the once rare black variety became dominant, still today a textbook example of Darwinian evolution in action.

During the deadly Great Smog of London of December 1952, a peculiar weather pattern prevented the circulation of air over the city for five days, with the result that the smog concentration became intolerable. People reported not being able to see their hands in front of their faces during the daytime or their feet as they tried to reach home walking, all transport having stopped due to lack of visibility: not even a flashlight held in front of a bus to guide the driver at walking pace could pierce the smog. Cinemas limited seating to the first three rows only—any farther back and the screen disappeared. The pea soup penetrated inside homes through every opening, so much so that people could not see the opposite wall in a room. An eyewitness recounts the nightmare invasion: "I went into the hall and there saw the smog pouring like water through the open letter box. I shut it and coughed. There was a small sea of smog in the hall."[10] Over ten thousand people are thought to have died as a consequence of the Great Smog.

Pictures of London during that fateful December justify the Great Smog's reputation: swirling black clouds plunged everything into ominous darkness; lampposts lit in daytime were barely visible. Even the Sun was blotted out, as were the Moon and stars. London's nightmare receded into the past after the UK government passed the Clean Air Act in 1956 in response to the attendant public health crisis. Unfortunately, many other cities today experience dangerous levels of air pollution, with twelve out of the fifteen most polluted cities being in India, according to the World Health Organization.[11] Human-produced pollution is turning large swaths of our planet into toxic worlds without stars.

LOVE AND SULFURIC ACID

To see nothing but clouds for a time or in a certain place is unsettling and can be debilitating, but I needed to up the challenge if I wanted a realistic blueprint for a world without stars. What about a whole planet covered in clouds?

The closest place to turn to for inspiration was the brightest object (bar the Moon) in the night sky: Venus. A director of the Smithsonian Observatory, astrophysicist Charles Abbot, argued in 1922 that "its high reflecting power seems to show that Venus is largely covered by clouds indicative of abundant moisture, probably at almost identical temperatures to ours."[12] The second planet from the Sun is a close twin of the Earth as far as its size and rocky composition are concerned, so much so that at the beginning of the twentieth century it was believed that it could harbor life under its impenetrable clouds, despite being much closer to the Sun than we are. Some scientists, including Nobel Prize–winning chemist Svante Arrhenius, took speculation one step further by describing the surface of Venus as a hot, tropical jungle on steroids, where everything is "dripping wet."[13] Such a scenario became the ideal setting for Ray Bradbury's haunting 1950 short story "The Long Rain," in which four stranded space travelers are driven to madness by the unrelenting rain as they search in vain for the Sun domes that will mean heat, light, and safety from fast-encroaching fungi and hostile, ocean-dwelling Venusians.[14]

Science fiction, itself a genre that owes its existence to the stars, has taken Venus as the theater of many an interplanetary adventure, and the planet's clouds often feature as one of the defining characteristics of its biosphere. Venus's surface being hidden from the gaze of our telescopes gave authors in the late nineteenth and early twentieth centuries greater imaginative license than Mars, another favorite location. Novels and short stories depict Venus as inhabited by dragons, dinosaurs (molded on Earth's prehistory), sentient plants, living colors, sea-dwelling monsters, winged, tentacled, or blue-skinned humans, angelic creatures, and (in observance of the association of Venus with the goddess of love) beautiful women.

Rarely, the clouds' role in civilization building is explored. Amtorians, as the inhabitants of Venus are called in Edgar Rice Burroughs's *Pirates of Venus*, believe that the stars occasionally glimpsed through a rip in the clouds are sparks from the sea of lava that surrounds what they think of as a disc-like planet.[15] A more intriguing account is found in a curious manuscript published in 1891 under the pseudonym "Antares Skorpios" by the Reverend James Barlow, father of Irish novelist Jane Barlow. Scarred by the painful divorce of his parents and obsessed with the thought of his mother suffering eternal damnation after her suicide, Barrow describes Venusians as godless immortals who contract and terminate marriages by mutual agreement. Since what he calls "Hesperians" have no children, their breakups cannot hurt their youth.

Hesperians puzzle about their Unknown Maker and have no conception of the vastness of the universe above the clouds. The impenetrable cloud screen above their heads means that they are "also debarred from reaching Him through the medium of His works," by which Barlow presumably means the heavenly bodies.[16] With great effort, the Hesperians at last manage to reach the summit of a twenty-mile-high mountain peak, which pierces the clouds and reveals to them the stars, a view of the distant Earth, and the fiery, rising Sun. Unlike with Isaac Asimov's Lagashians, the sight of the stars unleashes not madness in the Hesperians but a deep spiritual crisis. They rapidly work out celestial astronomy, quickly surpassing nineteenth-century Earth knowledge, and their initial awe rots into estrangement. They realize that in such a grand cosmos, they are "no more than insignificant specks in its unfathomed deeps. In the vast profusion of worlds they felt themselves lost. If their Maker had charge of that vast universe, he might well have forgotten them altogether."[17] The unexpected experience of the sublime breaks their spirit, untempered by early sight of the stars.[18]

Had they existed, the Hesperians would be an invaluable counterexample of intelligent life in a world without a view of the sky. But it wasn't to be. At the dawn of the space age, the discovery that the atmosphere of Venus is 96 percent carbon dioxide—a powerful greenhouse gas, as we know all too well today—tilted the odds away from the lusciously wet tropical paradise toward "an arid planetary desert," as Carl Sagan concluded in 1961.[19] The final nail in the coffin of the nonexistent Venusians was the realization that the main layer of clouds consists of sulfuric acid, rendering it millions of times worse than London's infamous pea soup, and that pressures at Venus's surface exceed those at the bottom of Earth's oceans. Any spacecraft sent in reconnaissance under the clouds was quickly dispatched by the lead-melting heat and crushing pressures—our last emissary, a Soviet probe in 1985, lasted less than an hour.[20]

WHEN A CLOUD WON'T DO

I needed to look farther afield, among the several thousand exoplanets discovered since 1992: worlds circling distant stars, offering a stupendous range of possibilities far exceeding anything in our own solar system. Planets with ferocious winds ripping at six thousand miles per hour; planets with rains of molten glass; planets the color of Vantablack, among the darkest substances known (and much loved by artist Anish Kapoor); planets covered in lava with average temperatures that would melt iron. And planets shrouded in pink haze and, in some cases, clouds.

The first confirmation of clouds on an exoplanet came in 2014, when a team led by Laura Kreidberg used the Hubble Space Telescope to analyze the starlight traversing the atmosphere of an exoplanet forty light-years away, as it passed in front of its star. They concluded that the planet is likely covered by a thick layer of clouds at high altitude, albeit not the water cirrus clouds of Earth but possibly clouds made of potassium chloride (the white powder often used in plant fertilizers) and zinc sulfite, a phosphorescent substance.[21] A planet with glow-in-the-dark clouds! This would have delighted Henri Poincaré, who had in jest imagined precisely this possibility to save his Jovians from darkness: "I know well that under this somber vault [of clouds] we should have been deprived of the light of the sun [had we lived on such an overclouded planet], necessary to organisms like those which inhabit the earth. But if you please, we shall assume that these clouds are phosphorescent and emit a soft and constant light. Since we are making hypotheses, another will cost no more."[22] Reality sometimes beats imagination.

While planets without stars might well be lurking in deep space, so far we have no clue as to the existence of life there, let alone intelligent civilizations. But the depths of Earth's oceans and underground habitats like caves offer examples of alien starless environments right under our noses. Life there has evolved clever evolutionary adaptations: the sunless abysses of the oceans are populated by creatures with enormous eyes, bizarre shapes, and bioluminescent bodies, fantastical beings that remain mysterious, as they are difficult to study and do not survive in laboratory conditions. Entirely dark caves are homes to animals that have lost their sight, compensated for by other means of navigating space, like sonar in bats.

A few humans have spent several months in a cave, investigating the bodily and psychological reactions to long-term isolation and sequestration from the natural cycle of sunlight. Unsurprisingly, dwelling in a space with only artificial light interferes with our circadian rhythms. Without clocks, natural or mechanical, the sleep pattern shifts and the perception of the passage of time is warped. Josie Laurel, a midwife who in 1965 volunteered to spend eighty-eight days alone in a cave, said of her time underground, "I knitted, and knitted some more, and looked forward to the time when I would finally see the sun."[23] An Italian sociologist, Maurizio Montalbini, who spent in total over three years of his life by himself in a cave (at one point enduring a whole year underground), was shaken by the experience: "I'm not going back in there. I need the sun. I used to dream about the dawn."[24]

These experiments were primarily concerned with the physiological and psychological impact of isolation in a secluded, constrained space, mimicking the conditions of long-duration space travel. More extreme sensory-deprivation studies also demonstrate that our grasp of reality quickly deteriorates when stimuli to the brain are reduced or eliminated altogether—for example, in total darkness—so much so that this has been employed as a form of torture. Extreme anxiety, paranoia, and hallucinations have been observed after volunteers spent only a few hours by themselves in a soundproof cubicle with minimal sensorial input.

As fascinating and revelatory about the human condition as such studies are, they do not offer much help in picturing a world without stars. The world I was looking for, the one that alone could bring to the fore the role of the stars in the cultural, spiritual, and psychological evolution of humankind, is not one of total darkness—rather, it is one where a single element, the clear view of the heavens, has been selectively removed. As such a world does not exist, at last I resorted to the powers of imagination to make up my own.

Picture a world where stratocumulus undulatus gives way to nimbostratus praecipitatio, chased off by stratus nebulosus, in a perennial incantation of gray hues that never reveal a corner of blue sky, never part to let through starlight, never even show the disc of the Sun or the changing shape of the Moon.

Welcome to Caligo, a world without stars.

WHERE CLOUDS NEVER PART

Caligo (pronounced *ka-lee-go*) is an imaginary alternative Earth, identical to ours in all respects but for a tiny, hugely consequential detail: like

Venus, it is completely shrouded in clouds. Unlike Venus's, Caligo's clouds are harmless water vapor.

Caligo, Latin for fog or mist, is still used in the Triestine dialect with this meaning; in Latin it also stands for darkness and, by extension, for the inability to see matters clearly. On any Caligo day, the cloud blanket glows softly with a diffuse light of unknown origin that moves across the sky, then disappears to the west, to return the next morning from the east. Pitch-black nights like the one I experienced on the Isle of Lewis are regularly and mysteriously graced by a faint gleam, growing stronger and then fading away in a cycle of 29.5 days. Nobody on Caligo has ever seen a star.

As a physicist, I wondered what kind of unscientific assumptions I would have to make to build the Caligo I wanted. Is an actual Caligo even possible, I asked myself? Although the complex physics of cloud formation is not fully understood, the large-scale patterning of clouds on Earth is the consequence of our planet's rotation, the tilt of its axis (which changes the solar irradiation across latitudes), and the shapes of landmasses, oceans, and mountain chains. A similar striped cloud pattern is seen in the atmosphere of Jupiter, only more clearly as the gas giant's rotation is faster than the Earth's. One could more easily obtain a total cloud cover if one supposed the surface of Caligo to be almost entirely oceans, then tweaked the chemistry of the atmosphere to keep the planet habitable despite the clouds reflecting away a larger fraction of sunlight, thus cooling the climate. But on such an aquatic Caligo, land-based primates might never evolve, and if we are to learn something about the influence of the stars on the human condition, we must at the very least start from a world that hosts humans similar to us. From the physics point of view, such a planet is not simple to engineer.

Turning to biology, the cloud screen on Caligo could be expected to modify the trajectory of evolution in many ways. Fortunately, photosynthesis—the basis for the pyramid of life on our planet—would work in the diffuse light of Caligo, as confirmed by a study of the ten-kilometer asteroid hit that wiped out the dinosaurs sixty-six million years ago. In one scenario, Earth was enveloped by a layer of soot spewed into the upper atmosphere by the massive wildfires caused by the impact. Shrouded in darkness, the planet cooled by sixteen to eighteen degrees

Celsius for a decade, while sunlight reaching the surface was reduced by about 80 percent for up to two years—approximately the same effect of very dark storm clouds.[25] Earth immediately after the dinosaurs was a pretty good model for Caligo.

Even under such dim lighting, photosynthesis remained effective, except in deep waters. On Caligo, we could expect broadleaf trees and oceanic algae to survive better, as their low-energy photosynthesis is more efficient, while warmth-loving grasses of the kind found in the great plains of North America would struggle. This would have direct implications for humans, as the staple foods of our diet come from the grass family, like maize, rice, and sorghum. But, of course, evolutionary adaptations of plants to low-light conditions would create a novel ecosystem with entirely new and hard-to-imagine characteristics. If certain cyanobacteria can survive exposure to space for over a year, withstanding vacuum, radiation, and extreme temperatures, we can be sure that evolution would find a bespoke solution to the problem of dim light on Caligo.

An animal kingdom shaped by evolution in an environment without direct sunlight might borrow and expand on adaptations we see in nocturnal animals, cave dwellers, and ocean-floor inhabitants on Earth: larger eyes to make the most of the fainter light, photoluminescence to chase away darkness, sensitivity to other wavelengths of light, exquisite hearing, exceptional smell and tactile perception, echolocation, and navigation by geomagnetism. The same evolutionary pressures would redirect the primate line that, on Earth, eventually led to us, selecting generation after generation for creatures that are supremely adapted to living under the clouds.

The inhabitants of Caligo might not suffer as we do when deprived of sunlight, as this would be a natural condition of their existence. If they did have our biology, apart from SAD, another consequence of life under permanent clouds would be a lack of vitamin D, an essential hormone produced by the skin when exposed to ultraviolet light and also found in foods like egg yolks and cod-liver oil. A lack of vitamin D prevents the absorption of calcium and phosphorus, thus weakening bones and hazarding a serious deficiency disease known as rickets, particularly in children. In 1890, Scottish medic Theobald Adrian Palm observed that rickets was prevalent in coal-mining districts of the country with soot-filled air and lacking in sunshine. As we saw earlier, the fumes of Industrial Revolution England conjured up an extremely unhealthy version of Caligo; on a planet naturally shadowed, humans would probably have evolved another way of getting their fix of vitamin D—or else gone extinct.

It is not even granted that we would exist at all on Caligo, under any recognizable guise. Biologist and evolutionist Stephen J. Gould pointed out that evolution is not a linear progression toward increased complexity of lifeforms and that we suffer from a "parochial focus" on ourselves as the self-proclaimed pinnacle of a process that started with unicellular life in a murky pond some four billion years ago. Contingency—fortuitous and unpredictable events, such as the dinosaur-killing asteroid strike 66 million years ago or the likely cometary impact that is believed to have contributed to the megafauna extinction of 12,800 years ago—plays as much a role in determining winners and losers in the lottery of life as random genetic variation and selection of the fittest do.[26] According to Gould, we must "entertain the strong possibility that *H. sapiens* is but a tiny, late-arising twig on life's enormously arborescent bush—a small bud that would almost surely not appear a second time if we could replant the bush from seed and let it grow again."[27] If *H. sapiens* is indeed unlikely to reappear even in identical evolutionary conditions, we must conclude that our species wouldn't exist—certainly not in a shape that we would recognize—if the bush of life were seeded under the altered skies of Caligo: clouds would influence evolution sufficiently to prune entire branches and supercharge twigs into boughs never seen on Earth.

Physics, chemistry, and biology together screamed that my thought experiment could not be performed in nature. Stretching clouds all around Earth would change the initial conditions of life so much that the history of our world would skid out of control too quickly for humans to arise. I had no choice but to follow Poincaré: since we are making hypotheses, another (or two or three) will cost no more. I shall therefore—unscientifically—pretend that Caligo's geological history, prehistory, biochemistry, and evolution are identical to those on Earth up until the point when our story begins—that is, fifty thousand years ago. Physics is not unfamiliar with thought experiments, often used as thinking devices to help grasp the salient properties of a system. Mathematicians even use counterfactuals to prove theorems: one starts by assuming that a certain statement is false and then shows that this leads to a logical contradiction—hence the initial statement must be true.

The stage is thus set for my thought experiment, and it is mostly a familiar one, featuring all the actors from our own prehistory, except for the clouds that are about to change everything. As we follow humankind's path from prehistory to today, Caligoans will depict one among infinite alternatives. While we explore how the stars helped us to build the modern world and, just as importantly, our understanding of who we are, the "Caligo Tales" present a fictional counterpoint, not as an actual counter-example to our own trajectory but as a provocation.

On that bleak February day in Calanais, I took refuge in the car, shivering—and not only because of the wind. I looked at the clouds once more. I looked at the standing stones: their enduring arrangement rested, I realized, on a delicate beam of starlight making it to the ground. I started then to appreciate how the flower of *Homo sapiens*, one bud on the unpredictable bush of life, had bloomed in the night—opened by the light of the stars.

THE CALIGO TALES

Freshwater's Tale

Cloud-Watcher had returned to her place in the circle, and Shepherd had risen once again to pick the next Rememberer, when a shadow emerged from the darkness. We all got to our feet, blades at the ready, but we soon recognized the figure that staggered into the light of the fire.

Beehive and I rushed to help Freshwater, but she didn't pay us any attention. She didn't seem to be hurt: she just stood there, her arms limp, her eyes gazing into the flames, seeing nothing. Beehive and I exchanged a look.

Shepherd, too, understood at once what had happened. He took Freshwater by the shoulders and gently led her to the middle of the circle. In the voice of the Cloud, he spoke to her thus:

"Freshwater, we welcome you back from the Dark! Join the circle of Remembering, and tell us—what did you see? Tell us so we may remember it!"

I knew what Shepherd was trying to do: to use the power of the Cloud to claw her back from the Great Blackness. But Freshwater seemed already too far gone, her body in front of us as empty of life as a bison's skin. So I was astonished when she spoke. It was only a whisper, a slow monotone barely audible above the hissing of the fire.

Thus spoke Freshwater:

"I salute you, Shepherd. Two glows ago I left the cave, following the footprints of aurochs into the woods. I searched all glow long for the spring that I knew the aurochs drink from—our stream is almost dry these days. When I reached it, the glow had almost gone and I knew then that I would not find my way back. Where the spring is, I no longer can tell you."

She paused, as if trying to remember something important. Then she continued.

"There is no point. With or without the spring, the Dark is inside us all . . . There is no point. Only darkness and no point. I can no longer . . . can't . . . I . . ."

She trailed off, then turned and walked out of the circle of light.

I moved to stop her, but Shepherd's hand on my shoulder broke my impulse. As I had suspected, the Great Blackness had taken Freshwater, like it had taken Skinner and Spear-Maker last bat awakening, and many others before. When the Great Blackness enters somebody, there is no saving them. Sooner or later, they wander off, never to be seen again, or, as with Skinner, they are found at the bottom of the Drop, their eyes eaten by the crow.

We turned back to the circle, and Shepherd commanded again, "Once-Upon-A-Glow, you have been told about the greatest dangers to our people! Remember the Battle with the Tricksters!"

Once-Upon-A-Glow stood, and thus began his story.

4

THE WEIGHT OF STARLIGHT

Pronaque cum spectent animalia cetera terram,
os homini sublime dedit caelumque videre
iussit et erectos ad sidera tollere vultus.

While other animals look downwards at the
ground, he [Prometheus] gave human beings
an upturned aspect, commanding them to
look towards the skies, and, upright, raise
their face to the stars.

—OVID, *Metamorphoses*

A PALEOLITHIC FACE-OFF

The Human Evolution section in the East Wing of London's Natural History Museum was, as usual, sparsely populated. The torrent of visitors and children pouring in through the Exhibition Road entrance forks into two streams, one toward Hintze Hall and its suspended blue whale skeleton, the other to the second-floor volcanoes via an escalator plunging through a reconstruction of the Earth's interior bathed in an infernal red gloom. It is easy to overlook the room where tongueless, eyeless skulls are neatly arranged in a pyramid: at the base, *Homo habilis*, small and monkey-like; alongside it, *Homo erectus* and *Homo rudolfensis* with its forward-facing cheekbones; higher up and later on, *Homo heidelbergensis* and its cousins; then, *Homo neanderthalensis* with its huge cranial capacity and prominent brows. At the top, grinning back without irony, *Homo sapiens*, its smooth, egg-shaped dome and empty orbs seen in many a production of *Hamlet*, classroom anatomy displays, and Halloween decorations.

But I have come to stand face-to-face with more than just skulls. A noisy class of kids in purple and black uniforms and orange high-viz jackets breezes through, giggling at the full-frontal nudity of the Neanderthal man I am there to meet. Once the kids' screams have faded into the distance, I find myself alone with the lifelike model that in my head I have begun to address as "Fred." Fred has an intensity about him, a presence that the wax reproductions of celebrity and historical figures in Madame Toussauds lack entirely. Perhaps it's the aura of a lineage that stretches back hundreds of thousands of years or the faint sorrow that also surrounds the dodo taxidermy a few rooms down. Much of it is in the pose: Fred is standing, naked, his left leg slightly bent, like a prehistoric David, his hands clasped behind his back, in the stance of a primitive philosopher. His bare, muscular chest and shoulders are tattooed with V-shaped parallel lines the color of burned-up meteorites, their pointy ends meeting between his pectorals. Thick bundles of muscles bulge from his short, powerful neck, partially hidden by a bushy beard and thick hair. As an average-height Sapiens, I am a little taller than he, despite the low plinth he stands on.

It seems as if Fred is contemplating me as intensely as I'm contemplating him, his strong brow creased with deep lines. I turn to follow his gaze, and that's when it dawns on me: across the narrow passage, another display case holds a second model, an older male, slimmer, taller, and of slenderer build: one of our own ancestors—a Sapiens. His creators, the Dutch twin paleoartists Adrie and Alfons Kennis, have recreated in their prehistoric diorama the face-off that played out fifty thousand years ago. The two models stare at each other in challenge, like primitive cowboys ready to draw their guns in a fight they know all too well only one of them will survive. Except, when Sapiens and Neanderthals shared the Earth, the only weapons they could rely on were stone darts and points and the ideas that formed, silently, in the space between their ears.

As I turn to consider Fred and his befuddled look one final time, I wonder how I have found myself outside the glass case rather than in it. The pyramidal arrangement of the skulls, with our species at the vertex, betrays the underlying assumption that Sapiens represent the apex of a progressive improvement, with earlier, less accomplished attempts ruthlessly snuffed out by evolution. But perhaps matters aren't so straightforward, our supremacy not as guaranteed. As archaeologist Rebecca Wragg Sykes writes, "Evolution didn't follow an arrow-straight Hominin Highway leading to ourselves."[1] We may be the last-standing representative of the genus *Homo*, but the skeletons now arranged in a neat (and misleading) evolutionary arrow pointing to ourselves were once serious contenders, and in some cases direct competitors, for the crown of master of the Earth. We are not even so different from some of our now-extinct cousins as we'd like to think: genetic analyses have shown that modern Europeans and Asians have between 1 and 2 percent Neanderthal genetic material, while Australian Aboriginals' DNA contains 3 to 6 percent of DNA from Denisovans, hominins that lived in Asia between two hundred thousand and fifty thousand years ago.[2]

Among our long-lost relatives, Fred's kind has especially fascinated us ever since Neanderthal 1 was discovered in the Feldhofer Cave of the Neander Valley in Germany in 1856 and recognized as the first fossil hominin species in 1864. Neanderthals were not, as it is still often believed, a bridge between us and apes, from which our common lineage diverged over six million years ago; recent discoveries in paleoanthropology show that they embodied an alternative way of being human, with mental capabilities similar to our own and superior physical strength. They were "state-of-the-art humans, just of a different sort," in the words of Wragg Sykes.[3] That we carry inside us traces of their DNA means that our ancestors met and, in some cases at least, mated with them.

Neanderthals had been at home over a vast territory, spanning Europe and the Middle East, for hundreds of thousands of years before our own ancestors, modern Sapiens, migrated out of Africa around seventy thousand years ago. By the time they met Sapiens, Neanderthals had overcome several ice ages, ebbing south with the rising glacial tide, then flowing back north when the temperature rose again. Their flat, large noses may have been an adaptation to the cold air, warming it as they breathed. They were peerless hunters, naturally endowed with a bodybuilder's muscles supported by a skeleton thicker than ours that could better withstand injuries; Olympic-caliber spear throwers, they had sharper than 20/20 vision and eyes 30 percent larger than ours, giving them improved night vision. Analysis of the genome of fossil microbes found in Neanderthals' teeth suggests that far from being exclusively meat eaters, they had a starch-rich diet, including perhaps wild barley porridge—ingestion of cooked tubers and plants might have helped support the growth of their brains, larger than ours, as early as six hundred thousand years ago.[4] They were confident nomads controlling vast territories from which they plucked not just food but also stones for tools, wood for burning, hides for warmth, medicinal plants for health, and pigments, shells, and raptor claws for adorning their bodies. They likely buried their dead.

Fifty thousand years ago, the dispassionate punter might have given Neanderthals the edge over the last kids on the block of hominid life, Sapiens. Both Neanderthals and *Homo sapiens* are equipped with oversized brains for their bodies, a precision grip for handling tools, and stereoscopic vision; both are masters of fire, builders of tools, skilled creators of objects, and keen devisers of marks that we would today call art. But the old guard were better athletes—better, seemingly, at the very mechanics of survival—with several ice ages under their collective belt.

And yet, forty thousand years ago they all but disappeared, leaving only well-hidden traces of their existence deep inside our cells and in bones at the bottom of caves.

What, then, condemned Fred and all of his kind to surviving only in museums, while we underdogs stand alone? True, during the ice age leading up to the Neanderthals' extinction, the climate had become bitterly cold as the ice cap slid south—winter temperatures reaching minus sixteen degrees centigrade in what would become London—but this was nothing that the Neanderthals hadn't experienced before. Perhaps the rapid succession of climate swings between forty-five thousand and forty thousand years ago altered the availability of food resources too quickly for Neanderthals to adapt, by which time ruthless competitors had trotted out of Africa. But any berries or aurochs the Sapiens snatched from under the large nose of the Neanderthals weren't secured by brute force alone: a Neanderthal's grip could crush a human hand. Our ancestors certainly didn't club the Neanderthals to death.

If it wasn't strength, then perhaps it was brains. With Neanderthals and Sapiens competing for scarce resources in an unpredictable environment, the tiniest novelty could have tipped the balance in our favor. Take the humble eyed sewing needle: fashioned out of bones (of birds and, in some cases, Przewalski's horses), it first appeared in Asia around forty thousand to forty-five thousand years ago, a Sapiens technological innovation. Such an apparently insignificant implement made a massive difference when Sapiens found themselves facing the recurrent advances of glaciers and fighting off longer and longer winter seasons. The chances of surviving in a world battered by snowstorms and covered in ice improved when they could wrap up in well-fitted furs and hides. The eyed needle pierced a path through multiple layers to stitch garments better able to fight windchill; to fashion out of hides and horse sinews hardy and durable shoes and bags for carrying tools, infants, and food; to sew together tents for shelter. The needle also launched prehistoric fashion: purely decorative embroidery and adornments meant new ways for establishing and signaling status withing a group.[5]

We know about prehistoric needles because they are made of durable material that survived the injury of time. But Sapiens may also have employed to their advantage a kind of knowledge that would have left no fossil record: that of the sky.

During the dramatic climate shifts that defined the time of the Neanderthals' demise, both their and Sapiens' ingenuity was tested to the extreme. To reach a grove of nutritious wild berries ahead of other animals (often equipped with better sensory organs) at the right time of the year could make all the difference between a baby nursed toward adulthood and a pile of little bones buried in a cave. To locate the right kind of stone, wood, bones, or shells meant crafting complex tools, from deadly flint-pointed spears to hand axes. To find water, shelter, and firewood while traveling was to avoid being mauled to death by predators. As seasons shifted within a single lifetime, those who knew how to be at the right place at the right time had an edge. And it wasn't only about material survival: social networks appear to have been stronger and wider among Sapiens than Neanderthals, judging by how far stone tools were being carried from their place of origin. Gatherings of bands of Sapiens required means of communication and coordination: to agree on, transmit, and remember a common location and time to meet. Passing knowledge and experience down the generations in a society without writing meant devising mnemonic devices that could be accessed by anybody at any time and that could be relied on in perpetuity, beyond the span of a single lifetime.

In meeting all these challenges, the rising of specific stars, the light of the Moon, the imagined shape of constellations in the sky, and the annual cycle of the Sun offered potentially crucial assistance to those who could decipher, remember, and predict them.

MEAT FROM THE MOON

What the prehistoric mind of Sapiens apprehended in the majestic beauty of the star-studded night, we will never know. The artifacts and scant remains that do exist, even when scrutinized at the atomic level with the powerful tools of modern science, cannot tell us of a hand raised to shield one's eyes against the glare of the setting Sun, looking for the first slice of the crescent Moon; they cannot reproduce a chant greeting the reappearance of the Pleiades. And yet, there are clues.

Perhaps the best place to start investigating how an intimate knowledge of the sky could have helped prehistoric Sapiens is with our satellite. The Moon circles the Earth every 27.3 days with respect to the distant stars, its appearance changing as the Sun's light illuminates it from different angles: it's dark when it's new and aligned with the Sun on the dayside of the Earth; it's round when it's fully lit, when it rises precisely as the Sun sets. As the Moon circles the Earth, our planet rushes along its yearlong orbit around the Sun, so that when the Moon has completed a full circle around the Earth with respect to the stars, after 27.3 days, it still needs to glide a little longer to line up with the Sun in the next new Moon phase. This is why the period between new Moons is longer and averages 29.5 days.

As used to artificial light as we are, we may not appreciate the dramatic difference moonlight made in a world otherwise almost entirely dark.

Even though the full Moon is four hundred thousand times fainter than the Sun, its cool glow would transform a landscape of menacing shadows into navigable terrain for Sapiens equipped with poorer night vision than many of their prey, predators, and competitors—including Neanderthals. The cool of night would also be preferable for traversing areas that would be scorching during the day, like the Sahara Desert—nomad populations have practiced this until very recent times, only traveling when the Moon's light is favorable. And because a full Moon doesn't occur every night, Sapiens would have had to learn how to keep track of its cycle, to plan ahead for their movements and their most fruitful hunts.

When man became a hunter, the moonlit night offered the most favorable conditions. The bigger the prey, the more difficult it was to slay on the spot. Much easier and less dangerous to wound it, then stalk it until it fell exhausted—something our ancestors excelled at, thanks to stamina that would allow them (and us) to outrun, over a long distance, even a healthy horse, and something that Neanderthals, with the higher energy requirements of their more powerful build, may have been less good at. What better moment for such a long-duration hunt than the end of the day, when the light of the setting Sun allowed for a clear sighting of the prey while the cooler part of the day beckoned? Should the chase spill over into the night, moonlight would help the excited hunters defy the lengthening shadows and the danger that their coveted mammoth, deer, or bison would collapse in a dark crevice, unseen and lost to its pursuers. The G/wi Bushmen of the Kalahari still follow the ritual of throwing bones from a slain animal toward the first sliver of the new Moon as it appears, asking for her help in the hunt: "There are bones of meat, show us tomorrow to see well that we do not wander and become lost. Let us be fat every day!"[6]

With nothing less than their survival at stake, the best hunters would not have failed to notice what most of us never have: the difference between the waxing and the waning Moon. The waxing Moon, with its cradle-like crescent oriented toward the Sun in the west, hangs from the sky right after the Sun has set, its faint gleam holding darkness at bay. Night after night, the growing crescent strengthens its light, and each successive sunset finds the Moon higher up in the sky. The hunters' searchlight would grow brighter and last for longer after the Sun had gone; if need be, the hunters' pursuit could stretch into the small hours of the night as their prey's strength dwindled. The full Moon marks the apex of this cycle, with its brightest light lasting the whole night and being most favorable to the hunt. In a version of the foundational myth of Egyptian culture, as recounted by the Greek historian Plutarch, it is precisely while out hunting "on the night of the full Moon" that Seth discovers the body of his brother Osiris, whom he had murdered.

But after the full Moon, the cosmic table would turn on the hunters: during the waning phase, the weakening Moon does not rise until sometime after sunset, lagging fifty minutes behind each night. The unwary hunter who hadn't laid hands on his prey yet and the hapless traveler who hadn't found shelter before sunset would find themselves caught out by total darkness before a diminished Moon could come to their rescue. The propitious times were, until the Moon was new again, over.[7]

A higher degree of safety from predators, swifter nighttime passage, and a better chance of procuring a nutritious meal would be the rewards of those Sapiens who knew how to read the Moon's cycle. At the beginning, Sapiens' abilities may have been entirely analogous to the navigation and orienting skills enjoyed by other animals. The spectacular migration of monarch butterflies relies on the Sun as a compass, while at least one species of moth has been found to orient using the Moon and the stars. Several nocturnal migratory songbirds use the center of the stellar rotation as a compass, and the humble dung beetle keeps the helm straight while transporting a dung ball away from its source with the help of moonlight, starlight, and even—in one species—the orientation of the Milky Way. Obviously the dung beetle doesn't have the faintest clue that the glow of the Milky Way is produced by hundreds of billions of distant suns. But through evolution's trial and error, dung beetles that "learned" to use the Milky Way as an arrow in the sky were able to transport their dung balls farther and with less effort than their less sky-savvy competitors. This would have meant greater food resources, better survival probability, greater strength and reproductive fitness, and thus a larger number of offspring. Within a few generations, the astronomer dung beetles took over the species.

Like the Milky Way–savvy dung beetle, early unconscious Sapiens astronomers would have thrived at the expenses of those who did not know the sky—including, perhaps, Neanderthals. Our obsession with the stars thus likely began with a little help here, a small advantage there. As with the whales that feel their way at the bottom of the ocean thanks to the invisible magnetic lines emanating from the Earth's iron core; the dung beetle that rolls its smelly parcel along tracks painted by the Milky Way; the starling that knows to keep the North Star to its rear as it travels to warmer climates at the end of the summer—that may have been how it started, an intuition that mysteriously provided larger, more frequent meals, kept dangers at bay, and generation after generation increased Sapiens' numbers compared with our less sky-savvy cousins, the Neanderthals. But crucially, and differently from the dung beetle, the knowledge about seasons, Moon phases, cardinal directions, and star guidance that could do so much to improve chances of surviving and increasing the number of one's offspring was passed down not through our genes but via a much swifter, more flexible, and more powerful means to transmit information: language.

LUNAR TALLIES

How language arose remains shrouded in mystery: some argue that it was enabled by a sudden neurological or genetic modification that kicked in between 70,000 and 50,000 years ago, when Sapiens fanned out of Africa; others posit that it was a gradual process that started as early as 150,000 years ago. Archaeologist Francesco d'Errico has suggested that cultural innovation rewired the brain without the need for any underlying genetic and morphological changes.[8] In this view, culture was the driving force of recent human evolution rather than a by-product of it. What seems certain is that complex language was both the prerequisite for and the consequence of the symbolic social organization of early humans, characterized by abstract thinking, artistic expression, bodily ornamentation, music, and cooperation in hunting large game. Part of the answer might lie in the social function of language and the close cooperation that it enabled: those individuals who could better grasp and work the dynamics of their social group would be preferred as sexual partners, leading to rapid evolutionary strengthening of cognitive abilities—a process that the anthropologist Brian Hare has dubbed "survival of the friendliest."[9]

In the process of bootstrapping culture from nothing, stars might have projected a sort of soft power onto those with the cognitive ability to wield it. Political leaders from the pharaohs to Louis XIV sought to justify the divine origin of their earthly power with the stars. Something similar might have been at work in the earliest forms of prehistoric human societies. Those who could infallibly lead their tribe to shelter by following the stars, or reliably provide meat during a full Moon, or tell when the days would begin to lengthen again would assume higher status within the group and thus rise to become tribe leaders. Literally, the stars would have been the first kingmaker.

In the view of anthropologist Chris Knight, the entire early social organization of Sapiens might have revolved around the lunar cycle: not only did it likely regulate hunting behavior, as we saw, but it also paced sexual and reproductive practices and mandated ritual feasts to coincide with the full Moon, when the hunting cycle came to its climax, perhaps in coincidence with women's fertility.[10] Knight sought to explain the mysterious, exact coincidence of the Moon's phases with the average length of women's menstrual cycle—a unique occurrence among primates (with the possible exception of gorillas and orangutans, who have cycles of around thirty days). And there is controversial evidence that the cycle of the Moon was carefully followed by Sapiens already thirty-five thousand to forty thousand years ago.

In the 1960s, journalist and science writer Alexander Marshack caused a stir in the archaeological community with a new explanation for markings seen on thousands of Paleolithic artifacts: lunar notation. Marshack believed that markings on etched pebbles, cavern wall paintings, and engraved reindeer bones and mammoth tusks, some as old as thirty-five thousand years, weren't just decorative doodles, as others had surmised, but bore evidence of sophisticated tracking of the lunar cycle, detailing in some cases the phases of the Moon, the passing of seasons, and even the location of the Moon's setting over the horizon.[11] This was an explosive hypothesis: How could prehistoric people possibly master counting and even arithmetic when they had no writing and an immature capacity for abstraction?[12]

Prime among the much-debated specimens collected by Marshack, a baboon right fibula (one of the two bones connecting knee and ankle) discovered in 1973 in the Lebombo Mountains, at the border between South Africa and the Kingdom of Eswatini, is remarkable for being carved with a sequence of twenty-nine regularly spaced notches and for its heavily polished surface, suggestive of continuous, longtime handling. The series of notches is interrupted by a break in the bone, so we cannot know whether the tally grew larger originally. A total of twenty-nine or thirty notches could have represented a full lunar cycle (whose exact length, 29.54 days, falls between twenty-nine and thirty)—or a complete menstrual cycle. The fibula has been dated to forty-three thousand years ago: around the time the last Neanderthal bowed off the stage of life, Sapiens might have been tallying the revolutions of celestial bodies. We can't know whether the Lebombo bone was indeed one of the first lunar or menstrual calendars, but a microscopic analysis reveals that the notches were carved in four sets, each with a different tool and at different times—suggesting that its creator was interested in adding and storing numerical information on it. Once finished, the bone could be consulted in the dark by feeling the notches, a protracted fondling of a cherished and perhaps shared device that would explain its smooth surface.[13]

The Lebombo bone is unique, but hundreds of prehistoric "Venus figurines" found throughout Europe, from Britain to Italy, from Spain to Ukraine, bear witness to a connection between lunar phases, the feminine cycle, and humankind's cultural development. The Venus figurines are carved statues or bas-reliefs representing naked women with large breasts, ponderous hips, and full, pregnant bellies, a celebration of fertility and the life-giving power of women, named for the Greek goddess of love.[14]

The most intriguing among them is the Venus of Laussel, a forty-six-centimeter-tall figurine carved into the limestone of a cave in the Dordogne region, near Bordeaux, France. Her left hand rests just above her belly button, perhaps caressing her unborn child, perhaps pointing toward her genitals. Her right hand holds up a bison horn, its pointed end facing upward. Her face is missing, but from the position of her head, turned to look over her right shoulder, it appears that she is contemplating the horn. If we follow her twenty-five-thousand-year-old invitation to look at it closely, we can discern thirteen regularly spaced vertical marks on the horn. The number is significant: there are usually twelve full Moons in a solar year, approximately one per month, but since the lunar cycle takes 29.5 days, the twelve Moon cycles add up to 354 days, 11.25 days short of a full solar year. Therefore, every two and a half years, a thirteenth full Moon will slot into a solar year—a so-called blue Moon.[15] Keeping track of the number of full Moons is crucial to sync a lunar-based calendar with the solar year. Thirteen is also the number of nights elapsing between the reappearance of the crescent Moon after the new Moon phase and the moment when it seems full (which happens a day or so before the actual full Moon, as it's difficult to distinguish a nearly full Moon from an actual full Moon).

The carving still bears traces of its original ochre paint: the same red as menstrual blood, in keeping with the figurine's exaggerated sexual traits and perhaps referring to the connection between menstruation and the lunar cycle. If Chris Knight is right, the Venus of Laussel can be considered to symbolize the link between the menstrual cycle, the cosmic cycle (the crescent Moon represented by the bison horn), and mathematical abstraction (the number thirteen linking lunar cycles and biological ones), and perhaps the seed of what would become a full lunar calendar. As we saw above, the full Moon brought prehistoric hunters a banquet rich in meat, and the attending celebration might have coincided with peak fertility for the women of the group. The cosmic cycle of the Moon therefore set the rhythm around which early prehistoric societies organized themselves: not just hunting and traveling but also ceremonies, lovemaking, and reproduction were paced by the rhythm of lunar phases.[16]

The number of days between full Moons—uncannily similar to the average fertility cycle—may thus have been among the very first tallies humankind kept, one of the greatest importance not only for individual survival but for the entire social organization of prehistoric life. Since women's fertility cycle was, individually or in a group, either symbolically or literally reflected by the Moon's appearance in the sky, it is plausible that it was women who kept tabs on the Moon and therefore were responsible for the first lunar calendars.

"Women were the first observers of the basic periodicity of nature, the periodicity upon which all later scientific observations were made," wrote social philosopher William Thompson, highlighting the importance of the lunar cycle not just as a basis for calendars but as a prototype for an inquisitive approach to the natural world.[17] Thompson's suggestion rings true when we consider the first named author in history, the Akkadian princess Enheduanna, daughter of Sargon the Great. In the twenty-third century BC, Enheduanna wrote poems that were copied over and over for thousands of years, and she was celebrated as a minor deity herself for centuries after her death. Her name reflects her role as the earthly spouse of the Moon god, Nanna—a male deity—to whom she was conjoined in a ritual sacred marriage: *en* for "high priestess," *hedu* for "ornament," and *anna* for "of heavens." She lived in the Sumerian city of Ur, in modern-day Iraq, where it was her duty to "measure the heavens above" and declare each new month commenced when the Moon's first waxing crescent was spotted above the western horizon.[18] In classical antiquity, the Romans worshipped Diana as the goddess of the hunt, of fertility, conception, and childbirth, and of the Moon. Her name may have come from *dius*, meaning "daylight": at full Moon, when the night was almost turned to day, the goddess propitiated the good hunt, and the ensuing celebrations ended with men and women congregating.

Whether or not Chris Knight's and Alexander Marshack's hypotheses are correct, it is suggestive that to this day, the same ancient root *me-*, meaning "to measure," links together in a faint echo of long-lost associations the words *month, meal, menstruation*—and *moon*.[19]

SONGLINES AND PEEPHOLES

A group of men crouching in tense ambush under the full Moon; an elder standing on a ridge awaiting the rising of the Pleiades before dawn; a chant guiding a band's steps across a pathless tundra under the stars—these do not leave fossils. The few that have survived the dusty march of the eons— skulls, fragments of food in fossilized teeth, the etches laboriously made in bones, the magnificent paintings adorning caves such as Lascaux, the

Venus figurines with gloriously large, fecund bellies—are but a shadow of the lived world of our forgotten ancestors.

In looking further for hints of evidence about the helpfulness of the sky, let us leave paleoanthropology and turn to ethnography. While modern hunter-gatherers are separated from prehistoric peoples by as many eons as the rest of us, they have in some way overcome similar challenges under similar conditions to our ancestors, acquiring complex wisdom from the heavens in doing so. This does not mean that Paleolithic Sapiens apprehended the sky in the same manner, but the study of the importance of stars for hunter-gatherers who lived in historical times is at least suggestive of the role stars might have played for our ancestors so long ago.

Among the best sources of ethnographic evidence are First Nation Aboriginal people, nomadic hunter-gatherers who, until the eighteenth century, had lived in almost complete cultural isolation in Australia since their ancestors colonized it around forty-five thousand years ago. From 70,000 BC, early Sapiens spread from central Africa through the Middle East, then followed the coast of India and China and crossed over from Papua New Guinea to reach the continent of Sahul, which then united in a single landmass (due to reduced sea levels) today's Australia, New Guinea, and Tasmania. Insofar as is known, no Neanderthal ever set foot in Australia.[20]

Aboriginal culture continued unbroken until the British invasion in 1788, which decimated many Aboriginal communities through famine, disease, and in some cases genocide: the Aboriginal population crashed from about three hundred thousand in 1788 to just ninety-three thousand in 1900. However, some groups have been able to maintain their traditions, language, and culture to the present day. Only recently has their ancient knowledge been "discovered" by ethnographic researchers, having been largely dismissed and overlooked for centuries. The Aboriginal people offer many striking examples of how an intimate knowledge of the sky was central—and helpful—to their traditional way of life.

The extent and detail of Aboriginal star lore is staggering: some elders can name nearly all of the approximately three thousand stars visible to the naked eye and know legends and stories associated with each. Star knowledge is considered sacred and is imparted to boys by the elders as part of their initiation into manhood. One of the most important features in Aboriginal astronomy, shared by many groups, is Gawarrgay, the "Emu in the Sky," not a constellation in the Western sense but two dark dust clouds obscuring the Milky Way, associated by their shape with the eponymous large, flightless bird native to Australia. "When the old people saw this shape [the emu] in the sky they knew that the emus were laying their eggs, so the old people would go out into the bush to find the eggs for food," recounted an Aboriginal participant in an ethnographic study.[21] When the head of the emu appears in February, the Kamilaroi people leave their summer camp; when the legs become visible in April, it's time for them to reach their winter camp. The arrival of certain stars just before dawn also heralds the change of seasons, announcing the imminent flocks of migratory birds, the flowering or fruiting of plants, or the availability of eggs and nuts. The sighting of the Pleiades before dawn tells the Pitjantjatjara of the Central Desert that the dingo-breeding season has begun, bringing the opportunity to feast on their pups and perform fertility rituals.[22]

In orienteering, too, the signposts in the sky played a significant role, especially when the men have been out all night hunting: Bill Yidumduma Harney, senior Wardaman elder, explained that "each night where we were going to travel back to the camp . . . the only tell was about a star. How to travel? Follow the star along."[23] When crossing vast distances, the starry sky was studied in preparation for the journey: both the Euahlayi and the Wardaman peoples memorized the route by associating individual stars with specific markers along the way—a river crossing, a waterhole, a bend in the road, a stone arrangement, a marked tree. The imagined path connecting the stars represented the road the traveler would tread on the ground, but only loosely so, more as an aide-mémoire than a star map in the Western sense. Each journey had its own songline, which unfolded along the track connecting the wayfinding points among them, a ritual set of directions that created a chanted, living bridge between land and sky. Some songlines—also called "Dreaming tracks," as they relate back to the Dreamtime, the ancestral creation of the world by the primal beings in Aboriginal culture—stretched across the entire continent, even changing language according to the territory being traversed. The eagle hawk songline of the Euahlayi people, for example, ran from Alice Springs, in the middle of Australia, to Byron Bay on the East Coast, thousands of kilometers away. European settlers followed the First People's paths, and roads were laid along songlines; that's where settlements sprang up, which grew into towns and cities. Some of today's highways, including Victoria Highway in the north and the Great Western Highway, are said to trace the Dreaming tracks of Aboriginal people. When zooming along in air-conditioned comfort across Australia, we should remember the ancient star lore that, even today, guides our path.

At the opposite end of the globe, in the unending whiteness of the Arctic Circle, stars also played an important—and often overlooked—part in traditional Inuit culture, despite surprisingly impoverished stargazing conditions. At the Inuit hamlet of Igloolik (meaning "there is a house here"), located at seventy degrees north and inhabited for thousands of years, the Sun never rises above the horizon between the end of November and mid-January. But in the long winter night, even in the absence of clouds, stars are often obscured by blowing snow, ice fog, moonlight, the aurora borealis, and their own light reflected on the ice. In spring and summer the never-setting Sun and its attending twilight mask the stars from mid-April to the end of August, so their movements can only be followed for three months a year. Given the extreme latitude, many heavenly sights familiar to those living in more temperate and equatorial regions remain all but invisible to the Inuit: for example, the Milky Way and the constellations of Scorpius and Sagittarius are never fully observable. For many hunters in the extreme North, the pole star is too close to the zenith to be a useful directional marker. The Inuit of northwestern Greenland don't even have a name for it.[24]

All in all, the Inuit knowledge of the starry heavens is greatly diminished compared with that of Australian Aboriginals, who enjoy much more favorable stargazing conditions (not to mention less frigid nighttime temperatures). The Inuit recognize only thirty-three individual stars, compared with three thousand, of which just six or seven have individual names: two star clusters (the Hyades and the Pleiades), the Orion nebula, and the Milky Way. Even in conditions that begin to resemble those of Caligo, the Inuit still make use of individual stars and constellations as wayfinding aids, complementing their remarkable knowledge of local landmarks, snowdrifts, and other orienting features that would completely escape a less expert eye. The Igloolik elder Noah Piugattuk explained that, on long trips "using one star for direction we would travel for some time . . . then we would start to move to the left of the star" to compensate for the movement of the stars in the sky over time.[25]

Another essential use of the stars, common to both Aboriginals and Inuit, was as timekeepers: "We didn't have a watch in those days. We always followed the star for the watch . . . Emu, Crocodile, Cat Fish, Eagle Hawk, and all in the sky in one of the stars," said Bill Harney, remembering his youth in northern Australia in the 1940s.[26] As Harney was nighttime hunting during the Australian summer with the stars above as a watch, the Inuit of Igloolik also relied on the stars to announce the time to rise or go to bed during the long, sunless Arctic winter. Igloos had dedicated peepholes for monitoring the stars: Tukturjuit (the Big Dipper), tracing the hours around Polaris, was seen as representing the shape of a caribou, and when it appeared to stand on its hind legs, its head rearing high, midnight was approaching. And beside the stars, the Moon had a universally central role in keeping track of time throughout the year, as we shall investigate further below.

THE LOST SISTER

While almost all modern hunter-gatherers have some kind of astronomical system—including the Inuit living in the nearly starless Arctic Circle— accounts of contemporary practices can be at most only suggestive of what prehistoric Sapiens might have seen in the heavens. Culture and knowledge have undoubtedly evolved and changed among hunter-gatherers in the course of time, and so have the heavens themselves. As we have seen, the shapes of constellations are deformed over the course of several thousands of years by the proper motion of stars, while the precession of the equinoxes knocks stars and their accompanying seasons out of sync. This means that, whatever knowledge of the sky proto-Australian Sapiens might have possessed forty thousand years ago, it was necessarily different in content from any oral tradition passed down to us.[27]

But given that sky lore plays such a central role for contemporary hunter-gatherers, it stands to reason that it was also prominent in the lives of our prehistoric ancestors, who might have used the stars to their advantage as wayfinding aids, timekeepers, and seasonal markers—perhaps even as inspiration for the first myths. The valuable (and often secret) star lore required to find one's way home from floating ice packs or a long trip in the outback had to be gathered, piece by piece, generation by generation, and passed down in oral form for thousands and thousands of years before the invention of writing. Legends, myths, stories, and songs might have been just the mnemonic device needed to encode the wealth of knowledge about stars, seasons, planets, the Moon, the Sun, and their complex relationship with all that prehistoric Sapiens needed to survive. Inuit elders today maintain that "stars [can] be remembered by the legends associated with them."[28]

There are indications that Sapiens' fascination with the stars stretches back as far as our gaze can pierce the pit of history. Aboriginal arrangements of stones are found throughout Australia, some rivaling in size the megalithic structures of northern Europe, and are thought to have had, at least in some cases, a ceremonial function. In particular, linear stone arrangements in New South Wales appear to be preferentially aligned with the cardinal directions, reflecting perhaps the interest of their builders in the sky. Other sacred sites in southeastern Australia, known as *bora*, were used for male initiatory ceremonies and consisted of two circles of stones, one smaller than the other, connected by a path. The stone circles are thought to mirror the two black clouds depicting the Emu in the Sky: since it is the male emu that raises the chicks, the bird was a fitting symbol for male initiation, when the elders turned boys into men. The beginning of ceremonial time was marked by the Emu in the Sky "coming down to drink," which is the time of the year when its head reaches the horizon.[29] In this case, the stars gave the clue for when the ceremony ought to take place; the stars guided attendees over hundreds of kilometers of trackless land to the ceremonial place; and the stars provided a heavenly sign for the ritual, thus bridging sky and Earth, time and space, reality and symbol.[30]

Back in Europe, a hint that Sapiens took notice of the sky as early as thirty thousand years ago is offered by the orientation of caves and shelters in the Dordogne's Vézère Valley in France, a complex that includes the world-famous Lascaux and the Venus of Laussel caves. The entrances of undecorated caves have random orientations, while those featuring cave art are more frequently oriented toward the direction of the rising Sun at winter solstice and the setting Sun at summer solstice.[31] It seems as if the prehistoric dwellers of those caves picked out the sites for their most expressive art in accordance with special points in the path of the Sun in the sky, thus anticipating the sophisticated alignments of Stonehenge and some Egyptian pyramids by thousands of years.

And a singular, tantalizing clue might just take us all the way back to one hundred thousand years ago. It is one of the most spectacular sights in the night sky, a tightly knit group of blue stars that shine like a box of jewels, resting lightly on the shoulder of the Bull, the constellation Taurus, where they "cluster like golden bees upon thy [the Bull's] mane," according to nineteenth-century American poet Bayard Taylor.[32]

The Pleiades are not particularly bright, but their closeness makes them unmissable; indeed, people all over the world throughout history have recognized them and often given them special significance. Their appearance at dawn opened the sailing season among the Greeks (that's probably how they got their name, derived from the verb *to sail*) and heralded the approaching harvest for the Romans; their rising in the winter sky signaled a time of scarcity and hunger for those living in the Northern Hemisphere, while their disappearance at the end of April brought forth the rainy season—and with it, lean weeks without game—for the Barasana people of Colombia. Known to Hindus as the Krittikās, they may have originated Diwali, the Festival of Lights.[33] The Pleiades appear in numerous works of literature and poetry, from the Bible to Milton, from Pliny to Shakespeare, from Homer to Tennyson, who much admired them:

> *Many a night I saw the Pleiads, rising thro' the*
> *mellow shade,*
> *Glitter like a swarm of fire-flies tangled in a silver*
> *braid.*[34]

In China, the Pleiades were first recorded over four thousand years ago; in Japan they have been known since ancient times as Mutsuraboshi, meaning "six stars," or, more recently, Subaru, meaning "gather together" (and serving as namesake and logo for the carmaker Subaru, born out of the merger of six companies).[35] It has even been claimed that the Pleiades might be represented in one of the prehistoric paintings of Lascaux, dating to between ten thousand and seventeen thousand years ago. Next to a magnificent depiction of a bull, and in a spatial relationship suggestive of the relative location of the Pleiades with respect to the constellation Taurus, an arrangement of six black dots is conspicuous.[36]

Their fame is hardly surprising, given how stunning they appear in the night sky. The mystery is, however, why in many cultures worldwide the Pleiades are described as *seven* sisters, or seven women, when the unaided human eye can normally only perceive six stars in the cluster. "The Sister Stars that once were seven / Mourn for their missing mate in Heaven," wrote English poet Alfred Austin.[37] It appears that in historical times and before the invention of the telescope, most people only ever saw six stars. In the third century BC, the Greek poet Aratus explained, "Seven are they in the songs of men, albeit only six are visible to the eyes. Yet not a star, I ween, has perished from the sky unmarked since the earliest memory of man, but even so the tale is told."[38]

Numerous legends describe how and why the missing Pleiade was lost: she withdrew in sorrow, was banned or saved by settling down as the seventh star of the Big Bear (which is similar in shape to the Pleiades, hence often associated with them), was struck by lightning, was abducted, faded away while ascending to the sky singing, hid her face in shame for falling in love with a mortal, fell back to Earth to become the Great Mosque, or was transformed into a comet.

Among those legends, uncanny parallels appear between the story of the seven daughters of Atlas and Pleione told in ancient Greece and that of the seven young girls recounted in countless versions among Aboriginal people throughout Australia. According to the Greek myth, when Zeus forced Atlas to carry the heavens on his shoulders for eternity as punishment for waging war against the Olympian gods, his seven daughters were left defenseless against the unwanted advances of Orion, the mighty hunter. Zeus, out of compassion for their father, turned them into stars and placed them in the sky, where Orion still pursues them.

At the other side of the planet, Aboriginal peoples—who have had no contact with classical Greece—passed down versions of the same essential story: the Pleiades are seven sisters or young women who are chased by a man or men in Orion, who desire them against their wishes. In some cases, the sisters are protected by their dingoes and shielded by a helper (sometimes an elder sister) identified with the horns of Taurus.[39] Variations of this basic myth have been documented across Australia, and at least until the 1970s they were still being enacted as part of the Seven Sisters Ceremony, an initiation ritual performed by Andagarinja women. During the ritual, the female ceremonial leader acts the part of Njuru (Orion), who captures and rapes one of the sisters. She subsequently dies of shame from the incestuous nature of the act, while the remaining six flee to the sky.[40]

If no star has "perished unmarked since the earliest memory of man," where do these myths come from? If we dare to take a bold leap, these parallel tales might be whispering to us from the depths of time: sprouted from a root first imagined when Sapiens walked, stalked prey, and lay in the darkness of the African savannah one hundred thousand years ago—when, astronomical evidence suggests, the seventh sister actually went lost.

Today, the seventh star is too close to one of the other six to be distinguished. In the very best observing conditions, those with exceptional eyesight may see more stars, but almost never seven—if a seventh is seen, so are a few more.[41] A modest binocular reveals a dozen more stars bathed in a blue-glowing gas cloud (Galileo saw thirty-six with his telescope), a fraction of the hundreds of members of this spectacular cluster. But one hundred thousand years ago, due to the relative motion between the two stars, the seventh star was three times as far away from its neighbor, which would have put it at sufficient distance to be clearly distinguishable by a human with average eyesight. It is conceivable, as Ray, Cilla, and Barnaby Norris have suggested, that the origin of the Seven Sisters myth predates the great migration that took Sapiens out of Africa and eventually to Europe, Australia, and beyond.[42] If so, the story of the lost sister might encode the long-vanished memory of our common ancestors' fascination for the stars. "The world-maker gave human beings an upturned aspect, commanding them to look towards the skies, and, upright, raise their face to the stars," sang Ovid—and it may well have been so from the very moment Sapiens first craned their necks toward the heavens.

THE SKY-SAVVY SAPIENS

Our early survival, cognition, language, and culture were all tended by the stars, perhaps decisively so. Two hundred million years ago, "the weight of a petal . . . changed the face of the world," as anthropologist Loren Eiseley has put it, when the angiosperms covered the Earth with flowers, fruit, grass, and herds of animals that would one day feed the calorie-hungry brains of Sapiens.[43] Fifty thousand years ago, the weight of starlight lifted the human mind upward.

It is plausible that the night sky was for our ancestors both a cognitive gym and a mnemonic device. A cognitive gym, as the mental stretch required to grasp the abstract notion that a group of stars could be a representation of something else, say a hunter pursuing seven sisters, or that their appearance would announce another event, without any evident causal connection, elevated our ancestors' thought from the material (a dark cloud is coming; rain will follow) to the symbolic (the Moon reappears regularly; we may one day be similarly reborn). Perhaps the need to keep track of celestial cycles—particularly the Moon phases, as they were so central as markers both of natural phenomena and of the burgeoning social life of the tribe—even sparked that most useful of ideas: numbers.

It certainly was the basis for the first recorded calendars, as we saw earlier. After the observations of lunar phases provided a shared measure of time, most cultures devised their own peculiar way of coming to terms with the 11.25-day discrepancy between twelve Moon cycles and a solar year and thus harmonizing their lunar calendar with the recurring seasons, which follow the Sun. The Babylonian, Egyptian, Jewish, Chinese, and Greek calendars were all lunar, and so was the ancient Roman calendar. Politicians colluded with priests to insert intercalary days (to bring the lunar calendar back into step with the solar year) whenever it was convenient, a practice that Julius Caesar nipped with his reform of 45 BC, which fixed the civil year to 365 days, plus a leap year every

four. Our modern twelve-month calendar still reflects the Roman arrangement, and even the names of the months from September through
December are inherited from their Roman designation (*septem*, meaning
"seven," as it was the seventh month in the Roman calendar, and so on
until *decem*, or "ten").[44]

The Egyptians played a game similar to Snakes and Ladders, called
senet, where a board was divided into thirty squares that followed the lunar
phases. The prize at the end was resurrection from the dead.[45] In Meso-
america, the time-obsessed Maya came up with a complex calendrical system that could determine heavenly cycles with the equivalent of five-digit
precision. Among contemporary hunter-gatherers, Australian Aboriginals
use digging sticks carved with notches representing their age in lunar
months, while the Inuit calendar, featuring up to thirteen named Moons,
was the regulator of virtually all aspects of that people's traditional way of
life before modernity set in.[46] Many religious festivals today remain tied
to the lunar calendar in all the major monotheistic religions: the Jewish
holiday calendar follows lunar months (with adjustments to avoid it getting too out of sync with the solar year), and so does that of the Muslims,
for whom the period of Ramadan starts and ends with the sighting of the
first lunar crescent of the ninth month. Christian Easter for the Western
denominations falls on the first Sunday after the first full Moon following
the spring equinox, while Eastern Orthodox Christians apply the same
formula but follow the old Julian calendar, according to which the first day
of spring falls on April 3.

The night sky has also long served as a mnemonic device, like the
Aboriginal songlines. Once a name, a story, a song, or the association with
an important natural event became embedded in a star or constellation, its
persistence beyond the life of the individual required that it be passed on
to new generations—and culture was born. I imagine prehistoric Sapiens,
from Greece to Australia, gathered around the fire, just like we do while
camping today, awaiting the Pleiades' rise so they could tell each other and
their children the story of how the seventh sister got lost from the blue
necklace of stars.

Why, then, did Fred the Neanderthal end up inside the glass cage at
the Natural History Museum and I, a Sapiens, outside? Time was on their

side: Neanderthals had a head start of three to four hundred thousand years to develop language, after which (if our experience is a valid blueprint) astronomy, religion, mathematics, and, in time, space technology and artificial intelligence could follow at an ever-increasing pace. Yet they failed to do so. While archaeologist Steven Mithen's characterization of Neanderthals' time on Earth as "a million years of technical monotony" is an exaggeration, it is true that their basic technology saw no breakthroughs, only incremental improvement.[47]

Taken as individuals, the evidence suggests, Neanderthals were in no way inferior to us, debunking their depiction by prominent anthropologists in the 1930s as "an earlier and inferior type" of human.[48] But they tended to live in more isolated groups, with fewer exchanges between bands, which meant that any technical innovation or advantageous piece of knowledge spread more slowly, if at all, to other groups. Smaller overall numbers of Neanderthals also meant smaller potential for innovation and discoveries by individuals. For hundreds of thousands of years, this was no obstacle to their survival. But when the fateful climate swings of forty-five thousand years ago upped the challenge, a new, more social kind of hominid came on the scene, and unlike Neanderthals, they saw potential in a secret weapon hidden in plain view right above their heads. The stars not only helped Sapiens bands in foraging and hunting, as we saw, but likely played a role in gathering groups together for ceremonial purposes—during which time tools, stories, and knowledge would be traded. This superior networking and exchange capability might well have been the crucial advantage of Sapiens over Neanderthals when the going got tough.

Here, too, the practices of Australian Aboriginal people are suggestive. The night sky provided not merely the guidance needed to travel to the ceremonial gathering place but also the key to synchronize the meeting, so that various tribes gathered in the same location at the same time. Among Australian Aboriginals, the phases of the Moon were used to time ceremonies, sometimes months or even years ahead of time. The invite took the shape of a message stick: a painted piece of bark, a carved stick, or an animal bone, sometimes with some of the sender's hair tied to it as a means to confirm identity (a practice strangely shared by traditional

Norse tribes). The content of the message was represented in pictograms: in a message stick from the nineteenth century, one side of a piece of quinin tree shows the location and a stylized picture of the sender and the route taken by the messenger across various rivers. An empty *C* shape indicates that the message was dispatched at the new Moon. On the reverse side, the invitee is shown together with the sender at the location of the meeting, at the intersection of two rivers. A full Moon indicates when the meeting will take place.[49]

A few more hints add to the picture: insofar as is known, Neanderthals never ventured outside Europe and the Middle East, while Sapiens migrated all the way from Africa to Australia, crossing the open waters separating Sahul from Papua New Guinea. There are indications that the colonization of Australia by proto-Aboriginals was intentional, as multiple sites appear to have been populated within the short time span of just a couple of millennia, something that the "pregnant woman adrift on a raft" theory of fortuitous arrival cannot explain.[50] Were those superior exploratory capabilities in some measure tied to the stars? It's impossible to know but difficult to deny—navigation on the open sea being crucially dependent on stars, as we shall see in Chapter 6.

It is true that generally Neanderthals lived at more northerly latitudes than Sapiens, at least before Sapiens migrated north into Europe, where stargazing conditions were less favorable. Could this have contributed to a lack of interest in the stars? The counterexample of the Inuit tells us that even in the difficult conditions of the Arctic, star lore remains important, if diminished in extent and usefulness with respect to what peoples living farther south enjoy. Perhaps the clear sky and balmy nights of the African savannah gave our Sapiens ancestors an advantage over Neanderthals living in less favorable climates—and they took this heavenly knowledge with them out of Africa, as the Seven Sisters hypothesis postulates.

From the far side of tens of millennia, we have yet to find a single material trace of Neanderthal interest in the stars.[51] The Sapiens' archaeological record is scant enough, to be sure, and the future might bring new findings about Fred's ancestors—but there are hints. Perhaps some individual Neanderthals gazed upward, but whatever they saw in the stars seems to have died with them.

Not with us. As we leave the depths of the Paleolithic to tread onto the more solid ground of recorded history, we will discover the role the stars played in helping us locate ourselves in both space and time, two inextricably linked concepts that are addressed in the next two chapters on clocks and navigation.

THE CALIGO TALES

Once-Upon-A-Glow's Tale

Once-Upon-A-Glow stood, and thus began his story.

Many glows ago—so many no one can remember—our people were fashioned out of the Cloud and the Lightning gave them the gift of life. Back then, game was aplenty, fruits bountiful, and mushrooms carpeted the forest's floor.

But one glow, our people found themselves in the middle of Tricksters. There were only a few at first. They had flying blades that could kill an auroch from far away; they got to the berries before our people could touch them; they ate our mushrooms and took our caves. Everywhere our people went, the Tricksters had gotten there before them: embers left smoking in a shelter, bones picked clean, flakes of rocks on the ground.

Our people fought them, but the Tricksters were quick. They had fire, and let it run wild among the dry grass to keep our people from following them. They stole the white fox's fur and wrapped themselves in it to escape far away, among the snow. They ran from place to place, carrying their tricks and their babies on their shoulders. We chased and chased them; yet they kept coming back.

For many bat awakenings the fight raged on. The ice tongues came and went, as many times as fit on the fingers of two hands. When the ice came down from the mountain, the Tricksters melted away. But they always returned, and our people went hungry.

Then, one glow, the woman we now call Trickster-Basher told our people how to stem the tide: they had to wait in the Darkness, until the Tricksters showed up.

Our people lay among the bushes, sleeping during the glow and clutching their axes during the Darkness, eyes wide and feet silent. The Great Blackness got hold of many of our heroes during the long wait, and yet Trickster-Basher didn't allow them to return to the fires in the cave. At long last, the Tricksters walked among the bushes, eating our mushrooms.

A terrible Rumble reached the Cloud as Trickster-Basher jumped out from behind the bushes and smashed the first Trickster's head. His fellows tried to fight back, but their flying stones were no use to them then, their babies on their shoulders a burden in combat. When it was over, the forest floor was red with Trickster blood, and the hyena gnawed at their white bones.

Trickster-Basher led our people back, the head of her first quarry held high up on a pole, his pouch full of tricks in her hand. The whole forest admired our strength. The Cloud was pleased: the Dark was kept away by Lightning so strong no one had ever seen anything like it before; nor have they since. Our people danced then in the rain and Lightning until the next glow.

And this is why, when the Lightning comes, we welcome it with the dance that Trickster-Basher taught us so many glows ago. We dance, and we show our strength, so the Tricksters may never come back.

5

CELESTIAL CLOCKS

*You must run through the tunnel faster than
the wind. You have just twelve hours.*

—Epic of Gilgamesh

ON A CLOCK FACE

"And what about now?" I asked my son, Benjamin, turning with my finger the small hand until it was midway between the four and five, then pointing the big hand down toward the number six, written large and green on the face of the wooden clock. A cloud of concentration came over his face. I could almost hear the synapses in his brain crackling as new connections were established, information retrieved, and a pattern reinforced. Then he brightened and exclaimed, "Half past four!"

It was a triumph for a four-year-old. I told him he was right and was about to move the hands of the clock for another go, when he struck me with one of those "why" questions grown-ups have mostly forgotten how to ask: "Daddy, why are there twelve hours on a clock?"

I paused. With the clear eyes of childhood, he had spotted a thread running deep into history.

"You see, Ben, it's all because of the stars . . ." I started, taking him into my lap.

Once upon a time, in the ancient city of Uruk, there lived a mighty but cruel king, who cared for nobody but himself. His name was Gilgamesh. Gilgamesh was incredibly strong and a terrific hunter, and the gods sent a wild man, called Enkidu, to fight him. But after they got into a terrible brawl, Gilgamesh and Enkidu became the best of friends and decided to share adventures together. One day, the goddess Ishtar saw Gilgamesh and Enkidu kill the monster Humbaba, the terrible guardian of the Cedar Forest, and immediately fell in love with the mighty Gilgamesh. But Gilgamesh didn't want anything to do with Ishtar, who, enraged by his refusal, convinced the gods to kill Enkidu as punishment.

Half mad with grief, Gilgamesh went on a journey to find immortality, looking for his ancestor Utnapishitim, who had been granted eternal life by the gods after the Great Flood. Gilgamesh crossed the Twin Peaks to find the entrance to the tunnel that traverses the Earth, through which the people of Uruk believed the Sun plunges when it sets. At the other side of the tunnel lies the garden of the gods, full of trees loaded with jewels and coral clusters hanging like dates. But getting there wouldn't be easy, warned the scorpion-man who guards the entrance:

"Ever downward through the deep darkness the tunnel leads. All will be pitch black before and behind you, all will be pitch black on either side. You must run through the tunnel faster than the wind. You have just twelve hours. If you don't emerge from the tunnel before the sun sets and enters, you will find no refuge from its deadly fire."[1]

As the sun rose, Gilgamesh plunged into the tunnel and ran in darkness. He ran and ran: at the eighth hour, fear overwhelmed him; at the ninth hour, a breeze caressed his face; at the twelfth hour, he emerged into the garden of the gods just as the sun was peeking through the entrance behind him, barely avoiding being burnt to a crisp. Gilgamesh faced many more scary adventures, but in the end he returned to Uruk as a wiser man, even though he failed to conquer death.

But did you notice the instructions given by the scorpion-man? Gilgamesh had only twelve hours between sunrise and the moment when the sun would traverse the tunnel to the underside of the Earth and burn him to a crisp. Twelve hours.

TWELVE DIAMONDS IN THE SKY

The Epic of Gilgamesh is a Sumerian poem whose roots can be traced back to 2700 BC. Its protagonist was perhaps inspired by the actual historic figure of the King of Uruk. It is the oldest epic known and a source of inspiration for many other heroic sagas (and stories for curious sons). Why twelve appears in the epic as the number of hours of nighttime, we don't know for sure. And the poem is silent on whether the same subdivision applied to daylight time.

We do know that the Sumerians and later the Babylonians had highly sophisticated mathematics: their number system was sexagesimal—that is, based on the number sixty (whereas ours is based on the number ten, which feels perhaps more natural as it is literally at your fingertips). Babylonians invented the place value—namely, the idea that a symbol's value depends on where it is placed in the representation of a number. In our decimal system the representation "11" stands for the number $1 \times 10 + 1$, where the left digit stands for a group of ten and the right digit for a single unit. The same representation in the Babylonian system would stand for the number $1 \times 60 + 1 = 61$. Babylonians knew that the diagonal of a square of unit length is equal to the square root of two some thirteen centuries before Pythagoras enunciated his general theorem; they could solve linear, quadratic, and some cubic equations and used a form of integral calculus to compute the position of Jupiter in the sky.[2]

The number sixty as a base was handy because it could be divided by many numbers (two, three, four, five, six, ten, twelve, fifteen, twenty, thirty), and this helped in computing fractions. The Sumerian fondness for fractions is embodied in Gilgamesh, who, as the son of a goddess and a mortal king, is not simply half divine and half mortal but rather "two-thirds god and one-third human," though how this peculiar division was arrived at is never explained.[3]

Maybe the twelve night hours reflected the usual number of lunar cycles in a year, or perhaps twelve was picked as one of the divisors of sixty, the numeric symbol of the father of all gods, the sky god Anu. The Moon god Nanna was linked to the number thirty, a reference to the length of the lunar cycle in days: at every new Moon, Nanna sat in judgment of the cases the gods brought him. The Sumerian calendar was dictated by Nanna, with each month announced by the sighting of the waxing Moon by dedicated priests and priestesses such as Enheduanna, with consequences for all aspects of society. For example, wages were paid after a lunar month, a practice continuing to this day. An intercalary month was added when the priests deemed it necessary to keep the lunar and solar calendars in sync. The Babylonians, who inherited the Sumerian calendar, introduced the *uš* (pronounced "oosh"), the length of time the sky takes to turn by one degree, corresponding to four of our minutes. One complete revolution took 360 *uš*—the division of the circle into 360 degrees that we still use today. Much like ancient Babylonians, astronomers still employ time units to measure distance on the celestial sphere: the hour angle (fifteen degrees), the arcminute (one-sixtieth of a degree), and the arcsecond (one-sixtieth of an arcminute).

By 2100 BC, the Egyptians had adopted the same subdivision for the hours between sunrise and sunset. The symmetry between night and day, between the upper world where the Sun shone and the netherworld that it visited each night, led to two periods of twelve hours each, with the whole twenty-four-hour cycle called "the double hours." Time was measured by the stars at night and by the Sun during the day; water clocks with twelve-segment scales took over when the sky was overcast. As the Sun traveled across the sky, the twelve slices of daytime were ticked off by the moving shadow of a vertical stick, called a gnomon (from the Greek for "indicator"): a sundial.

The earliest known sundial stems from the thirteenth century BC and was found in the workmen's area in the Valley of the Kings: it consists of a flat piece of limestone, the size of a salad plate, with a hole to hold the gnomon, which would have cast its shadow onto a half circle painted around it. The half circle is divided into twelve sections, like the slices of a cake, of about fifteen degrees each. Midway between the sides of each slice, a dot was used to indicate half hours. Its discoverers have suggested that the sundial might have been a portable clock, by which workmen could establish the time of their midday break, while the half-hour segments were perhaps used to delimit more strenuous activities.[4]

In the meantime, with the Egyptians the twelve-hour subdivision of the night had gained considerable sophistication—perhaps the only matter in which Egyptian astronomical lore rivaled the Babylonians'. The Egyptians worshipped Sirius, the brightest star in the sky, as connected to the goddess Isis. The new year began in Egypt when Sirius became visible just before dawn on the eastern horizon, because the arrival of Isis announced the flooding of the Nile, Isis's tears for the murder of her husband, Osiris. The flooding filled canals and basins with irrigation water and deposited fertile silt onto the land, both indispensable for agriculture.

Night after night, as the Earth traveled farther in its orbit (unbeknownst to the Egyptians), Sirius would rise four minutes earlier and move farther west, until it plunged below the western horizon just behind the Sun. For the next seventy days, Isis remained invisible: she had died and was cleansing herself in the Embalming House of the netherworld of Duat. In time she would be reborn from the east to raise the fertile power of the Nile once again.[5] As Isis/Sirius was found farther and farther away from the eastern horizon when the Sun rose each morning, a new star would take its place at dawn. A coterie of thirty-five stars, each about ten degrees apart, each rising ten days after its predecessor, lined up in a silent procession behind their leader, Sirius. Spaced regularly on the sky vault like diamonds around the face of a clock, the thirty-six decans (a name stemming from the ten-day interval between them: "It happens that one dies and another lives every ten days") ticked away the night.[6] Because of the twilight at sunset and dawn, it was convenient to use only twelve of the eighteen decans that could theoretically be seen, and the twelve-hour clock was born.

The procession of the decans was beautifully painted as a diagonal diagram inside the coffin lids of the pharaohs in the Valley of the Kings near Luxor, a personal clock for the illustrious dead in the afterlife. Performing rituals at the right time of night required interrogating the star clock, an important task that warranted the creation of a dedicated profession: hour watcher. The job description of these early astronomers, carved on a statue celebrating one of them, included some duties that we would recognize and some that have largely disappeared: "knowing the time of the rising and setting of stars, especially Sothis (Sirius), the progress of the sun towards north or south, the proper length of the hours of daytime and night, and the proper performance of rituals, as well as charms against scorpions."[7]

The Egyptian hours varied in length with the duration of daylight through the seasons. At summer solstice, one "hour" lasted sixty-nine minutes in Luxor, while at winter solstice it lasted just fifty-one minutes. It wasn't until the second century BC that the great Greek astronomer Hipparchus invented "equinoctial hours" (i.e., of equal length) "for determining with accuracy the hour of the night and for understanding the times of lunar eclipses and many other subjects contemplated in astronomy."[8]

The Greeks applied their mastery of geometry to the sundial, not limiting themselves to flat horizontal and vertical ones, but also inventing conical, cylindrical, spherical, and even roofed sundials, which told the time not by a shadow but by a beam of light piercing through a carefully devised hole in the roof. They also combined the Egyptian hours with the Babylonian number system based on sixty subdivisions, which astronomer, geographer, and mathematician Claudius Ptolemy found so superior for it saved him from "the embarrassment of Egyptian fractions."[9] In the second century AD, Ptolemy enshrined the Babylonian approach in his great treatise the *Almagest*, in which he summarized all that was known about astronomy up to his time. The *Almagest* contained a catalog of over one thousand stars, whose position above the horizon Ptolemy gave in hours, perhaps following Hipparchus—a system still used by astronomers today. Our units for measuring time, subdividing an hour into sixty minutes and a minute into sixty seconds, are another remnant of the ancient Sumerian infatuation with the number sixty.[10]

THE HEAVENLY CLOCK

The day's twenty-four divisions stemming from Sirius and its retinue of decans, stretching back all the way to King Gilgamesh's time, tumble down through the millennia to land on the face of one of the oldest clocks in existence—and the oldest working machine of its kind.

As they tottered from one market stall to the next, haggling over the price of sweet-scented figs and pomegranates the size of cannon balls, the Italian matrons filling Piazza dei Signori in Padua never glanced at the tower surmounting the archway on the east side of the square. They streamed past me, their shopping bags occasionally bumping my legs, as I looked up, transfixed.

Twenty-four Roman numerals are arranged around the dial that dominates the fifteenth-century clock tower of the Capitanio Palace. The clock face is older than the tower and designed by a doctor and astrologer named Jacopo de' Dondi, who, on his own tombstone, reminds us, "My art was medicine, and to know the sky and the stars, whither I now proceed, released from the prison of my body. . . . Yet, indeed, dear reader, know that it is my invention that, from afar, shows at the top of the lofty tower the time, and the changing hours which you count."[11]

Jacopo was rightfully proud of his brainchild, a "most ingenious clock, which in addition to striking and indicating the hours, indicated the days of the month, the course of the sun through the twelve signs of the zodiac, the days of the moon, the aspects of the same with the sun, and its phase," as an admiring citizen wrote in the seventeenth century.[12] Jacopo's clock was considered such a marvel when it was unveiled that thereafter his family was called *de gli Horologi*, "of the Clock."

On that mild November afternoon, as I stood amid market stalls, I could imagine why the citizens of medieval Padua were so enraptured. The series of twenty-four numerals, carved in the stone, starts with the number one positioned at around four o'clock on the outermost, fixed ring: that's when the first hour started, half an hour after sunset, the time for the evening Hail Mary prayer. Moving toward the center of the clockface, a second ring depicts the sphere of the fixed stars, golden on a deep blue background, followed by three further rotating rings. One shows the zodiac with golden figurines, the next the months of the year, and the innermost one the days of the month. The funny thing is that there are only eleven zodiac signs: the extralong Scorpio stretches to occupy the space that ought to be taken up by Libra. A popular myth, repeated with gusto by tour guides, is that the goldsmith commissioned to craft the figurines removed Libra, the symbol of justice, as he hadn't been paid all the money he was promised for the job. The real reason is simply that the zodiac follows the tradition of ancient Greece, when the region of the sky now occupied by Libra was seen as the claws of the Scorpio.

At the center of the dial a sphere stands in for the Earth, around which revolves a circular window cut into the face of the plate, showing the Moon with its current phase represented in elegant white and black. The clock's sole hand carries a resplendent Sun pointing outward toward the time of the day, while a gloved hand indicates on the inner rings the month and day. It is as magnificent today as it must have appeared to the citizens of Padua when first unveiled in 1344.

But the real marvel, the ticking heart of the clock, is the mechanism hidden inside the tower. Of the original, only the cage survives, a time-battered brass affair resting on elegant curving feet and topped by four wooden spheres. The complex system of interlocking toothed gears—some the size of a truck's tire, some as small as a teacup saucer—and weight-driven pulleys is a replacement built in 1434 by popular demand after the earlier clock was destroyed during a battle for control of the city in 1390.

The mechanism's constant demands for attention were sated thanks to the life sacrifice of a *temperatore*, who lived with his family in minuscule quarters at the top of the tower, looking after Jacopo's invention. The weights driving the clock needed to be hoisted up to nine times a day, and the *temperatore* slept in short fits to attend to his duties during the night, too. The clock was set against the time determined from observations of the Sun made by the university's astronomers using a sundial during the day and the stars at night—a practice that continued until the nineteenth century, as mechanical clocks weren't sufficiently accurate.

The clocks of medieval Europe were built more as a proof of human ingenuity than as devices to tell the time, and indeed simple, inexpensive sundials endured on churches to indicate the time of prayers. In the Islamic world, the times of the five daily ritual prayers (*salah*) are determined by the position of the Sun at the location of the observer, and from the thirteenth century onward, a specialized observer, the *muwaqqit*, regulated praying time. Already from the eighth century, the astrolabe—an astronomical instrument consisting of a plate often made of brass, inscribed with celestial coordinates, and featuring a rotating star map and an aiming line—was used both to set prayer times and to find the *kiblah*, the direction to face while praying. The astrolabe could carry out complex astronomical and calendrical calculations (medieval instruction manuals list more than forty different uses), in many ways replacing the need for a mechanical clock in the Islamic world.

But the early clock of medieval Europe had an immense impact on the development of the scientific method, for when, in the late seventeenth century, the need arose for new, more precise experimental instruments, there existed a class of artisans who had by then been constructing intricate mechanisms for centuries. With every tick of Jacopo's clock, the gears invented to reproduce the movements of the heavens morphed a little further into the levers, assembly line, and huge-toothed wheels of Charlie Chaplin's *Modern Times*. Physicist and historian of science Derek de Solla Price has observed that "given . . . the clockworks of the sixteenth century, one could proceed in reasonably continuous historical understanding to the advanced instruments built by Robert Hooke for the early Royal Society [one of the first of the scientific learned societies, founded in 1660 in England], and from that point by equally easy stages to the cyclotrons and radio telescopes of today's physics laboratories and also to the assembly lines of Detroit."[13]

The journey from astronomical clock to American Cadillac is shorter than one might think. Every high school physics student today remembers Robert Hooke for his eponymous law describing the force needed to deform a spring; largely forgotten are his obsession with sundials and his role in perfecting—if not inventing—one of the key components of cars and many other types of machinery.

As the first Curator of Experiments for the Royal Society, the multitalented but cantankerous Hooke was the first professionally employed scientist in history. From 1662, he was tasked with delivering three or four experiments and demonstrations a week to enliven the otherwise often dull meetings of the Royal Society, whose "natural philosophers" came together to debate the scientific breakthroughs of the day.[14] Among the curious devices Hooke demonstrated in front of the not always appreciative fellows, we find a whale-shooting engine, an air gun, a springy saddle, a lens-grinding engine, and an instrument to measure humidity in the air moved by the beard of a wild oat, which reportedly even aroused the interest of the king.[15] His friend John Aubrey described him as "the greatest mechanick this day in the world."[16]

Hooke roamed London's insalubrious coffeehouses, spending time with mechanics and artisans, looking for inspiration for his weekly demonstrations at the Society. His lifelong passion for sundials began as a boy when, growing up on the Isle of Wight, he reportedly built a sundial out of a round wooden platter without instruction.[17] This came in handy when he was short of new ideas: on two occasions he presented a device made of two shafts, connected in such a way that when the first shaft was rotated in steps of fifteen degrees (the angle by which the Sun moves in the sky in one hour), the second shaft would go through the uneven hours' division on a sundial's face. His "sundial delineator" simulated the workings of a sundial, thanks to the mechanical property of the joint, which transformed the uniform rotation of the first shaft into a nonuniform rotation for the second shaft.

It is likely that Hooke had "borrowed" this invention from his clockmaker friends, for a similar joint, needed to transfer rotational motion from a driving shaft to a second one at an angle with the first, had been in use for hundreds of years. Hooke realized that by coupling two such "universal joints," the resulting nonuniform rotation could be exactly canceled out, thus enabling the transmission of rotation across two shafts at any relative angle.

With the Industrial Revolution, Hooke's double joint became the essential component of a hugely diverse range of machines: from cotton mills to locomotive drive shafts, from the steam engine to the constant-velocity coupling that transmits the engine power to a car's wheels even as they bump up and down the road. Today, this scion of the sundial serves us every time we drive to the shops.

FALLEN ANGELS

Back in Padua, by 1364 Jacopo's son Giovanni had surpassed his father, so much so that his friend Francesco Petrarca, the poet, described Giovanni as "easily the leader of astronomers" in his last will and testament.[18] Giovanni de gli Horologi, or John of the Clock, designed and built, after sixteen years of labor, a machine considered "one of the wonders of the world," according to physician Michele Savonarola: the astrarium.[19] When Duke Gian Galeazzo Visconti bought the instrument in 1381, he gave it pride of place in the library of his castle in Pavia, near Milan. German mathematician and astronomer Regiomontanus brimmed with praise after visiting in 1463, reporting that "in order to see it, innumerable prelates and princes have flocked to that place as if they were about to see a miracle, and indeed not without cause."[20] Leonardo da Vinci spent hours studying and working in the ducal library during his 1489–1490 stay in Pavia; details of the astrarium's gears are among the most technical of his sketches.[21]

John of the Clock's wondrous machine was encased in a brass frame featuring seven sides, with paw-like feet identical to the ones on his father's clock cage, topped by seven dials, each the size of a car's steering wheel and engraved with exquisitely minute scales, some featuring a sky coordinate system and all provided with a moving hand. Each of the moving hands precisely showed the position of one of the "planets" in the sky: the Sun, the Moon, Mercury, Venus, Mars, Jupiter, and Saturn. The motions were executed by a complex set of gears, powered by a weight, and based on the Ptolemaic geocentric model in which the Earth was at the center of the universe and the Sun, the Moon, and all the other planets revolved around it. The astrarium had a dial for the points of intersection between the Sun's and Moon's orbits; gave the times of dawn and sunset in Padua; and featured an annual calendar indicating the fixed and movable feasts of the church, the saint of the day, the length of sunlight, and the day of the week. As a bonus, it also gave the time.

The clock is, in the words of de Solla Price, "nought but a fallen angel from the world of astronomy": a marvel brought into existence to duplicate the motions of the heavens, pared down over the centuries to simply tell the time of day, with hands moving in a "sunwise" direction, having been designed to mimic the horizontal sundials of the Northern Hemisphere.[22] Had they been invented in the Southern Hemisphere, our clocks' hands would likely turn in the opposite direction.

Even a time-telling clock was never just a clock. If the astrarium could faithfully predict the "the maddeningly near-uniform motion of the planets" in the sky, in the words of de Solla Price, was there any limit to the reaches of technology or to the explanatory power of science?[23] It is no coincidence that physics, powered by mathematical astronomy, was the first of the sciences to take off in the seventeenth century, as we will discuss in Chapter 7. The metaphor of "the clockwork universe" had already been deployed in 1377 by French philosopher Nicole Oresme, who, a mere three decades after Jacopo of the Clock's masterpiece, described the world as "a regular clockwork that was neither fast nor slow, never stopped, and worked in summer and winter."[24] After providing the inspiration for the astronomical clocks that imitated them, the heavens became synonymous with the very mechanisms they had spawned, a radical shift of perspective that would have immense consequences for civilization.

THE HEAVENS IN A SHOEBOX

The many admirers of John of the Clock's astrarium considered it a unique specimen, "a magnificent work, a work of divine speculation, a work unattainable by human genius and never produced in generations past," according to a friend, waxing lyrical in 1388.[25] But John of the Clock's admirer was wrong: even as he was writing, an artifact yet more astounding than the astrarium had been rotting away forty meters under the surface of the Mediterranean Sea for some fourteen centuries.

It would take five more centuries before the Antikythera mechanism, as it is now known, was retrieved by unsuspecting sponge divers in the spring of 1900. Between the islands of Kythera and Crete, in a busy and dangerous shipping passage connecting the eastern and western Mediterranean, Captain Demetrios Kondos and his storm-tossed crew discovered at the bottom of the sea a fifty-meter-long Roman ship, replete with tumbled amphoras from Kos, some still containing olive pits, clay wine jars, kitchen pottery, blue and brown glass bowls from Alexandria, gold coins and a gold brooch with pearls, a lamp from Ephesus, and two magnificent larger-than-life bronzes that were what attracted their attention at first. The lucky find earned the divers headlines in national newspapers, $30,000 compensation each in today's money, and, some say, an ancient curse—one died during salvaging operations, while Captain Kondos was crippled.[26]

But the real treasure of the first underwater archaeological find in history sat overlooked for eight more months in the backyard of a museum in Athens, encased in a lump of debris in a cage of unidentified fragments, while restorers were busy piecing back together the bronze and marble statues that had been recognized as interesting—but which had not been the ship's most arresting cargo. Drying up after two millennia underwater, one of the lumps of debris eventually split open, revealing astounding mechanical entrails: multiple interlocking bronze gears, some with graduated lines etched all around the rim, and a backplate with an instruction manual for the machine. What is now known as the Antikythera mechanism is not only the oldest known instrument bearing graduated marks in history but is also the first mechanical computer—a machine that could simulate the aspects of the heavens. Derek de Solla Price, the first to realize the importance of the discovery, described it as "as spectacular as if the opening of Tutankhamen's tomb had revealed the decayed but recognisable parts of an internal combustion engine."[27]

Roman orator, philosopher, and politician Marcus Tullius Cicero attributed the creation of a similarly prodigious artifact, in the form of an artificial sphere of the heavens, to Archimedes, the Greek inventor and astronomer who lived in Syracuse, Sicily, in the third century BC. When Syracuse fell to the Romans in 212 BC after a long siege, the Roman general Marcellus brought back home a single item from the city's fantastically rich booty: a geared planetarium allegedly invented by Archimedes himself. A century and a half later, Cicero claimed he saw Archimedes's planetarium demonstrated in the house of the general's grandson. A self-avowed admirer of Archimedes (whose forgotten tomb he identified in Syracuse, thanks to the inscription of a sphere and a cylinder, and had restored), Cicero describes the sphere with awe, a testament to the genius of the great old man of Syracuse: "For when Archimedes fastened on a globe the movements of moon, sun and five wandering stars [the planets], he, just like Plato's God who built the world in the Timaeus, made one revolution of the sphere control several movements utterly unlike in slowness and speed. Now if in this world of ours phenomena cannot take place without the act of God, neither could Archimedes have reproduced the same movements upon a globe without divine genius."[28]

Archimedes's sphere and similar globes Cicero says had been fashioned by his contemporary Posidonius had been considered by historians likely literary exaggerations if not outright fabrications—until the discovery of the Antikythera mechanism, the proof that a highly sophisticated (and star-inspired) technology did exist at the time of Cicero. De Solla Price even speculates that the Antikythera mechanism might have been part of the goods being sent back home by Cicero himself after his stay on Rhodes from 79 to 77 BC.

Archimedes's sphere, if it ever existed, has been lost, but decades of study have teased out the Antikythera mechanism's secrets. Thanks to X-ray images of the parts still encased in debris, de Solla Price uncovered and followed a trail of prime numbers (numbers that are only divisible by themselves and one) in the gearing's teeth. Among them, a section of a wheel likely to have counted 127 teeth struck him as peculiar: for 127 multiplied by two gives 254—the number of times the Moon circles around the Earth in nineteen years, swelling to a full 235 times, and returning at the end of the cycle to almost exactly the same position and phase it had at the beginning. At the end of this nineteen-year cycle, discovered by the Greek astronomer Meton of Athens in the fifth century BC and named after him, the Moon, Earth, and Sun are arranged exactly in the same way as they were at its start. This is a key observation for synchronizing a lunar calendar with the solar year, used to this day in setting the date of the Christian Easter, as we saw, and in keeping the Jewish festivities from drifting through the seasons—so important that each year's position in the Metonic cycle is called its "golden number."

Instead of representing the motion of the celestial bodies on a sphere, the unknown inventor of the Antikythera mechanism managed to compress its gearing into a volume no bigger than a shoebox and to represent its output with hands rotating on dials on the front of the machine. Imagine an exquisitely decorated wooden box, with highly polished dials on its front and a removable crank handle at the back: upon the turning of the handle, the interlocking wheels inside stir into motion, while the "golden little sphere" (as the instruction manual inscribed on the back says) of the Sun glides around the zodiac at the front, the black-and-white sphere of the Moon zips on its faster orbit, changing its appearance along the way, and a clever little pin advances on a spiral-shaped calendar showing the 235 lunar months. Other hands showed the position of the planets, complete with their retrograde motions, according to a recent reconstruction, and what has been dubbed a "dragon hand" may have warned of a coming eclipse—a reference to the Sun being swallowed by a monster, as in Chinese mythology.

Whoever captured the motions of the heavens in a shoebox in the second century BC, perhaps building on the example of Archimedes, pioneered technology the likes of which would not be seen again for over a thousand years. Some of the gearing solutions inside the Antikythera mechanism, like the differential gear that today allows car wheels to spin at different rates when navigating a bend in the road, would not be rediscovered until James Watt of the Industrial Revolution came along. Unlike the bulky astrarium, the size of a small fridge, the Antikythera mechanism was a miniaturized computer, its more than thirty gears, pins, and spindles so cleverly fitted together that we can't fathom how they could be crafted with the tools available in ancient Greece. The technology packed into it might have been dispersed to the east, preserved in Islamic geared astrolabes, and adopted in the monumental water clocks of imperial China, before eventually cycling back to Europe, where it resurfaced in the wondrous clocks of Jacopo and John of the Clock some fifteen centuries later.

Unlike the Antikythera mechanism, the astrarium didn't last. A delicate machine, it needed the constant attention of a master clockmaker to stay in good order; over the decades the "custodians of the clock" succeeded each other in the duke's payroll at an ever-increasing pace, some staying for just a year, like babysitters tiring of looking after a child whose tantrums have worsened with age. It fell into disrepair, its more than eighteen hundred gears wasted by rust, the name of John of the Clock forgotten. Its remains caught the eye of Holy Roman Emperor Charles V during a trip to Italy in 1530: he ordered it repaired, but the task proved impossible. In the fruitless attempt it was dismantled, and to assuage the emperor, a copy was made, which ended up in a convent on the border between Spain and Portugal, where it languished until 1809. When the invading

French army burned the convent down, the last traces of the astrarium went up in flames with it.

PING! A STAR GOES BY

"Daddy, do I need to remember all this stuff to read a clock? And I'm scared of scorpions!" said Benjamin.

"Don't worry, Ben—all you need to remember is how to read the big and the small hands. And the only scorpions you'll see are the tiny ones that creep in our garage in the summer, and they won't hurt you. But there's another little beast that was used to tell the time until not so long ago: spiders!"

Spiders supplied a key component of the technology that provided the best timekeeping until the advent of the atomic clock in the 1950s: the finest silk threads. By the fifth century BC the Greeks had realized that the Sun just isn't a particularly good timekeeper. The time between two noons, defined as the moment when the Sun reaches its highest point in the sky, isn't constant throughout the year. It varies by as much as sixteen minutes, as a consequence of the Earth's orbit around the Sun being an ellipse rather than a perfect circle and because of the tilt of Earth's axis. The difference between the mean solar time (as indicated by a clock) and the apparent solar time (as shown by a sundial), today called the Equation of Time, was already tabulated in Ptolemy's *Almagest*. The Greeks didn't care much about precision timekeeping, but if one needs high precision, a sundial simply won't do.[29]

Distant stars, by contrast, don't suffer from this irregularity. The time it takes for any given star to return to the same position overhead is almost exactly the same throughout the year (tiny variations are due to irregularities in the Earth's rotation speed). But this so-called sidereal day (from the Latin *sidus*, "star") is about four minutes shorter than an average solar day: this is because as the Earth spins on its axis, it also moves along its orbit around the Sun. The Sun is thus "left behind" by about one degree per day in the sky with respect to the starry background and takes four minutes more to come back to its exact same position in the sky with respect to an earthly observer. To measure precisely the passage of a sidereal day, astronomers needed specially designed telescopes to determine when a certain star had returned exactly overhead.

The so-called transit telescope was invented by Danish astronomer Ole Rømer in 1704, in his vain attempt to measure a change in the position of distant stars. A transit telescope is aligned with a meridian—a circle of constant longitude on the surface of the Earth—and built in such a way as to only swivel up and down. Undoubtedly, the most famous such instrument is the Airy Transit Circle, designed by Astronomer Royal George Airy and built in 1851. It can still be seen at the Royal Greenwich Observatory, where its axis, as we shall see in the next chapter, defines the prime meridian in the world—the line of zero degrees longitude from which all east- and westward distances are measured.

That's where spiders come into the picture, or more precisely, into the eyepiece. Astronomers needed a means to equip their eyepieces with extremely fine crosshairs in order to point the telescope precisely at the star they were interested in. Astronomer William Herschel complained in 1782 that human hair was unsatisfactorily thick: "I have attempted in vain to find hairs sufficiently thin to extend them across the centres of the stars so that their thickness might be neglected."[30] An English amateur astronomer named William Gascoigne had been inspired when he found a spider web inside his telescope tube in 1639, but the difficulty of stretching the filament across the eyepiece discouraged him. American surveyor, astronomer, and clockmaker David Rittenhouse is credited with introducing spider webs as crosshairs in 1785, when he reported, "I have lately with no small difficulty placed the thread of a spider in some of my instruments, it has a beautiful effect, it is not one tenth of the size of the thread of the silkworm, and is rounder and more evenly of a thickness."[31]

By 1824, William Herschel's son John was employing spider web–equipped eyepieces to measure the position of stars to a hitherto incredible precision of one second of arc, or 1/3,600th of degree. "Spider ranches" sprung up to provide astronomers, surveyors, and the military (for use in bomb and gun sights) with a large variety of spun-to-order spider threads, five times stronger than steel and three times tougher than Kevlar: from extraheavy (1/5,000th of an inch), to extrafine (1/50,000th of an inch), to even the nearly invisible premium silk of week-old baby spiders

(1/500,000th of an inch). The Californian spider rancher Nan Songer, whose hundreds of spider "workers" supplied the US government during World War II, favored the silk of venomous black widow spider females, which she "milked" by tickling them with a soft camel-hair brush. "Some of the regular producers get as docile as old milk cows, particularly the Black Widows," she reported.[32]

Every night, when one of the many "clock stars" circling in the sky moved across one of the spider silk wires inside the eyepiece, the astronomer knew that, at that exact moment, the Earth had come around full circle—a sidereal day had just finished. Ping! The clock star goes by. Ping! Another one. And that's how they set their clocks, just like the ancient Egyptians did.

"Ah," said Benjamin, unconvinced. "But when I learn to read my watch, I won't need to worry about the stars anymore, right? My watch knows what time it is all by itself!"

Quartz watches keep excellent time, and even Benjamin's—with its garishly colored wristband and tiny flashlight that projects scary dinosaur pictures in the dark—is better than John Harrison's masterpiece, a marine chronometer that we will encounter in the next chapter. But the standard of time today comes from atomic clocks: the second itself is defined as the duration of 9,192,631,770 periods of the electromagnetic radiation emitted as a cesium atom undergoes a particular type of quantum mechanical transition. In a sense, we have merely replaced the old pendulum with atomic-scale oscillators and the clocks' hands with a beam of light. These atomic clocks are so precise that they keep time to better than a lag of one second in thirty thousand years.

"So the special clocks made of light tell us what time it is, then?" asked Benjamin, hopeful to have gotten to the bottom of the matter at last.

"Not quite, buddy. The atomic clocks are almost perfect, but the spinning of the Earth isn't. Mostly, it slows down over time: when the dinosaurs were around, a day was only twenty-three hours long! To keep our clocks matching the actual time it takes for the Earth to do a full spin on itself, every now and then the timekeepers adjust the atomic clocks' time by adding or taking away a second."

"And where do they find all these seconds, Daddy?"

"In the sky, of course! Only nearby stars move around too much to be useful for this, so astronomers look out almost to the end of the visible universe, where radio galaxies are as good as pinned. The astronomers fetch the extra second from those distant galaxies. Ping! The radio galaxy goes by, and an extra second is added.[33] You see, deep down, we still live by the time of the stars."

Benjamin had gotten rather more out of his question than he asked for. But the relationship between timekeeping and the stars is enmeshed with other stories of our debt to astronomy. An accurate determination of time is today, like centuries ago, absolutely necessary for a precise measurement of location. The building of colonial empires, the shipping of goods from one end of the globe to the other, and the drawing of faithful maps of the Earth required the ability to determine one's position in the middle of the ocean or on the shore of a strange, new land. And that meant good clocks—or, at least, a clear view of the skies.

THE CALIGO TALES

Shepherd's Tale

As Once-Upon-A-Glow ended his story, we all jumped up and stomped on the ground with all of our strength, shells rattling around our necks, to show any Trickster what they would face if they ever came back.

When our cries had died down, dust swirling in the air, Shepherd rose again. He poured more water onto a fresh batch of coals and spoke thus:

> Daughters and sons! The Tricksters came and went, but one thing stays the same: the Cloud rolls from one end of the Disc to the other, glow after glow, wave after wave, gray after black after gray. And so live our people too: Cloud-Watcher follows Elder-Cloud-Watcher, Once-Upon-A-Glow follows Elder Once-Upon-A-Glow, Shepherd follows Elder-Shepherd. Like the Cloud, we follow one another, each different but always the same.
>
> Now that Freshwater has gone, Freshwater-the-Young will take her place in the circle. This is the way of the Cloud.
>
> But be warned! Other peoples don't heed the way of the Cloud! They don't honor Lightning by

112

cutting their skin in its shape every time it visits! They don't Remember, glow after glow, like we do, calling in the power of the Cloud against the Dark!

Daughters and sons! I saw them with my very eyes: they call themselves "People of Foucault," and their cave is but one glow away for a fast walker. You have to descend through a narrow hole, slippery with moss, then crawl in Darkness along a narrow passage, warm with the breath of the Earth. At the very end, a vast cave opens, its roof so high that fire cannot reach it. That's where I saw it.

A rope hangs from darkness, its far end lost to the eye, thick as a strong man's arm. How it got there and why, the People of Foucault themselves don't remember. At the bottom of the rope, a huge rock hangs, its pointed bottom just scraping the floor.

The rock swings back and forth, tracing a line in the sand on the floor. Two of their people push the rock, one on each side, gently, their eyes empty. For many bat awakenings they have been waiting their turn to push the rock. When they start, they no longer eat, they no longer drink, they no longer sleep, always watching the rock—until each falls, their face in the dust, and another takes their place, pushing the rock.[34]

Daughters and sons! I saw them with my very eyes! As the rock swings back and forth, the line in the sand moves by itself! Such is the power of their Dark! They say it's the Disc itself moving under the rock, and swear that if the rock should ever stop, the Disc will end. They say that keeping the rock swinging is the only task worth doing. I saw them push the rock until they could no more, then fall with their eyes open into the dust, smiling.

But we know that the Disc cannot move. It is the Cloud that rolls and flows and keeps being itself even as it changes, glow after glow. The People of Foucault come from the Dark! Should you ever meet one, their arms uncut by Lightning, draw your blade and chop off their tongue!

So spoke Shepherd, warning us against the people of the Dark, those wretched people who do not follow the way of the Cloud.

In the silence that followed, Shepherd spoke again: "The People of Foucault are not the only strange folk we Remember. Way-Finder, tell us about the Big Swap and what we brought back when we last went!"

Way-Finder stood and began his story.

6

TRIPLE BRONZE AND OAK

Whose light, among so many lights,
Was like that star on starry nights
The seaman singles from the sky,
To steer his bark for ever by!

—Thomas Moore, "The Light of the Haram"

NAVIGATING BY THE STARS

Among the vast collection of artifacts preserved at the British Library in London, one bears witness to the brief, extraordinary encounter of two approaches to navigation: both relied on the stars but in radically contrasting ways. The artifact is a map, drawn in a neat hand on paper gone yellow after over two centuries, showing seventy-four labeled islands and the four cardinal directions, with north at the top. You would scour contemporary atlases in vain looking for this same arrangement of islands: as depicted on the map, they do not exist. The map is a chimera, in the sense of both an impossible aim and an entity containing a mixture of genetic material. It is also an epitaph, of sorts, marking the moment of apparent quiet before the almost total destruction of the way of life of one of its makers.

The story of Tupaia's Map (as it is today known) will take us to the heart of the art and science of navigating by the stars. In brief, the pursuit of a rare astronomical phenomenon brought James Cook's *Endeavour* to Tahiti, the first European ship to intentionally sail to a mid-Pacific island, where he found the unexpected help of the Polynesian high priest and navigator Tupaia. The meeting of these two master navigators, each the pinnacle of his respective culture's ability to find its way in the middle of the high seas, distills the course of hundreds, perhaps thousands, of years of voyaging by the light of the stars: Lieutenant James Cook, with his instruments and lunar tables provided by the king's finest astronomers and infused with the best science the Western Hemisphere had to offer, and Tupaia, who had never heard of Polaris but knew every star in the southern skies and could sail between specks of land scattered across the Pacific with no maps or instruments—only the ancient lore he carried between his ears.

Leading up to their encounter, we will see how getting a ship safely and swiftly to its destination eventually became dependent, in the eyes of Europeans, on solving the problem of reliably measuring time—with the solution to both questions provided by the stars. On the other side of the globe, the Polynesian way of navigation had far more in common with the traditions of star knowledge we saw in Chapter 4: those of the Australian Aboriginals, who relied on songlines and star lore for thousands of years when traveling overland, and the Inuit, who did so in the even more forbidding conditions of the Arctic Circle, where stars are less reliably present and the topography of ice literally moves underfoot. Navigation by the stars was important on the high seas, especially when sailors ventured into unfamiliar waters and were deprived of local knowledge of winds, currents, swells, and other subtle signals that might help them locate themselves within the blue stretching from horizon to horizon. The world's greatest masters of noninstrumental navigation were—and still are—Pacific Islanders.

Having grown up in landlocked Switzerland, I am no mariner, but my one experience of the high seas filled me with awe. Invited as the astronomer aboard a cruise ship touring the Caribbean, I led a stargazing session one night. On my request, the captain ordered all lights except those essential for navigation and safety to be extinguished, and when I emerged on the top deck, the bejeweled sky above was overwhelming, sparkling with an intensity I had never experienced before—nor have I again since. Black depths surrounded us on all sides, which the fourteen-deck ocean liner under me traversed—I knew—guided by satellites, cocooned by radar, and propelled by a two-hundred-thousand-horsepower engine. I shivered at the thought of facing this immensity on a wooden ship tossed by the waves, at the mercy of the winds and with not so much as a chart to steer by. And yet, this was precisely the act seafarers, explorers, and voyagers chose to undertake for millennia. As Horace sings,

> *Triple bronze and oak encircled*
> *the breast of the man who first committed*
> *his fragile bark to the cruel sea.*[1]

The Phoenicians are credited as having been the greatest mariners of the ancient world—perhaps misguidedly, as we shall see. Around 1500 BC, Phoenician territory stretched along the coast that is today Lebanon, Syria, and northern Israel, but the "purple people" (as they were called by the Greeks, in reference to the red-purple color they used to dye their robes) soon established trading posts along the coast of the entire Mediterranean Sea. Sailing on ships adorned with a horse's head in honor of their god of the sea, the Phoenicians crisscrossed the Mediterranean, ventured as far as Britain, and even circumnavigated Africa in the seventh century BC at the behest of Pharaoh Necho II.

The Phoenicians knew no navigational instruments. They relied on keeping the coast within sight, on their knowledge of winds, currents, and coastal landmarks, and on the position of the Sun during the day. At night, they steered by the constellation of the Little Bear (today also called the Little Dipper in North America and the Lesser Wain; to them, it was the Guiding One), keeping it to their left when sailing eastward and to their right when going west.[2] We know of this because Thales of Miletus, who was of Phoenician descent and whom we encountered earlier as the author of the first successful total solar eclipse prediction, "was said to have measured out the little stars of the Wain, by which the Phoenicians sail" in the sixth century BC.[3] Before then, the Greeks, who used the same method, made reference to the constellation of the Great

Bear instead (also known as the Big Dipper in North America, the Plough in the United Kingdom and Ireland, the Wagon since Mesopotamian times, and, somewhat confusingly, Charles's Wain, a medieval association with Charlemagne), presumably because they didn't recognize the Little Bear. As Homer tells us in the eighth century BC, Ulysses is instructed by Calypso to keep the stars of the Great Bear "hard to port"—that is, to his left-hand side, to head back east, to Ithaca and to his faithful wife, Penelope:

> *The wind lifting his spirits high, royal Odysseus*
> *spread sail—gripping the tiller, seated astern—*
> *and now the master mariner steered his craft,*
> *sleep never closing his eyes, forever scanning*
> *the stars, the Pleiades and the Plowman late to set*
> *and the Great Bear that mankind also calls the*
>   *Wagon:*
> *she wheels on her axis always fixed, watching the*
>   *Hunter,*
> *and she alone is denied a plunge in the Ocean's baths.*
> *Hers were the stars the lustrous goddess told him*
> *to keep hard to port as he cut across the sea.*[4]

One may wonder why the Phoenicians and the Greeks didn't simply use the one star we have come to consider the epitome of navigation: the North Star. American Romantic poet William Cullen Bryant celebrates it as the lodestar by which humanity steers its course, both literally and figuratively:

> *On thy unaltering blaze*
> *The half-wrecked mariner, his compass lost,*
> *Fixes his steady gaze,*
> *And steers, undoubting, to the friendly coast;*
> *And they who stray in perilous wastes by night,*
> *Are glad when thou dost shine to guide their footsteps*
>   *right.*[5]

Shakespeare also adopts it as a metaphor for steadfastness, having Julius Caesar proclaim,

> *But I am constant as the northern star,*
> *Of whose true-fixed and resting quality*
> *There is no fellow in the firmament.*[6]

In fact, the North Star is not all "true-fixed" or "unaltering," and there is a simple reason why the Phoenicians, the Greeks, and even the Romans didn't recognize it as special: at their time, it wasn't yet in the right place.

FIFTEEN HUNDRED YEARS OF FAME

As we saw earlier, the distant stars often called "fixed" are not immobile. They just take hundreds of thousands of years to change their relative positions and so change the shapes of constellations. Over a much shorter period, another effect shifts the apparent position of all the stars in the same direction, as if the whole starry canopy had been rotated as a solid sphere. I say "apparent" position because the effect, called "precession," is not a real motion of the stars in space but a consequence of a shift in our perspective on them.

In the second century BC, the Greek astronomer Hipparchus compared the location of stars in the sky at the fall equinox, when day and night are each exactly twelve hours long everywhere on Earth, with records held at the Alexandria observatory from 150 years earlier. He discovered that all the stars had moved forward (i.e., east) by four times the diameter of the full Moon—hence the name *precession* (moving forward). Ptolemy, in the second century AD, attributed the phenomenon to a shift of the celestial sphere around a motionless Earth, a model that held sway until Isaac Newton and his theory of gravitation in the seventeenth century.

Far from being motionless, the Earth spins on its axis as it orbits the Sun. The Earth's rotation causes its equatorial regions to bulge out, like a squat orange twirling on an imaginary skewer that pierces it through its green navel, the North Pole. The skewer is tilted 23.5 degrees with respect to the vertical, and the direction in which it points indicates true north (the geographic north pole), which differs from the so-called magnetic north shown by the needle of a compass: the latter follows the Earth's magnetic field, which is constantly shifting, and is currently located about twelve hundred miles away from true north. Newton realized that because of the squat shape of the Earth, the gravitational attraction from the Sun and the Moon at the equator is stronger on the closer side than on the far side. This causes the skewer itself to rotate in a slow circle, like the wobbling axle of a spinning top not spun exactly on the vertical, which takes about twenty-six thousand years to complete and causes the phenomenon of precession.

When the Earth returns to the exact same location on its orbit around the Sun after a solar year, its rotation axis (the skewer) has moved by a fraction of a degree, so the equinox shifts by the same amount—one day every seventy-two years—with respect to the distant stars. This means that seasons cycle through the starry canopy every twenty-six thousand years, a span of time called a "great year." The position of the stars on the day of the equinox several centuries apart will show them all shifted to the east, something we will return to in connection with astrology in Chapter 9. As the tip of the skewer traces out an imaginary circle in the sky over the course of a great year, the point on the celestial sphere around which all stars appear to rotate changes over time. Fortunately for explorers and travelers in the Northern Hemisphere, for the past five centuries or so the skewer has pointed in the vicinity of an easily recognizable star: Alpha Ursae Minoris, the brightest star in the constellation of the Little Bear, or Ursa Minor in Latin.

Like a passerby finding themselves at the center of a great commotion and capturing on camera a historic event, Alpha Ursae Minoris grew in fame wildly once the wandering skewer that is the Earth's rotation axis leaned its way. In stellar tables drawn in 1492, it assumed for the first time the more genial nickname of Polaris, though at that time it was three times farther away from the celestial north pole than it is today. Around the same time, it became for the Arabs Al Kiblah, because it showed to every faithful Muslim the *kiblah*, the direction toward which to turn in prayer; for the inhabitants of Lapland, the "Nail of the North"; for the Navajo Native Americans, "the Campfire of the North";

for the Chinese, the Great Imperial Ruler of the Heaven, since all other stars circled it in homage; for the Italians, *la tramontana*, "the one who is beyond the mountains," hanging beyond the jagged alpine tops marking the northerly border of their territory.[7] Today, *perdere la tramontana* (to lose the *tramontana*) means in Italian to lose one's bearings and therefore one's temper.

At the time when Shakespeare's *Julius Caesar* is set, in 44 BC, what we now call Polaris was about ten degrees away from the celestial north pole, which lay unmarked by any star. Alpha Ursae Minoris rotated just like any other starlet, so Caesar himself could not possibly have used it as a metaphor for steadfastness.[8] Nor will Polaris's special status endure: as the skewer relentlessly moves on along the great year, Polaris's distance from the celestial north pole will at first shrink farther to less than half a degree in 2095 and then inexorably grow larger again. In about a thousand years, Polaris will cede the scepter of lodestar to Errai, in the constellation Cepheus. To paraphrase Andy Warhol, Polaris will have had its fifteen hundred years of fame—at least, until another great year has passed, and who knows what eyes, if any, will be turned to the skies then.[9]

THE POLYNESIAN MASTERS

While the Phoenicians, the Greeks, and the Romans hugged the coastlines of the Mediterranean Sea, Micronesian and Polynesian seafarers were tackling the Pacific Ocean, in one of the highest achievements of human exploration: the discovery and settlement of almost all habitable Pacific islands, many of which are less than a mile wide in a blue expanse totaling sixty-four million square miles, over four times the entire surface of the Moon. How exactly Polynesian people did this remains contested. Anthropologist Geoffrey Irwin argues that it was a systematic process of discovery proceeding from west to east, and then from north to south, over the course of more than two thousand years: starting in the second millennium BC from New Guinea to Fiji, Samoa, and Tonga, then to the Cook Islands and Hiva (the Marquesas), and from there to Hawaii. Aotearoa/New Zealand was, in this perspective, the last to be settled around the eleventh century AD.[10]

Until the fifteenth century, Polynesians regularly sailed the four thousand miles separating Aotearoa/New Zealand and Hawaii. This they achieved with no navigational instruments, no charts, and no written instructions, sailing on remarkable voyaging canoes dug out of tree trunks, held together by woven coconut fibers and sealed with sap from the bread tree.[11] Chilean sculptor Ignacio Andía y Varela marveled when he visited Tahiti in 1774:

> It would give the most skilful [European] builder a shock to see craft having no more breadth of beam than three [arm] spans carrying a spread of sail so large as to befit one of ours with a beam of eight or ten spans, and which, though without means of lowering or furling the sail, make sport of the winds and waves during a gale. . . . These canoes are as fine forward as the edge of a knife, so that they travel faster than the swiftest of our vessels; and they are marvellous, not only in this respect, but for their smartness in shifting from one tack to the other.[12]

One of the best documented voyaging traditions in the Pacific is that of the Gilbert Islands, a chain of sixteen atolls in Micronesia, midway between Hawaii and Papua New Guinea, the largest of which is about half the area of Manhattan. In the Gilbertese tongue, there is no word for "astronomer"; if one needs an expert on the stars, one asks for "a navigator."[13] The Gilbertese, like all other Polynesian and Micronesian master mariners, used all clues at their disposal to judge direction, distance covered, and position while voyaging: types of seaweed and fish encountered, ocean currents and swells, the refraction and color of the water, the flight of birds (indicating proximity to land), the shapes, types, and sizes of clouds, and, of course, the positions of the Moon, Sun, and stars. Polynesian navigators recognized over two hundred stars, and each was thought to belong to the island over which it passed when at the zenith (i.e., directly overhead). So when a guiding star was at the zenith, navigators knew they were at the correct latitude of the target island. The presence of land birds expanded the footprint of even a tiny island to about thirty miles, still a very small target in the immensity of the Pacific. An error of half a degree in judging the position of the zenith star translated into heading thirty miles off course at sea, sufficient to sail past the destination and on toward certain death.[14] Still, the stars were the most trustworthy guides, as shown by the Tongan aphorism "The compass may be wrong; the stars never."[15] The exactitude and reliability of the Polynesian method was witnessed and much admired by Andía y Varela:

> When the night is a clear one they steer by the stars; and this is the
> easiest navigation for them because, there being many stars not only
> do they note by them the bearings on which the several islands with
> which they are in touch lie, but also the harbours in them, so that they
> make straight for the entrance by following the rhumb of the particu-
> lar star that rises or sets over it; and they hit it off with as much preci-
> sion as the most expert navigator of civilized nations could achieve.[16]

The *Bounty* mutineer James Morrison agreed in 1788: "It may seem strange to European navigators how these people find their way to such a distance without the help or knowledge of letters, figures, or instruments of any kind but their Judgement of the Motion of the Heavenly bodys [*sic*], at which they are more expert and can give better account of the Stars which rise and set in their Horison [*sic*] then a European Astronomer would be willing to believe, which is nevertheless a Fact."[17]

Strange, indeed, and even stranger that no one took a closer interest in the eighteenth century, at the time of first contact between Europeans and Polynesians—though if they had, they would likely have been rebuffed by the Polynesian navigators: the Polynesian master mariners considered their knowledge sacred and secret, not to be shared with the uninitiated.

Sadly, under the onslaught of European colonization, much of it would be lost to all.

A MOONSHOT PRIZE

The European voyages of exploration—and exploitation—of the fifteenth and sixteenth centuries opened up transoceanic sailing to the maritime pow-ers of the day: Spain, Portugal, France, and Britain. In his historic voyage to the Americas in 1492, Christopher Columbus used little more than a compass (introduced already in the twelfth century) and dead reckoning, the often inaccurate estimate of a ship's position given its last known location accounting for its speed and direction of travel. The little astronomical navigation he used was wildly off, even by the standards of his day.[18]

Sailing with compass and dead reckoning was sufficient while keeping within sight of the coast, but it became hopelessly inaccurate in the middle of the Atlantic or Pacific Ocean: errors of up to ten degrees—equivalent to six hundred nautical miles at the equator—were common. A more reliable method was needed. By the time Ferdinand Magellan reached the Pacific in November 1520, in what would become the first circumnavigation of the globe, the Portuguese had pretty much solved the problem of determining latitude, the north-south position of the ship, by measuring the height of Polaris over the horizon and compensating for its circling around the true celestial north pole during the night—a feat that Polynesian seafarers had been practicing for centuries already with their zenith star method, which also worked in the Southern Hemisphere, where Polaris was useless. When Polaris was not visible, the Portuguese measured the height of the Sun over the horizon at noon, when it had climbed to its highest point in the sky, to estimate latitude once corrected for the seasonal variations of the Sun's trajectory (higher in summer, lower in winter).

Finding longitude, the distance east or west from a reference point, was the hard part. There simply was no reference point, as all the stars move from east to west, while the planets and the Moon drift along their own orbits. By the sixteenth century the problem of longitude had become an urgent matter of state. The Treaty of Tordesillas in 1494 had determined a demarcation line, 370 leagues west of the Cape Verde Islands, separating lands that could be claimed by the Spanish Empire (to the west until the antimeridian line on the opposite side of the globe) from those under the influence of the Portuguese Empire (to the east until the antimeridian line). The problem was that no one knew where the demarcation line ran in the Atlantic or where its antimeridian was in the Pacific. Uruguayan historian Rolando Laguardia Trías suggests that this uncertainty was to blame for Magellan's death in Guam in 1521: having set off with the intention of proving that the Moluccas, an archipelago east of

Indonesia rich in coveted spices, were in the Spanish zone, Magellan concluded (correctly) from his longitude estimates that they were actually in the Portuguese hemisphere, contrary to what he had assured the Spanish king before his departure. To avoid the humiliation of returning home and facing the king with such disappointing news, Laguarda Trías speculates, Magellan exposed himself to reckless risks in the battle in which he was killed.[19]

With longitude remaining elusive, maps of distant lands and the maritime charts needed to get there remained approximate at best. As Britain, France, Portugal, Spain, and the Netherlands vied for supremacy in the seas and over the riches of new colonies, accurate knowledge of their locations, coastlines, and islands and the distances between them was of paramount strategic importance. Countless ships, crews, and cargo were lost for the lack of longitude, having either disappeared in the uncharted vastness of the oceans or been sunk in poor weather by shoals they didn't know were there.

Desperate to find a workable solution, in 1714 the British Parliament promised the fantastic sum of £20,000 (equivalent to several million dollars today) to anybody who could come up with a "Practicable and Useful" method to discover longitude at sea.[20] This wasn't the first time a government offered a large sum of money as enticement to inventors: the Spanish Crown had done it twice, the States General of the United Provinces of the Dutch Republic had done it, and the provincial States of Holland and Westfrisia had done it. No one had ever been awarded any of these much-coveted prizes.

The longitude prize was a moonshot whose solution would provide decisive strategic advantage to the British. The winner could count on fabulous riches, fame, and an assured place in history. As such, the prize attracted any number of crackpot proposals, of which the most notable is surely the "powder of sympathy" suggestion: that an injured dog be carried aboard the ship and that the bandage previously applied to the wound be left behind in Greenwich and dipped every day precisely at noon into a pot of powder of sympathy, whose effect, it was claimed, was to instantly make the dog howl, wherever it may be. This would have told the ship's crew that it was then noon in Greenwich.

This proposal was "entirely tongue in cheek," according to historian of science Owen Gingerich; yet it did contain an ounce of wisdom: longitude could be found by determining the local time at the ship's location and comparing it with the time at a reference point, for example, the Royal Greenwich Observatory.[21] If the local time aboard the ship was 9 a.m. when it was noon at Greenwich, the ship was three hours west of Greenwich, which meant a longitude of forty-five degrees west (as one hour time difference corresponds to 15 degrees of longitude: it takes twenty-four hours for the Earth to make a full rotation of 360 degrees). The difficulty was to find a means to know when it was noon back at home. So the howling dog with its magic bandage was merely a satiric way to implement what would have been an actual solution.

In the early eighteenth century, none of the intricate mechanical clocks that had followed Jacopo de' Dondi's pioneering machines could be relied upon to keep time at sea, tossed by the waves and subject to extreme changes in temperature, humidity, and pressure—certainly not to the exacting standards mandated by the Board of Longitude, which stipulated that the top prize required an error in longitude no larger than half a degree over the course of a six-week voyage from England to the Caribbean. This meant a clock that didn't gain or lose more than three seconds per day, at sea.

Newton, a member of the original Board of Longitude tasked with judging entries, was skeptical that such a clock could ever be built: "By reason of the motion of the Ship, the Variation of Heat and Cold, Wet and Dry, and the difference of Gravity in different Latitudes [which changed the swinging of pendulums], such a watch hath not yet been made"—and perhaps never would, he suspected.[22] Newton had put his eggs in a different basket: as you might expect from the discoverer of the laws of gravity, his favorite method was to use the Moon as a watch. Any celestial phenomenon whose timing could be predicted with precision well in advance would do: mariners observed the event and recorded their local time, which only required a simple watch that could be reset each day at local noon. Then they would look up in tables prepared for the purpose the time at which the event was predicted to happen in Greenwich. From the comparison of the times, longitude would fall out.

The principle was simple enough, but the devil was in the astronomical detail. Even Newton couldn't predict the complex trajectory of the Moon with sufficient accuracy for the purpose. Lunar eclipses could be used, but they are too rare to be practical. Galileo Galilei had in 1612 attempted to convince the Spanish king that the moons of Jupiter were a suitable option, but given that Jupiter becomes invisible when it passes behind the Sun for ten weeks at a time—not to mention the difficulty of pinpointing tiny dots of light with a telescope from the deck of a waved-tossed ship—this too had been a dead end.[23]

By the time His Majesty's Bark *Endeavour* set sail from Plymouth on August 26, 1768, no scientific quest was more valued, or bedeviled, than the taking of longitude, and a battle for the prize had been raging for a decade. Yet Lieutenant James Cook and his eighty-five-man crew—including twelve royal marines, astronomer Charles Green, and naturalist Joseph Banks—had another mission on which the pride and luster of the empire depended: to determine once and for all the size of the solar system.[24]

FOR THE PRIDE OF THE EMPIRE

The distance between the Earth and the Sun had preoccupied astronomers ever since the first estimate in the second century BC by Hipparchus, who had it at a relatively meager 490 Earth radii, compared to the actual value of 23,400 Earth radii (or 150 million kilometers). By the early seventeenth century, the three laws of planetary motion announced by Johannes Kepler provided a way to obtain the relative distances from the Sun to all the planets in the solar system. If the Earth's distance to the Sun was assigned a value of 1 astronomical unit (AU), it followed that Mercury was 0.39 AU from the Sun; Venus, 0.72 AU; Mars, 1.52 AU; Jupiter, 5.20 AU; and Saturn, 9.5 AU (the outer planets would only be discovered later). But to obtain the absolute distances of all the planets, and thus determine the actual size of the solar system, one needed a way to measure the AU.

That's where Edmond Halley came in. In 1716, the British astronomer issued a call to arms to all "diligent searchers of the Heavens" to take advantage of the rare astronomical phenomenon of Venus transiting in front of the Sun to at last measure the Sun-Earth distance. Venus being the second planet in the solar system, its orbit around the Sun is smaller than the Earth's, and thus its period is shorter: the Sun, Venus, and the Earth line up in a straight line approximately every 1.5 Earth years. However, because of the relative inclination of the Venusian and terrestrial orbits, Venus is seen gliding across the face of the Sun (a so-called transit) only on the rare occasions when the alignment matches the intersection of the planes of the planets' orbits. A transit will happen twice 8 years apart, then not at all for 105 years, followed by another pair of passages 8 years apart, then another wait of 121 years, after which the cycle recommences. Halley rallied "natural scientists" to mount a worldwide expedition for the pair of transits expected in 1761 and 1769, which, if timed from locations as far apart as possible on the surface of the Earth, would give a means of measuring the solar parallax—the apparent shift of the Sun's position when seen from two observers half an Earth radius apart—and with it the distance between the Earth and the Sun.

When the 1761 transit came, over 120 missions in sixty-two countries went about timing it, the grandest international scientific effort ever attempted until that point. But the quality of the data collected turned out to be unsatisfactory to determine the Earth-Sun distance. A second chance—the last for more than a century—would present itself on June 3, 1769. This time, failure was not an option.

James Cook was chosen to lead the English expedition to observe the 1769 transit because of his reputation as the ablest navigator and cartographer among His Majesty's seamen. His orders were to take the *Endeavour* and its cargo of telescopes, clocks, and natural philosophers to the other side of the planet, cross half the Pacific Ocean, make landfall on a speck of an island twenty miles across (Tahiti), befriend the locals, and set up a high-precision observatory in good time for the transit so that he and Green, the astronomer, would "have leisure to adjust and try [their] instruments before the observations." And that was the easy part. Once the transit observation was completed—ostensibly the only mission of the *Endeavour*—Cook wasn't to relax in the tropical paradise of Tahiti but instead was "to put to sea without loss of time, and carry into execution the Additional Instructions contained in the enclosed sealed package."[25]

But first, Lieutenant Cook needed to get to Tahiti, and that meant taking lunars.

THE LORD OF THE LUNARS

Back in Greenwich, one of the instigators of the mission was also the man who had provided Cook with his latest, still experimental method for precision navigation and longitude determination: Astronomer Royal Nevil Maskelyne. Maskelyne had taken up Newton's suggestion of using the position of the Moon in the sky to determine the time difference with Greenwich and adopted the much-improved predictions of the genius German mathematician and astronomer Tobias Meyer. Whereas Newton's old method gave results that were off by up to one degree, Meyer's lunar tables achieved the hitherto unheard-of accuracy of an arcminute, sufficient to guarantee a measurement of the longitude to within half a degree. And that would bring the method of "taking lunars," as it was called, into contention for the Longitude Prize.

The problem was that computing Meyer's tables entailed the slow and painfully dull work of applying by hand, over and over again, complicated trigonometric formulas, with pen, paper, candlelight, and a set of seven-digit logarithmic tables as the only aids (we shall see in Chapter 8 how this very issue spurred Charles Babbage to attempt to build a mechanical general-purpose computer). So Maskelyne contracted out the job of crunching the numbers to a small army of human "computers," as they were called, making sure each calculation was carried out independently by two different computers to cross-check for errors. Once, when he found evidence of two of them copying from each other, he fired them both.

Maskelyne supplied Cook with the first edition of his *Nautical Almanac*, containing detailed tables of the angle separating the Moon and the Sun, and the Moon and a few reference stars, every three hours, for the next two years—a shorter period than Maskelyne would have liked, for the lieutenant was bound to run out of data before he could return. Crucially, Maskelyne had simplified the calculations needed onboard the ship to obtain a longitude estimate from the measured lunar distance from the Sun or a star, reducing it from four hours to half an hour. If Maskelyne hoped that Cook would vouch for his method of lunars, he wasn't to be disappointed, for Cook resoundingly endorsed "a method that we have generally found may be depended upon to within half a degree; which is a degree of accuracy more than Sufficient for all Nautical purposes."[26] More than sufficient, in fact, for Maskelyne to pocket the marvelous Longitude Prize, as Cook undoubtedly realized.

But while Cook was at sea, another contender for the prize was on the ascendancy at home, one so good that it would commandeer his allegiance on his second voyage. Still, en route to Tahiti, taking lunars was the method of choice, and comparisons of Cook's and especially Green's results with the actual longitudes of landmarks show remarkably good agreement. Green was so dedicated to the task that he coolly kept taking lunars even as the *Endeavour* narrowly escaped shipwreck in the treacherous waters around the Great Barrier Reef off the east coast of Australia.

When the *Endeavour* sailed into Tahiti's Matavai Bay in the early morning of April 13, 1769, in good time for the transit, the locals met the British with a friendly welcome, as recounted by Joseph Banks, a landed young gentleman with a passion for botany who had spared no expense to join the expedition: "Before the anchor was down we were surrounded by a large number of canoes, the people trading very quietly and civilly, chiefly for beads, in exchange for which they gave cocoanuts, bread-fruit both roasted and raw, some small fish and apples."[27]

Among the dignitaries whom Cook soon befriended was a high priest of the 'Arioi cast: Tupaia. No portrait of Tupaia has been found to date, but Banks judged him to be about forty-five years old, making him a contemporary of Cook. Driven away from his home island of Raiatea, 340 miles to the northwest of Tahiti, by a bloody invasion that saw him wounded in battle, the high-born Tupaia exiled himself to Tahiti, where he shed his previous name, Parua, and called himself Tupaia, "the beaten." There, he became the political advisor and lover of Purea, the mother of a high-ranking chief. Tupaia must have cut an impressive figure: only good-looking men and women could become 'Arioi, who wore only the finest bark cloth and often had their legs tattooed from thigh to heel. Dedicated to the worship of the war and fertility god 'Oro, they were master navigators and repositories of ancestral, sacred knowledge. They led rituals and sacrifices (including human ones) but would lose their sacred status unless their babies were killed at birth.

During the *Endeavour*'s three-month stay in Tahiti, Banks became close with Tupaia, who guided the Englishman around the island and proved himself a fine diplomat and gifted linguist, acquiring a good command of English. He demonstrated how to butcher and cook a dog in the traditional Tahitian manner, a meal that both Cook and Banks found delicious, much to the horror of the rest of the crew. When the day of the transit came, Cook and his crew stood at the ready, barricaded inside "Fort Venus": a fortified camp guarded by forty-five men and ten heavy guns to make sure their observations could proceed undisturbed by the locals. They were blessed with ideal conditions, as Cook reported in his journal: "This day proved as favourable to our purpose as we could wish. Not a Cloud was to be seen the whole day, and the Air was perfectly Clear, so that we had every advantage we could desire in observing the whole of the Passage of the planet Venus over the Sun's Disk."[28]

With the first part of his mission accomplished, it was time for Cook to turn to the sealed instructions he had been carrying from the Admiralty. In two pages marked "Secret," Cook was told to "proceed to the Southward in order to make discovery of the Continent above mentioned [Terra Australis Incognita] until you arrive in the Latitude of 40°, unless you sooner fall in with it."[29] This imaginary continent was not Australia, which Europeans had known about since the early seventeenth century, but a new landmass that was supposed to lie below forty degrees south. Should Cook locate it, he was to chart its coastline thoroughly and report on "beasts and fowls," trees, seeds, fruits, minerals, gems, and the "genius, temper and disposition" of any locals. The whole crew was to be put under strict orders to maintain absolute secrecy over any such discovery upon their return.

As the *Endeavour* prepared to venture farther into what, for the Europeans, was a vast, blank space on their maps, Cook realized that Tupaia's knowledge of those waters—and their islands—would be invaluable. Tupaia claimed to know the names, positions, and characteristics of nearly 130 Pacific islands, most of which no European had ever laid eyes on before. Even more astonishingly, he could pilot the ship from one island to the next and reach a safe harbor without the aid of any instrument or chart. Cook was impressed, as he confided to the ship's journal: "We found him to be a very intelligent person, and to know more of the Geography of the Islands situated in these Seas, their produce, and the religion, laws, and Customs of the inhabitants, than any one we had met with, and was the likeliest person to answer our Purpose."[30]

As the crew prepped the *Endeavour* for the next leg of her mission, Tupaia was keen to join them. He hoped to persuade Cook to help him take back his home island from the invaders who had driven him away and perhaps had reason to fear for his own life in Tahiti, as his loyalty to his British friends bordered on treachery in the view of the islanders' chief. Tupaia managed to convince Cook to take him and his young disciple Taiata along with the endorsement of his friend Banks, who seems to have felt genuine affection for Tupaia while at the same time regarding him as less than human, a living trophy, a pet. "I do not know why I may not keep him as a curiosity as well as my neighbours do lions and tigers at a larger expense than he will ever probably put me to. The amusement I shall have in his future conversation, and the benefit which will be derived by this ship, as well as any other which may in the future be sent into these seas, will, I think, fully repay me," he confided to his journal.[31]

When the *Endeavour* left Tahiti on July 13, 1769, Tupaia, according to Banks, "stood firm at last in his resolution of accompanying us, [and] parted with a few heart-felt tears, so I judge them to have been by the efforts I saw him make to hide them."[32] The anchor was weighed, amid a sea foaming with canoes filled with the islanders crying out their good-byes at the top of their lungs. Banks and Tupaia stood "for a long time" shoulder to shoulder on the topmast platform, waving at the canoes until they disappeared beyond the horizon. A new and unique chapter in the history of navigation—and in the conflicted relationship between European colonial powers and Polynesian people—was about to begin.

A CLASH OF CULTURES

Cook lost no time putting Tupaia to the test. For the next four weeks, the 'Arioi priest took over piloting the *Endeavour*, which he first sailed to the island of Huahine, where Tupaia took on the role of cultural liaison between the Englishmen and the locals and secured a friendly welcome, as well as supplies, for the visitors. From there, Tupaia guided the ship back to his home island of Raiatea, traversing a difficult and sacred passage through the reef to anchor in a sheltered bay in front of a ceremonial site, where he officiated a rite of welcome for the British. Banks was impressed by Tupaia's piloting skills:

> Tupia [*sic*] to-day shows us a large breach in the reef of Otahah, through which the ship might conveniently pass into a large bay, where he says there is good anchorage. We have now a very good opinion of Tupia's pilotage, especially since we observed him at Huahine send a man to dive down to the heel of the ship's rudder; this the man did several times, and reported to him the depth of water the ship drew, since when he had never suffered her to go in less than five fathoms without being much alarmed.[33]

Tupaia's fame grew quickly onboard, with the crew begrudgingly acknowledging his navigational prowess. Accustomed as he was to being treated with deference by his people in virtue of his sacred status, he came across as haughty, "proud and austere, extorting homage, which the sailors . . . were very unwilling to pay," according to Cook.[34] But his abilities were undeniable. At any time of day or night, he could point back in the direction of Tahiti without fail. "There was not a star fixt or erratic [a planet] but Tupaia could give a name to, tell when and where it would appear and disappear; and what was still more wonderful, could foretell from the aspect of the heavens the changes of the wind, and the alterations of the weather, several days before they happened," noted a sailor.[35] Banks, too, was intrigued by Tupaia's extraordinary method of predicting the weather, which was if "not infallible . . . far more clever than Europeans [*sic*]": Tupaia explained to him that he believed that the wind buckled the arch of the Milky Way, so that its shape would foretell the direction of the wind the next day.[36] Streams of air in the Earth's atmosphere do not change the shape of the galaxy, of course, but perhaps Tupaia was looking for a far subtler influence of wind or clouds on the appearance of the

Milky Way—something that only his trained eye could detect—similarly to how Australian Aboriginals use the twinkling of stars, an indication of high-altitude winds, to forecast the weather. Day after day, island after island, the *Endeavour* ventured farther west with Tupaia as its guide: "we again launched out into the ocean in search of what chance and Tupia [Tupaia] might direct us to," wrote Banks.[37]

When Cook questioned Tupaia about the southern continent he was after, Tupaia told him that his father had once spoken of islands to the south, which, however, he had never visited. Tupaia assured him that by continuing west, within ten or twelve days of navigation they would encounter plenty more islands, information which Cook had "no reason to doubt," based on how accurate Tupaia's directions had been thus far. But by mid-August, the lieutenant had tired of Polynesian island-hopping and was determined to follow his new orders. "I shall spend no more time searching for [the islands to the south that Tupaia had spoken of], being now fully resolv'd to stand directly to the Southward in search of a Continent."[38]

As the *Endeavour* searched in vain for the nonexistent continent the Admiralty expected Cook to claim for the king, Tupaia, Banks, the master's mate Richard Pickersgill, and Cook himself likely co-created the extraordinary cultural artifact that is Tupaia's Map.[39] Covering an area of the Pacific Ocean equivalent to the US landmass, the chart is, according to historian of science David Turnbull, an instance in which "two knowledge traditions encounter one another and become blurred in the representation."[40] No matter how the map is rotated, flipped, or warped, the seventy-odd islands represented thereupon do not match the geographical location of the archipelagos they purport to chart. Revealingly, the forty or so islands that were known to, or had been visited by, the British are in their "correct" location, according to the Western, bird's-eye-view mapmaking convention. There is no doubt that Tupaia understood how the British intended to set up the chart, with the cardinal directions indicated at the four sides of the sheet of paper, north at the top. But the Polynesian way of understanding navigation was profoundly different, taking as reference point either the voyager's canoe or the departure island. In the well-documented approach of navigators from the Caroline Island, an archipelago stretching over eighteen hundred miles in the Western Pacific, the canoe was thought of as stationary, with the islands moving backward with respect to the rising and setting points of the stars. The sky was called "the roof of voyaging" and endowed with a star compass comprising thirty-two stars, whose rising and setting locations on the horizon provided orientation. It was not an absolute reference frame but one relative to the observer's position.

Tupaia's stroke of genius was to override Western convention and place the islands of which he had exclusive knowledge in a relative (rather than absolute) position to the observer, understood as a voyager aboard a canoe. Hard evidence for this interpretation comes from the word *Eawatea*, found labeling the crossing of the cardinal axes at the center of the map. The actual Tahitian word, *avatea*, means "noon," or the position of the Sun at midday—the north direction in the Southern Hemisphere. So the reference point, for Tupaia, lies at the center of the chart, where the sea voyager always is to be found: at the center of their own journey, with the sea, the islands, and the stars flowing all around them.

Tupaia's Map, then, is a unique example of cultural mélange, highlighting the fundamental incomprehensibility of the Polynesian knowledge system for the Europeans, blinded by their patronizing and chauvinistic view that scientific Western knowledge was the only knowledge worth possessing.[41] This also explains why the astounding Polynesian astronomical and navigational lore, which the British saw and admired in action, wasn't investigated in more depth by Banks, Green, and Cook, even as a mere curiosity. To them, it simply could not have possessed comparable value.

Cook's secret mission appeared successfully concluded when, on October 6, 1769, land was sighted, and "all hands seem to agree that this is certainly the Continent we are in search of," as Banks jubilantly wrote in his diary.[42] It wasn't: the *Endeavour* had reached Aotearoa/New Zealand, where Tupaia found he could speak the local language and was feted by the Māori, who thought he was in charge of the ship. Tupaia had found the long-forgotten island his father had spoken about, centuries after voyaging between the Society Islands and Aotearoa/New Zealand had ceased. Despite his initial reservations about bringing Tupaia along, Cook was at last won over and recommended to the Admiralty that Tupaia should be sent back in for future voyages of exploration: "But should it be thought proper to send a Ship out upon this Service [exploring further the Southern Pacific] while Tupia [Tupaia] lives, and he to come out in her, in that case she would have a prodidgious [*sic*] Advantage over every ship that hath been upon discoveries in those Seas. . . . This would enable the Navigator to make his discoveries the more perfect and Compleat [*sic*]."[43]

But Tupaia wasn't to set eyes on the North Star. According to the expedition's artist, Sydney Parkinson, Tupaia became demoralized by his much-reduced usefulness after the *Endeavour* left Aotearoa/New Zealand behind, and with it people and languages he was familiar with. He became increasingly isolated and despondent, to the point that "he gave himself up to grief; regretting, in the highest degree, that he had left his own country."[44] Perhaps Tupaia grew disillusioned with the British, too, after he witnessed Cook and his men fire wantonly on the Māori on multiple occasions in October 1769, then later claim New South Wales for the king when they anchored at an island Cook called Possession Island on August 22, 1770.

Unaccustomed to the British ship diet lacking in fresh vegetables and fish, Tupaia ate poorly and became sick with scurvy, refusing medical help. He nursed himself back to health while the *Endeavour* was undergoing repairs in Batavia (in today's Jakarta), but like much of the rest of the crew, he and his adored disciple Taiata both succumbed to dysentery and tropical fever. When Taiata died, Tupaia wept desperately and called out his name for two days before laying down his own life on December 20, 1770.

When Cook returned to Aotearoa/New Zealand during his second voyage in 1773, a steady cortege of canoes paddled out to see him, full of Māori Cook had never met before. The strangers were eager to meet Tupaia, and they wept hard upon hearing of his demise. Cook concluded that "the Name of Tupia [Tupaia] was at that time so popular among [the Māori] that it would be no wonder if at this time it is known over the great part of New Zealand."[45] Stories of Tupaia's visit to Aotearoa/New Zealand (and perhaps some direct descendants of his) have reached down the centuries to today. His name lives on in a mistletoe species native to Aotearoa/New Zealand, *Tupeia antarctica*.

Cook was feted as a hero upon his return to Britain in July 1771 and promoted to commander by the king himself. The data on the transit of Venus he brought allowed Thomas Hornsby, the Savilian Professor of Astronomy at Oxford, to estimate the AU to 93,726,900 English miles—within 1 percent of the actual value.[46] In his absence, he also discovered, an acrimonious battle had reached its climax in London on the awarding of the Longitude Prize: as a reward not for the method of lunars that Cook had been so pleased with but for a new kind of clock, a high-tech mechanism that, if successful, could revolutionize Western navigation as well as the exploration, and exploitation, of distant lands.

OUR TRUSTY FRIEND, THE WATCH

Archimedes had allegedly translated heavenly cycles onto the surface of his sphere, and de' Dondi had done so onto the face of his astronomical clock. When it came to determining longitude onboard a ship, nothing could beat the convenience of a mechanical clock, provided it could reliably keep the time of the port of departure over the course of several weeks at sea. From the Antikythera mechanism onward, the returning regularity of the Sun, the Moon, and the stars above had inspired the creation of machines to imitate and reproduce them: it was now time for the starry time-givers to be wholly replaced by a mechanical surrogate. Deep down, navigating by mechanical timekeeping was still a gift of the stars.

Among the heirs of the clock-making tradition of Jacopo and Giovanni de gli Horologi, an English carpenter had risen to the threshold of claiming the fabulous Longitude Prize. Like James Cook, John Harrison was a Yorkshireman of humble beginnings, who through talent and tenacity had achieved what many thought impossible. By the time Cook set sail on his second voyage around the world in July 1772, Harrison had been locked in a spiteful fight with the Board of Longitude for over forty years. The bone of contention was whether Harrison's increasingly sophisticated marine chronometers had met the exacting standards of precision required for him to claim the top prize. His first "sea clock," the H-1, completed in 1736, was a brass affair with mechanical entrails exposed behind the four front dials. Its system of springs and swinging balances was designed to be unaffected by the ship's motion, to withstand temperature and humidity changes, and to forego lubrication. It consisted of fifteen hundred parts, including two meter-long cone-shaped pulleys, each made of two thousand pieces.

Harrison's fourth prototype, the H-4, which he nicknamed simply "the Watch," was much soberer: in appearance, nothing more than a hefty pocket watch in a steel encasement, with elegant black roman numerals on a white dial. All the magic was hidden in its miniaturized mechanism, which distilled over fifty years of research and experience by its maker. The H-4 had been sailed to Jamaica in 1762, losing only five seconds after eighty-one days at sea; then to Barbados in 1764, where it was found to be accurate to within forty seconds on arrival—well within the two-minute tolerance required to win the longitude prize. But the Board of Longitude was still not satisfied: one of its most influential members, Astronomer Royal Nevil Maskelyne himself, had convinced the board to confiscate all of Harrison's prototypes for further testing. Before he could be given the full longitude award, Harrison had been requested to disassemble his H-4 in front of a committee of watchmakers, to explain its functioning to them, and to build two more copies from memory.

With no end to the dispute in sight, Cook was tasked with testing Harrison's watch—not the H-4, deemed too precious to be risked on a circumnavigation of the globe, but an exact copy made by the watchmaker Larcum Kendall, called the K-1. This was pricey tech: the Admiralty shelled out £500 for the K-1, approximately a tenth of the purchase cost of the entire ship it sailed on. Armed with a chronometer that was supposed to keep Greenwich time to within a fraction of a second per day, Cook needed only to determine local noon from the position of the Sun to work out his longitude by comparing it with the time shown by the K-1. This simplified matters enormously, and the method worked even when the Moon wasn't visible.

Just as he had been enamored of the then novel lunar method during his first voyage, Cook was swayed by the new gizmo he brought along on his second. He reported to the Admiralty that "Mr. Kendall's Watch has exceeded the expectations of its most Zealous advocate and by being now and then corrected by lunar observations has been our faithful guide through all the vicissitudes of climates."[47] Thanks to his "trusty friend, the Watch," Cook was able to sail directly to Saint Helena and find it in the middle of the South Atlantic, short-circuiting the old, time-consuming practice of first reaching the latitude of the target island and then steering east or west until land was sighted. This was an impressive demonstration of the marine chronometer's potential to reduce time at sea, and therefore costs, for the British commercial fleet.[48]

When Cook returned home in July 1775 with his glowing endorsement of the Watch, the Harrison affair had already been adjudicated. Enraged by the board's continuing delaying tactics, John Harrison had petitioned the king himself, who personally ascertained the accuracy of the H-4 and thundered, "By God, Harrison, I will see you righted." In 1773, Harrison received £8,750 by act of Parliament on top of the £10,000 he had already been given, but he was never officially awarded the Longitude Prize—nobody was, in fact. Harrison died a wealthy man in March 1776. Maskelyne, who had stalled him for decades, held the office of Astronomer Royal until his death in 1811, and his yearly lunar distance tables weren't published until 1906. The Board of Longitude was disbanded in 1828.

As marine chronometers became cheaper and more reliable in the second half of the nineteenth century, ships were equipped with triplicate copies to guard against failure. The HMS *Beagle*, for example, carried in 1831 no fewer than twenty-two chronometers. Its mission: to travel around the world and correct charts of South America and the Pacific by means of more precise longitude measurements. Its captain, Robert Fitz-Roy, fearing the solitude of the high seas that had pushed the previous commander of the *Beagle* to suicide, sought the company of a gentleman during the voyage. He found one Charles Darwin, a twenty-two-year-old budding naturalist. The voyage took them to South America, the Galapagos, Tahiti, Aotearoa/New Zealand, Australia, Mauritius, and South Africa, then back across the Atlantic to Brazil to check some longitude measurements FitzRoy wasn't too sure about (they were fine).

Darwin returned in 1836 a changed man. The longitude measurements that were FitzRoy's main mission gave Darwin the opportunity to see firsthand the variety of natural forms that plants and animals exhibited on far-flung, isolated shores, thus planting the seed for his theory of natural evolution. Weeks after his return, Darwin's ideas would be further influenced by a star-inspired encounter at one of Victorian London's most dazzling salons, as Chapter 8 will recount.

TIME FOR A CHANGE OF TIME

With the rapid development of transportation links across the world's oceans, enabled by more and more reliable means of navigation, from the nineteenth century onward the need arose to standardize charts and location measurements. If the equator was the obvious choice as reference for the north-south location, the "meridian zero" dividing east from west was a matter of convention, and different nations each used their own reference meridian. Choosing a "meridian zero" also meant choosing a reference point in time, since chronometer-based longitude measurements were obtained from the difference between local time and time at the reference meridian. On a planetary scale, space and time had become inextricably enmeshed, decades before Hendrik Lorentz, Henri Poincaré, and Albert Einstein introduced the idea of four-dimensional space-time.

The International Meridian Conference was convened in October 1884 in Washington, DC, to decide on a common reference point for the zero of longitude and to standardize time reckoning around the world. By this time mariners were long used to Maskelyne's *Nautical Almanac*, whose tables were computed with reference to Greenwich: 65 percent of the sixty thousand ships in the world already adopted it, often in conjunction with or to double-check chronometer-based measurements. The prime meridian, the line separating east from west and from which time zones are computed, was therefore drawn through the axis of the great transit circle telescope at the the Royal Greenwich Observatory, much to the dismay of the French, who had used the Paris observatory instead. For the next ninety-four years, the French haughtily continued to use "Paris Mean Time, retarded by 9 minutes and 21 seconds" rather than utter the bitter words "Greenwich Mean Time." In 1978, they eventually moved to Coordinated Universal Time, which by definition is always kept within 0.9 seconds of Greenwich Mean Time.

Meanwhile, the advent of railways first and the telegraph later had dramatically increased the ease of movement and flow of information long-distance. This gave sudden urgency to the question of establishing a uniform time across a nation, thus overriding a patchwork of time-reckoning systems based on the local position of the Sun in the sky. In the United Kingdom, railway operators on different lines ran their trains according to the local times of whatever city they chose. Clocks at train stations kept local time, resulting in utter confusion for travelers. In the 1860s, the ever-resourceful dons of Christ Church College, Oxford, "solved" the problem by adding a second minute hand to the great clock on the college's main tower. One hand showed Greenwich time, the other Oxford time, five minutes behind.

The difficulty was magnified in the United States, where the time difference from coast to coast is more than three and a half hours. A traveler taking the train from Portland, Maine, could arrive at Buffalo, New York, at 12:40 p.m. local (Buffalo) time. While her own watch would indicate 1:15 p.m. (Portland time), upon disembarking from her train she would see the New York Central Railroad clock display 1 p.m. (New York City time) and the Lake Shore and Michigan Southern Railway clock show 12:25 p.m. (Columbus, Ohio, time).

To put a stop to the temporal chaos in the United Kingdom, the Royal Greenwich Observatory was tasked with distributing a uniform time to the nation—and that meant that astronomers needed to observe it first. Every night, Greenwich astronomers measured the transit of clock stars through the spider web threads in the transit circle telescope, then translated the sidereal time into solar mean time, as we saw. The exact time from the stars was then communicated by several means. From 1833 onward, a large ball was dropped along a mast at precisely 1 p.m. every day from the top of the the Royal Greenwich Observatory, so shipmasters on the Thames could set their watches by it, a practice that continues to this day. From 1892 to well into the 1930s, Ms. Ruth Belville, aka the "Greenwich time lady," had a chronometer reset to Greenwich time each Monday and then visited up to fifty watchmakers in London during the week to sync their watches with Greenwich. Each morning a messenger from the Admiralty carried a watch with the time to Euston train station, so that the correct London time could be taken to Holyhead, in the northwestern tip of Wales, and from there by boat to Dublin.

By 1852, a few select users could count on the hourly distribution of the exact time by electrical signals, which radiated out from the observatory and almost instantaneously synchronized clocks at the Central Telegraph Office, the General Post Office in London, and the iconic Westminster Tower clock, Big Ben. Daily signals were sent to provincial towns all over the kingdom, where the exact time was marked sometimes by the automated firing of a gun, the dropping of a ball, or the lighting of a light, or simply by putting right the time on another clock. Automatically and mostly silently, the time from the stars was spread all over the nation, changing people's habits and helping establish industrial society, with its railroad timetables, regulated trading hours, and strict factory shifts.

In the twentieth century, radio signals and later the satellite-based Global Positioning System (GPS) replaced marine chronometers, sextants, and lunars aboard ships and airplanes. Just as Harrison's H-4 would probably have not existed had the stars not inspired the long line of its predecessors, so would GPS not work today without Einstein's theories of special and general relativity, both of which owe much to the stars. The little dot that shows us our location on our mobile phones today triangulates radio signals coming from four satellites orbiting the Earth. But time flows differently at your location, on the surface of the Earth, and onboard the satellites, as a consequence of both their relative speed and the curvature of space-time induced by the Earth's mass. The resulting tiny time shift, approximately thirty-eight milliseconds per day, is sufficient to throw off the GPS's accuracy unless it is accounted for by using Einstein's theories of special and general relativity. General relativity, in part motivated by anomalies in the orbit of Mercury that Newtonian gravity could not explain, was first tested in 1919 by looking at the displacement of stars during a total solar eclipse. Even today, navigation as we know it wouldn't exist were it not for the stars.

Celestial navigation is making a comeback in the twenty-first century. The US Navy has reintroduced training in it as mandatory for its officers, amid concerns that its operations are becoming overreliant on GPS, whose satellite signals can be jammed.[49] The nearly lost Polynesian voyaging tradition is also being brought back to life, building on the wayfinding knowledge that survived colonialism. In 1980, Hawaii native Nainoa Thompson piloted a replica of a double-hull voyaging canoe from Hawaii to Tahiti and back, a round trip of four thousand miles, with traditional voyaging techniques—among them, using his hand in place of a sextant to judge the height of a star over the horizon, a method made popular by the animated Disney film *Moana*. "When everything is going right," said Thompson, "you have the star patterns in mind and you seem to know where you are even when the sky is cloudy and you can't see the stars."[50]

Finding patterns in the stars has meant many things over time, as we've already seen, but one particular kind of pattern seeking has influenced all aspects of our life: science. To fully understand how the stars have guided us to where and who we are today, we'll next retrace our steps to the roots of the Scientific Revolution. We need to revisit the seismic shift of perspective flowing from the Copernican heliocentric system and peer over the shoulder of Galileo as he points his telescope to the night sky; we need to follow Newton as he gets lost walking from his lodgings to the dining hall, en route to inaugurating a new conception of the cosmos and our place in it. To understand the modern world, in short, we need to understand the starry science at its foundation.

THE CALIGO TALES

Way-Finder's Tale

Way-Finder stood, and thus began his story.

I remember what Elder-Way-Finder told me—a story
he remembered from his Elder, who remembered
it from his Elder, and so on for many Elder-Way-
Finders, as many as fit on your two hands.

Many Elder-Way-Finders ago, our people used to
walk to the border of the Disc at each bat awakening, to meet other peoples on the shores of the Great
Water.

It was a long and difficult trip, and Long-Ago-
Elder-Way-Finder was the only one who remembered the way: down the river, over the great rock,
along the woodland's edge you walk, until after
three glows you reach the Great Oak—that's where
you turn into the forest, and walk another three
glows until the Meadow of the Big Swap lies in front
of you, the Cloud almost touching the treetops. To
get there, Long-Ago-Elder-Way-Finder remembered the markings in the bark, the carvings on the
rocks, the fording of streams—all of which has been
forgotten.

But one bat awakening, the Great Oak was gone. The Cloud had sent a wind that, like a huge hand, had flattened the woods. Long-Ago-Elder-Way-Finder could no longer find the turn into the forest, and from then on our people never went again.

At the last Big Swap before the way was lost, our people brought their best hides and blades, and brought back the finest needles, and two strong husbands.

From that last Big Swap we also received the story of the people from across the water. These strange folk told Long-Ago-Elder-Way-Finder that they had traveled to the Big Swap on a log that could grab the wind, though no one had seen it. The man who spoke for them called himself Cook, but he prepared no food, so everybody called him Big Sail. Big Sail said he had been on his log all around the Disc, seen fish the size of mammoth, and walked on sand so hot it burned his feet. People looked at him, their eyes wide, their mouths open, not sure whether to believe him.

Everybody knew he had a secret: a black rock that, it was said, had been sent to Big Sail in a ball of fire by the Cloud, and that told him which way to go.[51] They said the rock moved by itself. They said the rock was magical. Big Sail never said a word about it, but kept a pouch close to his heart.

When Darkness fell, Long-Ago-Elder-Way-Finder went for the pouch, but Big Sail was a light sleeper. They fought and Big Sail cried for help. Our people were ready, and drove their blades through the strangers' hearts.

Inside the pouch they did find a flake of black rock, thin and long, like a needle, and a fine thread of spider web. They looked at it for a long time, but the rock didn't move; nor did it show them the way. Now that Big Sail was dead, its magic was gone, said Long-Ago-Elder-Way-Finder.

They forgot about the rock and came back to the cave with the finest needles and two strong husbands.

7

FROM BEAUTY, ORDER

*I can never look upon the Stars without wondering why
the whole World does not become Astronomers.*

—THOMAS WRIGHT, *An Original Theory or
New Hypothesis of the Universe*

A GOLDEN FLOWER IN SPACE

As the first images trickled in from deep space, we were in the garden tracking down the Summer Triangle, an easy-to-spot reference pattern in the Northern Hemisphere's summer sky. The Triangle's vertices are among the brightest stars: Vega, in the constellation Lyra; Altair, in the neck of the Eagle; and Deneb, in the tail of the Swan. An ancient Chinese myth, I told the kids, says that the weaver fairy Zhinü, or Vega, and her husband, the cowherd Niuland, or Altair, are separated by an impassable river in the sky, the Milky Way. Once a year, all the world's magpies are said to flock to the sky, forming a bridge across the Milky Way and reuniting the lovers for one night.

The full Moon spilled silver light over the sunbaked tiles of the nearby farm, washing out the night and calling time on our observing. But I knew that somewhere high above the Moon and immune to its light, an unblinking, man-made eye was fixed on the dawn of time.

Later, with Emma and Benjamin asleep, I browse on my laptop the first image release of the James Webb Space Telescope (JWST), the successor to the hallowed Hubble Space Telescope.

The JWST is our most advanced telescope, the pinnacle of technologic progress in astronomy. What is not so progressive is its name, eponymous of a high-ranking NASA administrator in the 1960s who has come under criticism for his connection to a homophobic cull within the space agency.[1] James Webb may or may not have been openly homophobic, but he certainly was a white, cisgendered man—like almost all of the historical figures we are about to meet in this chapter. We cannot change the history of prejudice and gatekeeping in science, but there is much we can and should do today to change the future of astronomy, to make it more inclusive and diverse, and to celebrate the achievements of *all* of its members.

The JWST itself is still an incredible observatory. Comparisons of its images with those of its predecessor are merciless—at least as much of a leap forward as Hubble's were over those of ground-based telescopes in the 1990s. The crispness, detail, and resolution are stunning. On this evening, I gape for a long while at a cluster of distant galaxies, whose shape is distorted by the invisible hand of gravity as its light journeys through darkness to us. Some of my colleagues are already at work on these images to tease out the signature of dark matter, but tonight I am content to marvel. The delicately bent galaxy shapes, like glowing commas, remind me of the flourishes I once saw a Venetian glassblower magic out of molten glass. An image of the Carina Nebula is sculptural: young stars blaze in a blue haze of hydrogen, sharply bordered by a landscape of ochre dust three-dimensional in its shades, nooks, and edges. The professional in me knows that the framing, orientation, and color palette have all been selected for maximum aesthetic impact; yet my awe is real.

The first telescope pointed at the stars was a slim tube of wood and glass, made in seventeenth-century Padua by one Galileo Galilei. Four hundred years later, its descendant is a space telescope with eighteen hexagonal beryllium mirrors, engineered to unfold into the shape of a flower 1.5 million kilometers from Earth. JWST has a much larger mirror than the Hubble Space Telescope and is designed to see infrared radiation, largely invisible to Hubble, which enables it to see farther out in space and further back in time. The JWST's mirrors are gold plated, gilded like the domes of cathedrals used to be. It is designed to peer back in time to the dawn of stars, when gravity ignited the first atomic fires in huge hydrogen balls and, according to the Bible, "God said, 'Let there be light.'" Its spectrometer will be trained onto stars in our own galaxy, looking for the telltale signatures of chemical compounds in the atmospheres of their planets that can betray life there. In a slimy tidal pool on some unremarkable rocky planet hundreds of light-years away, a cosmic shift of perspective is perhaps brewing, awaiting the JWST to announce its existence to us Earthlings. To find life, however small, outside our pale blue dot would shatter any lingering doubts as to our uniqueness in the universe—a dethroning comparable to only one other moment in the history of humankind, when we found out the universe did not literally revolve around us.

Explaining the motion of planets, stars, and luminaries in the sky was always the stuff of myth, legend, and religion. But it also set off the Copernican Revolution and instilled the key insight that nature's laws are intelligible—and, later, exploitable—because they are written in mathematical language. The stars shaped, then changed, our view of the universe and of ourselves—and may do so again in our lifetimes.

To understand how this all came to be, Florence in 1610 is not a bad place to start.

FROM HUMBLE BEGINNINGS

The Museo Galileo is hosted in a magnificent palazzo from the eleventh century, its windows overlooking the right bank of the Arno River in Florence, around the corner from the world-famous Uffizi and the Palazzo Vecchio, which during the Renaissance was the residence of the de' Medici family. I had traveled to Florence to pay my respects to the ancestor of the JWST.

It looked like a toy, something an ingenious child might cobble together on a slow summer afternoon. It was slender and, measuring over a meter, longer than I had imagined. A nineteenth-century fresco depicts Galileo demonstrating it to the doge of Venice, mounted on an intricately decorated bronze stand, but no such stand was on display. The tube itself was made of strips of wood held together by red leather that had long ago turned brown; at each end, a snugly fitted thicker wooden cylinder, decorated in gold leaf, concealed its key technology: two glass lenses, each flat on one side and ground in the shape of a flattened dome on the far end and a shallow bowl on the near end, where—I imagine—Galileo's eye would have widened with amazement.

This was but one of the progenitors to today's telescopes. Only two of Galileo Galilei's creations survive, but the Renaissance inventor, astronomer, and polymath built over a hundred variations, refining his technique with experience. The invention of the telescope is credited to Dutch spectacle maker Hans Lippershey, who in 1608, according to a version of the story, observed that the weather vane of a nearby church appeared magnified when viewed through a certain combination of lenses that some children had been playing with in his shop.[2] Word of Lippershey's work reached Galileo in Venice in 1609, and his curiosity was tickled, as he recalled in 1623: "The news arrived that a Dutchman had presented Count Maurice with an eyeglass which showed distant objects as perfectly as if they were quite close. Nothing further was added. At this report, I went back to Padua, where I was then living, and set about thinking the problem over. The first night after my return I had solved it."[3]

His first *cannocchiale* (an Italian word invented in 1611 by fusing *canna*, "barrel," and *occhiale*, "glasses," to describe what had until then been called a *perspicillum* by its maker and "Galileo's tube" by everybody else) magnified a mere three times. Within six days, Galileo's rapid prototyping achieved nine enlargements. A month later, the *cannocchiale* was all the rage in Venice, where noblemen and senators hauled themselves to the top of a steeple to marvel at boats, two hours' sailing away and invisible to the naked eye, magically materializing in perfect detail at the other end of Galileo's tube. The shrewd inventor presented his miraculous device to the doge, who extended Galileo's appointment at the University of Padua for life and doubled his salary on the spot.

This seems to have been Galileo's plan: to be rewarded by the doge for an instrument that conferred him strategic advantage at war. Yet he was not just a practical inventor. He had an insatiable mind and by 1609 had been obsessed for decades with discovering how God had constructed the world and establishing its hidden structure. Now that he had built for himself the means of seeing farther than anybody ever had, the most revolutionary use of his *cannocchiale* wasn't war; nor were the most intriguing targets distant ships on sea or landmarks on land. The view he wanted was nothing less than that of God's own abode—the firmament that had attracted and resisted closer human scrutiny for thousands of years. Skirting blasphemy, Galileo pointed his *cannocchiale* at the heavenly spheres. In doing so he would shatter fourteen hundred years of belief.

A FOUR-PRONGED ASSAULT

At the turn of the sixteenth century, our understanding of our place in the universe was about to change forever. With the four-pronged assault of a reclusive Polish canon, an ill-tempered Danish nobleman, a meek German high school teacher, and our man Galileo, by the end of the seventeenth century the great Ptolemaic castle would fall to the flag of heliocentrism— the Sun shining at its center, the Earth and other planets revolving around it—raised finally by a Cambridge don who once played pranks with fake comets. Along the way, their aloof yet exacting general, the starry firmament, led them to victory.

When the revolution began, a single vision of the universe, the Ptolemaic model, had held sway for fourteen hundred years. Its captivating central tenet: everything revolves around us. Ptolemy perfected in the second century AD the model put forward in the third century BC by the great Greek philosopher Aristotle, who maintained that the Earth was spherical and stationary at the center of the universe. The Moon, Mercury, Venus, the Sun, Mars, Jupiter, and Saturn (in order of distance) all circled around the Earth, carried along invisible, solid crystalline spheres nested inside each other. The highest sphere was studded with the so-called fixed stars and marked the boundary of the universe. The ninth and last sphere was the seat of the First Mover, responsible for generating and transmitting movement to all the other spheres.

Ptolemy—a skilled observer, as we saw in Chapter 5, who compiled a catalog of over one thousand stars—realized that updates were necessary to make Aristotle's model accord with observations. Ptolemy knew that the Sun's apparent speed against the stars changed in the course of the year, in contradiction to Aristotle's model, in which the celestial bodies revolved along circles with constant speed, their perfect movements in keeping with their divine, incorruptible nature. The planets presented another headache: night after night they drifted eastward with respect to the stars but stopped at regular intervals, reversed direction, then stopped again and resumed their previous course. (*Planet* comes from the Greek word for "wanderer.") Ptolemy drew on previous theories from Apollonius and Hipparchus, while retaining Aristotle's key notions of the circular perfection of planetary orbits and a stationary Earth, and introduced the idea of epicycles to align the geocentric model with observations.

In the Ptolemaic system, each planet moves with constant speed in a circle (the epicycle), whose center is carried along a bigger circle (the deferent). The deferent's center is displaced from the position of the Earth. A second point, called the equant, lies on the opposite side of the deferent's center with respect to the location of the Earth. An imaginary observer located at the equant would see the center of the epicycle move at constant speed—a subterfuge to maintain the appearances of the Aristotelian uniform motion. By choosing the size and speed of epicycles and deferents for each planet, the Sun, and the Moon, Ptolemy was able to reproduce in his model the complex motions of the heavenly bodies in the sky.

The great astronomer Aristarchus of Samos, a contemporary of Aristotle, had suggested a different way to make sense of observations: a heliocentric model in which the Sun, not the Earth, was fixed at the center of the solar system. But it never caught on. Ptolemy had offered too convincing a translation of Aristotle's elegant theory. And had Aristotle not made it a matter of common sense that the Earth could not be moving or spinning around? If it were, he had said, an arrow shot vertically would be left behind by the movement of the Earth, which everyone could attest was not the case.

Another argument against Aristarchus's ludicrous idea was that if the Earth were orbiting the Sun, the position of distant stars should appear to change every six months, a phenomenon called *parallax*. Parallax is easily experienced by holding a finger up at arm's length and closing each eye in turn. The finger appears to jump left and right with respect to the background, due to the different perspective of each eye. In the same way, when observed from opposite ends of the Earth's orbit, stars should appear to move—by more the closer the star is to us. But no parallax was noticeable with the naked eye; its discovery would have to wait until 1838, as we shall see. This meant that either the Earth was fixed, as Aristotle dictated, or the stars were at such an unfathomable distance as to make their parallax too small to be seen, which seemed preposterous. The universe is simply vaster than anybody in antiquity was prepared to imagine.

Ptolemy's model and the numerical tables he created for predicting the positions of the planets worked beautifully for centuries: copied by generations of scholars, Ptolemy's books traveled eastward to Persia and India and then, like the planets whose motion they described, retraced their steps back into medieval Europe and into Latin, thanks to the work of Arabic astronomers. From Spain, Ptolemy's greatest work, the *Almagest*, resumed its eastward course to reach the new seats of learning—later called universities—sprouting up in Bologna, Paris, Padua, Oxford, and Cambridge.

In the meantime, the belief that the Earth was at the center of the universe had become cemented in common sense supported by the Judeo-Christian-Islamic worldview. The Old Testament or Torah revealed a geocentric universe: "[O Lord my God] who laid the foundations of the earth, that it should not be removed for ever."[4] In the Muslim world, the Qur'an described the Earth as "spread out like a carpet," covered by seven heavens sustained by invisible pillars, with the stars adorning the lowest sphere, where the Moon and the Sun "swim along, each in its rounded course."[5] Whether one believed in God, Allah, or Jehovah, the Earth was at the center of creation, and the Ptolemaic universe designed with humankind in mind was, almost literally, Gospel truth.

It would take more than a millennium for a few bold thinkers and sky watchers to prove Ptolemy wrong. But at the beginning of the sixteenth century, the Ptolemaic edifice was showing cracks. Plato's work, infused with the Greek philosopher's mathematical conception of the world and his vision of cosmic harmony and symmetry, was being studied and held in high esteem in European universities, and the Ptolemaic model fell short of the Platonic mark. For one thing, the equant construction was inelegant, and Ptolemy offered no insight into the relative distance of the planets and no explanation for their observed orbital periods. One of the scholars who studied Aristotle and Ptolemy in depth concluded that the geocentric model was composed of disparate, disharmonious parts that are "just like someone taking from different places hands, feet, head and the other limbs . . . not modelled from the same body and not matching one another—so that such parts would produce a monster rather than a man."[6]

This scholar was Nicolaus Copernicus, the son of a Polish merchant born in 1473, who took up astronomy while in Bologna to study canon law at the turn of the fifteenth century. To fix this Frankenstein of a geocentric cosmos, he dusted off the heliocentric universe. All of a sudden, the ugliness of the Ptolemaic system evaporated: previously Mercury and Venus had inexplicably shared in their epicycles the one-year period of the Sun; in Copernicus's formulation, Mercury and Venus orbited the Sun instead, and their true periods came out to be eighty-eight days and seven and a half months, putting them inside the orbit of the Earth, which takes one year to complete. This explained why Venus and Mercury always appear close to the Sun in the sky. The planets were thus arranged geometrically in order of their periods (Mercury, Venus, Earth, Mars, Jupiter, and Saturn), revolving in concentric orbits around the Sun at the center, for "who would place this lamp in another or better place than that from which it can illuminate the whole at one and the same time?" wrote Copernicus.[7] As for the puzzling retrograde motion of the planets, it was the obvious consequence of the combination of their orbital revolutions with that of the Earth.

The truth of his model, Copernicus believed, was to be found not in observational data, which could not distinguish at the time between the two options, but rather in its intrinsic elegance, "a wonderful commensurability . . . a sure linking together in harmony of the movement and magnitude of the orbital circles, such as cannot be found in any other way," he wrote.[8] Copernicus had beaten Aristotle, Plato's student for twenty years, at being a Platonist.

In 1543 Copernicus published his model in a book called *De revolutionibus orbium coelestium* (*On the Revolutions of the Celestial Spheres*). The story goes that he died clutching one of the first printed copies. But a later preface by the book's editor, Andreas Osiander, undermined its potency: it urged readers to take the Copernican model not as a description of reality but merely as a trick for computing more conveniently the orbits and positions of the planets. The nature of the solar system was not in question, wrote Osiander. And so Copernicus's model neither caught on nor raised the ire of the church—at first. The telescope would give his theory fangs.

THE CASTLE OF URANIA

Despite his being an accomplished astronomer, Copernicus's strength wasn't sky watching; his eye was directed to the beauty of theory. Enter Tyge Brahe, the scion of a noble Danish family who, at age thirteen, was hooked by the spectacle of a total lunar eclipse in 1560 in Copenhagen. From then on, astronomy and astrology combined in producing the greatest naked-eye observer in history. He latinized his name to Tycho (pronounced "Tee-ko") and set about producing horoscopes and nativities (descriptions of the astrological character of newborn babies) in the service of his patrons, first Frederick II of Denmark and then Holy Roman Emperor Rudolph II, whose court in Prague was renowned as a lusciously funded hotspot for alchemy and astrology. The short-tempered Tycho didn't take it lightly when made fun of during a betrothal celebration, allegedly for a wrong astrological prediction he had made. The quarrel escalated to a full-blown sword duel in the churchyard outside. When the other guests, alarmed by the ruckus, rushed outside, it was too late: Tycho's face was covered in blood, his nose neatly severed. The twenty-year-old Tycho was fortunate to survive, and for the rest of his life he wore an artificial nose that he fashioned in his alchemical laboratory. It was rumored that the nose was made of gold and silver, although analysis of his remains indicate it was a humbler brass. It's possible, though, that he did wear a gold nose on special occasions.

As a teenager, Tycho realized that both the Alphonsine tables, compiled three hundred years earlier on the basis of the Ptolemaic model, and the more recent Prutenic tables, based on the Copernican model, were inadequate to the task of predicting astrologically significant phenomena, such as the conjunction between Jupiter and Saturn he observed in 1563. As a young man of brains and means, he set out to rectify that. He convinced the king of Denmark to grant him the fiefdom of the isle of Hven, a windswept rock in the icy waters of the Øresund strait, twice the size of Central Park in New York City. By that time, he had already made his name as a first-rate astronomer by proving that the "new star" (a supernova explosion in the Milky Way, it turns out) that had appeared in 1572 in the constellation Cassiopeia was well beyond the lunar orbit. The immutable spheres of the Empyreans were not so immutable after all.

As the new lord of the island, he ordered the disgruntled peasants to dig up the common grazing field at the exact center of Hven, then had them build an extravagant cross between fairy-tale manor and observatory, surrounded by a geometric herb garden and arboretum planted with over three hundred varieties of trees. He designed his palace's four wings with musically harmonious ratios in mind and aligned with the cardinal directions. At its heart he placed a fountain with water sprayed all around by a rotating figure—a spectacular feature to impress his guests that not even Queen Elizabeth I could sport at her palace. Uraniborg, the Castle of Urania, muse of astronomy, was born.

To achieve the kind of observational precision Tycho demanded, limited only by the acuteness of human sight at a time when Galileo's *cannocchiale* was still three decades into the future, he set up an in-house workshop where he trained skilled craftsmen to build astronomical instruments the likes of which had never existed before, some taking six artisans three years to complete. In the workshop, he devised and built new, more exacting versions of his instruments, then tested them with repeated observations, in an unprecedented prototyping and development cycle. He later built a dedicated off-site observatory, sunken into the ground to shield it from the winds, which he called Stjerneborg, the Star Castle.[9]

After two decades of assiduous observation of the heavens, Tycho had accumulated a trove of uniquely high-precision data—which, however, he didn't have the means to interpret. A twist of fate put this data into the hands of perhaps the only person alive who understood their true significance: a math teacher with a penchant for mysticism, who had written a peculiar book explaining the planets' orbits with geometric figures and the Sun at the center.

THE MEASURE OF ALL THINGS

Born to a Protestant family of modest means near Stuttgart in 1571, Johannes Kepler did not enjoy the privileged upbringing of Tycho Brahe. When Kepler's father, a mercenary soldier, went missing in battle when Johannes was five, never to be seen again, the child helped out by serving in his mother's inn. After attending the local school and a nearby seminar, he studied at the University of Tübingen, where he was introduced to the Copernican model by one of the leading astronomers of the time, Michael Mästin. Abandoning his plans to become ordained, perhaps as a consequence of his not entirely orthodox religious views, Kepler became district mathematician in the provincial town of Graz (in present-day Austria), where his duties included producing a yearly astrological calendar with predictions for the region. For 1595 he foresaw an extremely cold winter, a devastating attack by the Turks, and a major uprising, all of which came true.

As a teacher in the local Protestant seminary school, Kepler taught classes in advanced mathematics, astronomy, ethics, history, rhetoric, and astrology. On July 19, 1595, Kepler was explaining the astrological pattern of Jupiter-Saturn conjunctions, the same event that had inspired young Tycho to become an astronomer decades earlier. Drawn on the zodiac, conjunctions occurring twenty years apart move by approximately 120 degrees, so that connecting three of them gives the shape of a triangle. The fourth conjunction, however, does not match the location of the first one exactly, so the next triangle is slightly rotated with respect to the first. Conjunction after conjunction, the vertices of the triangle sweep the zodiac, while the mid-points of its sides trace a smaller circle, exactly half the radius of the zodiac's circle.

When Kepler drew this pattern on the board for his students, he was struck by a new path to solving a problem he had been pondering for years: Copernicus had put the Sun at the center of the solar system but had offered no explanation for the distances between the planets and the Sun. Could the solution be geometry? Was it a coincidence that the diagram of great conjunctions between Jupiter and Saturn produces two circles, one half the radius of the other, while the periods of the two planets are in approximately the same proportion? Could geometry explain God's thinking in creating the solar system the way it is? "The delight that I took in my discovery, I shall never be able to describe with words," he wrote.[10]

Armed with this discovery, Kepler began looking for a geometrically necessary scheme that could not only underpin the relative distances of the planets in the Copernican system but also explain *why* there were only six of them altogether—a crucial step that prefigured the kind of reasoning that today is the hallmark of theoretical physics. Two-dimensional polygons (like triangles, squares, pentagrams, and so on) wouldn't do, he quickly realized, for by adding more and more sides, one can create an infinite number of such figures. But planets move in three-dimensional space, so perhaps solid figures, not flat ones, were needed. With a sense of having gleaned a page from God's very book of creation, he realized that there exist only five regular polyhedrons, the so-called Platonic solids, each composed of identical faces: the tetrahedron (whose faces are four triangles), the cube (six squares), the octahedron (eight triangles), the dodecahedron (twelve pentagons), and the icosahedron (twenty triangles). Five solids, one for each gap between the planets! And—lo and behold—when the Platonic solids were arranged in a certain order, a simple geometrical construction appeared to explain the planetary distances, as he argued in his 1596 book *Cosmographic Mystery*:

> The Earth's orbit is the measure of all things; circumscribe around it a dodecahedron, and the circle containing this will be [the orbit of] Mars; circumscribe around [the orbit of] Mars a tetrahedron, and the circle containing this will be [the orbit of] Jupiter; circumscribe around [the orbit of] Jupiter a cube, and the circle containing this will be [the orbit of] Saturn. Now inscribe within [the orbit of] the Earth an icosahedron, and the circle contained in it will be [the orbit of] Venus; inscribe within [the orbit of] Venus an octahedron, and the circle contained in it will be [the orbit of] Mercury. You now have the reason for the number of planets.[11]

We know today that Kepler's geometrical construction doesn't explain the makeup of the solar system—for one, it doesn't account for Uranus and Neptune, undiscovered at Kepler's time—though the definitive reason behind the spacing of planetary orbits remains elusive. But it was inspiration enough for Kepler, who relentlessly pursued a quantitative mathematical verification of his theory, which in 1600 led him to the man who had by far the best observational data ever gathered: Tycho Brahe.

The Brahe-Kepler alliance was never going to be an easy one. On the one side, the haughty Dane—brooding after having fallen from the graces of the young Danish king and living in self-exile at the court of Rudolph II—desperately sought to cement his legacy by having Kepler validate his idea of the solar system, a curious blend in which the Earth was fixed, with the Sun and the Moon revolving around it, while the other planets circled the Sun. On the other side, the staunch Copernican wanted to mine Tycho's data to prove that the Sun sat at the center of the universe. In the end their collaboration was brief: Tycho died suddenly in 1601, aged fifty-four, of a purported urinary tract infection (tests on his remains conducted in 2010 revealed high levels of mercury, raising questions of foul play that were later laid to rest).[12] His last words, repeated in a delirious state over and over again, were, as related by Kepler, "Let me not seem to have lived in vain."[13]

Kepler took Tycho's precious Mars observations and ran with them, borrowing them without permission from the Brahe estate. After years of dogged mathematical wrangling, he was "roused from a dream and saw a new light" when he discovered his first two laws of planetary motion.[14] Kepler worked out from Tycho's data that the orbits of the planets, including the Earth, are ellipses, with the Sun at one of the foci (his first law), and that a line joining a planet to the Sun sweeps equal areas during equal time intervals (the second law). Kepler's third law, which he formulated in 1619, relates the orbital period of a planet to the size of its orbit—or, more precisely, its semimajor axis. By proving the Copernican system right, Kepler—who by then had succeeded Brahe as imperial mathematician at the court of Rudolph II—ironically sounded the death knell for Tycho's own picture of the solar system.

THE SHAPES OF CYNTHIA AND OTHER WONDERS

As Kepler was publishing his findings in 1609, in a book grandiosely— but appropriately—titled *New Astronomy*, Galileo was busy improving his *cannocchiale*, to the point of reaching twenty magnifications. Soon afterward, on a January night, his efforts were rewarded:

> On the 7th day of January in the present year, 1610, in the first hour
> of the following night, when I was viewing the constellations of the
> heavens through a telescope, the planet Jupiter presented itself to my
> view, and as I had prepared for myself a very excellent instrument, I
> noticed a circumstance which I had never been able to notice before,
> owing to want of power in my other telescope, namely, that three little stars, small but very bright, were near the planet.[15]

Further observation revealed that the three "starlets" wandered on either side of the planet but without ever departing too much from it. At times he could only make out two "wanderers" around Jupiter, and he later noticed a fourth one.

Galileo had discovered the four largest moons of Jupiter, which he soon afterward attempted to use as a means to find longitude, as we saw. When word of his astonishing finding reached the British ambassador, he reported to King James I, "The author runneth a fortune to be either exceeding famous or exceeding ridiculous."[16] It was to be the former: Galileo named the moons "Medicean stars" in honor of Cosimo II de' Medici, grand duke of Tuscany, and the flattery worked. Within six months, Galileo had left the University of Padua and its heavy teaching duties for Florence, where he was appointed philosopher and mathematician to the duke.

Galileo's observations piled up night after night, and he realized that the four wanderers had to be circling Jupiter, each with its own period. At this point, his "perplexity was . . . transformed in amazement" when he at last understood what he was seeing, and its far-reaching implications:

> Here we have a fine and elegant argument for quieting the doubts
> of those who, while accepting with tranquil mind the revolutions of
> the planets about the sun in the Copernican system, are mightily disturbed to have the moon alone revolve about the earth and accompany it in an annual rotation about the sun. Some have believed that
> this structure of the universe should be rejected as impossible. But

now we have not just one planet rotating about another while both run through a great orbit around the sun; our own eyes show us four stars which wander around Jupiter as does the moon around the earth, while all together trace out a grand revolution about the sun in the space of twelve years.[17]

The Medicean stars were only one of the many wonders that Galileo's telescope uncovered. Under its magnified gaze, the Moon's celestial perfection became pockmarked by craters, valleys, crevices, and mountains, revealing a world not dissimilar to ours. Even the face of the Sun was blemished by spots, dark flecks that Galileo demonstrated in 1612 were part of the star itself by charting their rotation over time (today, we understand them as cooler patches on the Sun's surface, where magnetic field lines erupt).

In July 1610, Saturn astonished Galileo by appearing to be flanked by two other smaller stars that never changed position but vanished after a few months, leaving Galileo puzzled: "Have the two minor stars been consumed after the manner of the solar spots? Have they disappeared and suddenly fled? Has Saturn devoured his own sons?" he wrote.[18] Half a century later, Dutch astronomer Christiaan Huygens, equipped with a much more powerful telescope, recognized them for what they are: rings, rendered periodically unobservable when they present edgewise to Earth.

It was Venus that provided incontrovertible evidence, to the eyes of Galileo, against the Ptolemaic system. Galileo followed the planet as it morphed from a small, fully illuminated disc into a larger, thin crescent. He concluded that Venus orbits the Sun, appearing full when it is on the other side of the Sun with respect to us and as a larger crescent when on our same side of the Sun, thus much closer to us and lit sidewise. This explosive revelation he first concealed as a Latin anagram in a letter intended for Kepler written in December 1610—a common stratagem at the time to establish priority on a discovery without revealing it at first. Once unscrambled, Galileo's announcement read, "The shapes of Cynthia [the Moon] are imitated by the mother of love [Venus]."[19]

What was left of the unchanging, incorruptible perfection of the Aristotelian heaven? And how could the new sights be squared with the revelation of the sacred scriptures? Does the Bible not say, of the day when God fought on the side of Israel, "And the sun stood still, and the moon stayed, until the people had avenged themselves upon their enemies"?[20] A fixed Sun was a miraculous intervention of the Almighty, not the everyday order of things. Some prelates chose to simply not believe their eyes, attributing the never-before-seen phenomena described by Galileo to the deceptive power of the telescope itself.

When, in 1610, Galileo announced in *The Sidereal Messenger* his discovery of the moons of Jupiter and of countless stars invisible to the naked eye, he was at first feted by intellectuals and the clergy alike—even as a circle of staunch Aristotelians began a campaign to undermine him. On a trip to Rome in spring 1611, he was received as a celebrity: the Jesuit fathers praised his book in a ceremony; Pope Paul V bade him rise from his knees to stand during an audience; the Academy of Lynxes, one of the first learned societies in the world, held a splendid banquet during which he demonstrated his *cannocchiale*. Everybody was delighted, and the academy's president, Federico Cesi, invented a new word to describe Galileo's wondrous invention: from then on it would be called a *telescope*, from the Greek for "farseeing."

Galileo's fame spread like wildfire through Europe's intellectual circles. Upon receiving a copy of *The Sidereal Messenger* in Prague less than a month after it was published in Florence, Kepler rushed to compose a ringing endorsement in answer to Galileo's request to shore up his claims. Kepler's pamphlet comes across as fawning, though he admits to having "no support from [his] own experience" about the sights Galileo described. Galileo was grateful for the endorsement and wrote to Kepler, "I thank you because you were the first one, and practically the only one, to have complete faith in my assertions, even though you had made no visual observations."[21] Kepler was desperate to see for himself the marvelous sights announced by Galileo and begged the Italian to send him one of his telescopes. Galileo never did.

But powerful enemies lurked in the shadows, and all they needed was an excuse to pounce. In 1613 Galileo claimed in a widely publicized letter that matters of science should not be judged by religious authorities, who didn't possess the necessary competence—which attracted the attention of

Jesuit cardinal Robert Bellarmine, who considered the immovable Earth described in the scriptures a literal representation of reality.[22] Bellarmine summoned Galileo to Rome in 1616 and forbade him to speak of the Copernican system as a truthful model of the cosmos. Galileo's telescopic observations, he argued, didn't pass St. Augustine's muster for "clear and manifest" evidence against the literary truth of the scriptures.[23] The Holy Office concurred and also accepted Bellarmine's denouncement of *De revolutionibus*, putting Copernicus's book on the index of forbidden texts, where it remained until 1835. Bellarmine's was no empty threat: he had overseen the trial that led in 1600 to philosopher Giordano Bruno's being burned at the stake for refusing to abjure heretical ideas (among them that the universe is infinite and that it contains "infinite worlds").[24]

Intimidated, Galileo acquiesced, but in 1632 he reckoned that the storm had sufficiently abated for him to go ahead with publishing *The Dialogue on the Two Chief World Systems*, which he framed as a debate between the Aristotelian Simplicius, a proponent of the Ptolemaic view; Salviati, representing Galileo's own Copernican ideas; and the initially neutral layman Sagredo. With the *Dialogue*, Galileo followed the letter of Bellarmine's imposition not to teach Copernicanism as a true theory but introduced Simplicius as a thinly disguised caricature of Pope Urban VIII. Put under pressure by the heap of evidence brought forth by Salviati in favor of a rotating Earth revolving around the Sun, Simplicius confesses, "Truly, I am not entirely capable."[25]

Pope Urban was enraged to have been publicly humiliated in the *Dialogue*. Galileo had flouted his authority and therefore that of the Catholic Church, right when the Counter-Reformation was reasserting the primacy of doctrines to which Protestants were opposed. Urban summoned an almost septuagenarian and sick Galileo to Rome in the middle of a frigid January 1633. Over the next four months, the Inquisition put him on trial for disregarding Bellarmine's admonition from 1616 and forced him to abjure his views under threat of torture.[26] Galileo was condemned to indefinite incarceration, soon converted to house arrest in consideration of his age and frailty. He died in January 1642 in Florence and was buried without pomp or ceremony in a tiny chapel of the Santa Croce basilica. His soul, according to his last and most devoted disciple, Vincenzo

Viviani, at last "went to enjoy and inspect more closely those eternal and unalterable wonders which by way of a fragile instrument it eagerly and impatiently succeeded in drawing closer to our mortal eyes."[27]

Together, Copernicus, Brahe, Kepler, and Galileo had ushered in ideas and shored them up with evidence-based arguments that had blown away a fourteen-hundred-year-old universe—although the revolution they had set in motion would take another century to take hold. A baby born in an unremarkable British village the year after Galileo died, named Isaac after his deceased, illiterate father, would clamber onto the shoulders of these giants.

ISAAC'S DIALS

After an unhappy childhood spent largely in the care of his grandmother, the young Isaac Newton moved to the village of Grantham to attend grammar school. At Grantham, the majesty of the heavens had impressed itself onto the mind of the schoolboy who, as we saw in Chapter 2, played pranks on the villagers by flying lanterns on kites to imitate comets. When he wasn't scaring the villagers to death or building a model windmill or wooden cart, young Isaac was busy filling his lodgings with sundials. From his room in the local apothecary, "Isaac's dials," as they became known to family and neighbors, spilled over into the entrance hall and sprouted on every wall where the Sun shone. Pegs and strings marked the hours, and Newton learned how to tell from his sundials the day of the month and the exact time of equinoxes and solstices. Seven decades later, as an old man, Newton still looked at the shadows in a room to tell the time.

By 1669 Newton had become the second Lucasian Professor of Mathematics at Trinity College, Cambridge. The Lucasian Chair was one of the best endowed positions at the university and the only one connected with mathematics or natural philosophy. Quite how Newton managed to be appointed to it by his predecessor, Isaac Barrow, remains shrouded in mystery. We do know that for the twenty-eight years Newton held the position, he enjoyed a generous stipend on top of free lodgings and dining privileges in the college and almost no academic duties. He only ever had three pupils, and even after he had become England's most famous natural philosopher, whatever lectures he gave remained poorly attended: few went to hear him, and even fewer understood him, so that he often found himself lecturing to the walls and returned to his rooms and his work within a quarter hour for want of an audience.

This does not seem to have bothered him at all. Free to dedicate himself wholly to his studies, Newton threw himself into optics, mathematics, and physics. His many achievements include clarifying the nature and behavior of light, particularly colors; the invention of calculus, notwithstanding German mathematician Gottfried Leibniz's claim to the independent development of similar ideas (which led to decades of bad blood between the two); and, of course, devising the law of universal gravitation and the laws of motion that bear his name. He also spent a decade pouring over obscure alchemical texts, in the hope of synthetizing from them a new form of natural philosophy. He focused on his work to the detriment of everything else, "thinking all Hours lost, that was not spent in his Studyes," according to his assistant: he usually ate alone in his rooms, hardly received visitors, and did not mingle with the other fellows, leaving his lodgings only to give his lectures—short as they often were.[28] His assistant said he only saw him laugh once.[29] On the rare occasions when he elected to dine in the shared hall, his hair disheveled and his clothes untidy, he sometimes absentmindedly ended up in the street instead or forgot to help himself to food before the table was cleared away. Whenever a fresh layer of gravel was laid onto Trinity's garden paths, before long it would be covered by Sir Isaac's diagrams, which the other fellows were careful to step around.

In 1669 Newton invented the reflective telescope, which replaced the lenses of the Galilean telescope with an arrangement of mirrors that gave a much higher magnification for a given tube length and did not split the light into its component colors as lenses tended to do. Newton's telescope—which he fashioned with his own hands, mastering the difficult process of casting and grounding the mirrors from an alloy of tin and copper of his own invention—created a sensation when presented to the Royal Society, which swiftly inducted its inventor into its ranks in January 1672.

By 1680 Newton had published important papers on optics but none of his groundbreaking results in calculus or celestial mechanics yet. The results of decades of genius remained scattered among stacks of unfinished papers, buried under lengthy treatises on theology and alchemy. A new impulse was needed, a challenge that would concentrate Newton's immense power of analysis on a worthwhile problem and hold it steady until a breakthrough was achieved. The heavens obliged by providing Newton with not one but two comets. The first, during the 1680–1681 winter, was four times the diameter of the Moon with a tail over seventy degrees long, the largest ever seen according to John Flamsteed, the royal astronomer. Newton followed it closely with a purpose-built telescope and corresponded with Flamsteed about its trajectory and nature. As was his style when a problem captivated him, Newton read everything he could lay his hands on regarding observations of earlier comets, attempting to figure out the shape of their orbits.

Newton had already solved the greatest unanswered question in natural philosophy at the time: although he had not published his solution, he had demonstrated that the elliptical orbits discovered by Kepler were the consequence of the gravitational attraction between the Sun and the planets. He had done this by proving mathematically that the elliptical shape was a consequence of gravity diminishing with the square of the distance between the Sun and the planet. Where Kepler had discovered that planetary orbits are ellipses from his analysis of Tycho's observations of Mars, Newton had found a physical explanation for that shape. He had, in other words, derived Kepler's first law from dynamical principles. But he hadn't yet extended the concept of gravitational attraction to bodies other than the planets, and as he puzzled over the path of the comet, an idea, fuzzy at first like the tail of a broom star, began to crystallize: What if comets, too, followed the same law as planets? What if gravitation was a universal force in the cosmos?

Newton recalled an intuition that had planted itself in his youthful mind in the summer of 1666, when, in a relative's later account, "whilst he was musing in a garden it came into his thought that the power of gravity (wch [*sic*] brought an apple from the tree to the ground) was not limited to a certain distance from the earth but that this power must extend much further. . . . Why not as high as the moon said he to himself & if so that must influence her motion & perhaps retain her in her orbit."[30] Perhaps the same "power of gravity" applied to the comet too, Newton began to suspect. (There is no evidence that the apple actually hit Newton on his head, as the popular myth would have it!)

In 1682, another comet enflamed natural philosophers' discussions (the word *scientist* was still 150 years in the future). Newton once again systematically recorded its position, and when he met Edmond Halley, he quizzed the astronomer closely about his observations. But it took a challenge from Robert Hooke, whom we encountered earlier as the first salaried scientist in history, to truly focus Newton's thinking. In August 1684, Halley visited Newton in Trinity to consult him on a matter he had long been debating with Hooke and architect Sir Christopher Wren: If one supposed the force between the Sun and a planet to fall off as the inverse of the square of their distance, what shape would the planet's orbit take? This was the very puzzle Newton had cracked years before, without bothering to let anybody know: the shape would be, of course, an ellipse, exactly as Kepler had determined. When Newton informed Halley of the answer on the spot, per mathematician Abraham DeMoivre's account of their conversation, "the Doctor [Halley] struck with joy & amazement asked him how he knew it, why saith he I have calculated it, whereupon Dr Halley asked him for his calculation without any farther delay."[31]

Halley told Newton that Hooke claimed to have found a solution himself but was refusing to divulge it in full. That greatly irked Newton and led to an acrimonious dispute between the two (not the first or the last of its kind in Newton's life). Stung in his pride, Newton wasn't content to present to the world the calculation that had lain among his papers for years but gave himself up to pursuing fully the ramifications. "Now I am upon this subject," wrote Newton to Flamsteed in January 1685, "I would gladly know the bottom of it before I publish my papers."[32] For the next four years, Newton abandoned his alchemical experiments, worked through most of the night, and often forgot to eat.[33] The resulting masterwork, the *Principia* (from the first word of its Latin title, which translates to *Mathematical Principles of Natural Philosophy*), did far more than even Halley, who instigated it and later shepherded its publication through a thousand difficulties, could have ever expected.

In the *Principia*, Newton presented the laws of motion that today are named after him and remain one of the cornerstones of physics. With a mathematical tour de force that used the infinitesimal calculus he had invented decades earlier, the Lucasian professor demonstrated how the elliptical orbits of the planets followed from an attractive force, gravity, that falls off as the square of the distance; how tides are a consequence of the attraction of the Sun and the Moon; and how the motion of all heavenly bodies, including Jupiter's satellites and comets, is governed by the same law: universal gravitation. The section on comets was, in Newton's own opinion, the most challenging in the whole book, but also one of the most compelling proofs of his theory of gravitation.[34] Newton proved that the trajectory of the Great Comet of 1680–1681, which had so aroused his curiosity, had the shape of a parabola—a special case of a family of curves, including the circle and the ellipse, that could all be derived from the inverse square law. Comets were subject to the same laws as all other heavenly bodies: gravity was universal. The flash of insight from a falling apple two decades earlier had been transmuted into one of the greatest scientific discoveries of all time. Lord Byron's wry description of Newton rings true:

> *And this is the sole mortal who could grapple,*
> *Since Adam, with a fall, or with an apple.*[35]

The publication of the *Principia* in 1687 took the world of natural philosophers by storm, impenetrable as its advanced mathematics was for most. Some of the recipients of the first printed edition, a gift from Newton to other Cambridge dons and heads of colleges, candidly admitted that "they might study seven years, before They understood any thing of it," according to Newton's assistant.[36] In reviewing the book for the Royal Society, Halley recognized its epochal quality: "so many and so Valuable Philosophical Truths, as are herein discovered and put past Dispute, were never yet owing to the Capacity and Industry of any one Man."[37] When the prominent French mathematician the Marquis de l'Hôpital received a copy, he was so impressed by its "fund of knowledge" that he could scarcely believe Newton was human: "does he eat & drink & sleep, is he like other men?" he wondered.[38]

He wasn't. With the *Principia*, Newton showed that the book of nature was written in mathematical characters and that its pages could be deciphered to reveal universal laws. The cosmos, in all of its majestic parade

of stars, its cortège of wandering planets, and its retinue of apparently disorderly comets, followed rules that humans could decrypt. This sudden
enlightenment is pithily described in the epitaph that English poet Alexander Pope had intended for Newton's monument in Westminster Abbey
(where it was never engraved):

> *Nature and Nature's Laws lay hid in Night:*
> *God said, "Let Newton be!" and all was light.*

The monument itself is crowned by a weeping female figure: Astronomy,
the Queen of the Sciences.

The earliest surviving portrait of Newton is the one he commissioned himself from Godfrey Kneller, the leading artist of the time, two
years after the publication of the *Principia*—the gesture of a man who, at
forty-six, is assured of his place in history. In the full-scale oil, Newton's
visage is framed by a cascade of silvery hair, his high brow creased over his
aquiline nose and a delicate, feminine mouth. His gaze is fixed on a point
outside the frame, as if to dare us to follow him to his newly discovered
land. An otherworldly light, absent from the later, more formal portraits,
shines from his eyes: the gaze of a man who, in his obsessive quest to decipher the language of nature, has visited far, dark places and barely escaped
with his mind intact.

A FIERY RETURN

As a boy, Newton had been playing with kites fashioned to resemble comets, rare and special visitors to the Earth to which, Roman philosopher
Seneca wrote in AD 65, nature had "assigned . . . a different place, different periods from the other stars, and motions unlike theirs."[39] With the
Principia, Newton had instead shown that those apparitions that used to
terrorize emperors and kings meekly obeyed the tug of gravity's leash, just
like every other heavenly body and ripe fruit. But it was Edmond Halley
who perfected Newton's calculations on the orbit of comets, when, after
investigating twenty-four of them, he became convinced that the broom

star he had observed in 1682 was the same as the one reported in 1531 and 1607 (by Kepler, among others).

Newton had computed the trajectory of the Great Comet of 1680–1681 as a parabola, a curve that does not close upon itself, making the Great Comet a onetime visitor of the Earth. But Halley's calculation showed that the comet of 1682 followed a closed, elliptical orbit, which would bring it back toward the Sun after its long, solitary voyage beyond the far confines of the solar system. The case for the comet's return was strong, but Halley couldn't be sure, for he realized that its orbit would be affected by the gravity of Jupiter and Saturn. A small gravitational kick from either of those gas giants would be sufficient to increase the comet's velocity beyond the point of no return. What had started as a piece of intellectual bravado in 1705 ("I can undertake confidently to predict the return of the comet in 1758," he wrote) became a much more cautious claim by 1715 ("I think, I may venture to foretell").[40] He found out that other cometary passages, in 1305, 1380, and 1456, also fit in with the same approximate period of seventy-five and a half years. He predicted that the 1682 comet would return "about the end of the year 1758." Already an old man, he knew he would not live to see it but entrusted his fellow natural philosophers to verify his augury—just as he had done for the transit of Venus, which resulted in the Royal Society mounting James Cook's expedition to Tahiti half a century later.

By late 1758, the frenzy about the predicted comet's return reached its peak: gentlemen's magazines carried articles on the imminent end of the world, the will of the Almighty to be brought about by Halley's Comet. The highest-ranking astronomers of England and France anxiously scanned the skies, searching for the glittering snowball where their painstaking refinements of the comet's orbit guided them, keen to claim the honor of the first glimpse. On Christmas night, 1758, a German farmer was the first to spot the comet's return, its cold light dawning precisely on Halley's timetable, sixteen years after the astronomer's eyes last saw the stars. The preface that Halley wrote to Newton's *Principia* became, with the predicted return of the comet that would forever bear his name, a fitting tribute to Halley's own achievement:

> *When Mathematics drove away the Cloud.*
> *No longer doubting in the Mists we stray;*

> *Genius' high Summit grants to us the Way*
> *To reach the blessed Gods' Abodes and pierce*
> *The lofty Limits of the Universe.*

Seneca had concluded his musing on comets by foretelling, "Men will some day be able to demonstrate in what regions comets have their paths, why their course is so far moved from the other stars, what is their size and constitution. Let us be satisfied with what we have discovered, and leave a little truth for our descendants to find out."[41] His prophecy had come to pass.

THE TRACKLESS ABYSSES OF SPACE

Despite the church's efforts to stem the advance of Copernicanism, a radical reconceptualization of space and time was afoot. This wave of change had swelled from the natural philosophers' questioning of the heavens, but within two centuries it would grow to overturn our conception of the nature of our planet, and of life itself. It took a convicted heretic like Tommaso Campanella (who spent twenty-seven years imprisoned for his unorthodox views and practiced a kind of black magic, as we shall later see) to recognize the seismic shift in perspective that was coming before almost everybody else: "This news of ancient truths, of new worlds, new stars, new systems, new nations, etc are the beginning of a new age," he wrote to Galileo in the summer of 1632, upon reading *The Dialogue*.[42]

But putting the Sun in charge of the solar system was not the only, and perhaps not even the most fundamental, insight Galileo's telescope brought about. His narrow tube expanded the realm of scientific inquiry itself. The starry universe, previously an unreachable backdrop to life on Earth, became the object of quantitative scrutiny. And it was a mighty shock to realize, for the first time, its true vastness.

Wherever Galileo pointed his telescope, never-before-seen stars came into range: he counted thirty-six in the Seven Sisters, whereas only six appear to the naked eye, as we saw in Chapter 4. The Milky Way wasn't an atmospheric phenomenon, as Aristotle had said, or the abode of virtuous souls after death, as Cicero had it. "All the disputes which have tormented philosophers through so many ages are exploded at once by the irrefragable evidence of our eyes," reported Galileo in *The Sidereal Messenger*, for "upon whatever part of it [the Milky Way] you direct the telescope straightaway a vast crowd of stars present itself to view."[43] With this crowded cosmos new questions arose: Why had God created such a splendid firmament, only to then hide it from the contemplation of man until the advent of the telescope? Was the universe finite or infinite? And what was the true nature of the stars? Galileo yanked into the realm of empirical investigation what for millennia had been theological and philosophical questions.

If the Moon's surface appeared similar to the Earth's when seen through a telescope, could the stars not be made of the same fiery substance and be of the same size as our own Sun? Not even the most powerful of Galileo's telescopes, capable of thirty-two enlargements, could show anything but a pinprick of light in the place of a star. Stars, Galileo concluded, had to be incredibly far away, much more so than the planets, which instead looked like little discs through his *cannocchiale*. And could the philosopher and mathematician René Descartes be right in his supposition that "the matter that God shall have created extends quite far beyond in all directions, out to an indefinite distance"?[44] An infinite universe might just be conceivable, filled with a multitude of suns—yet in 1633 Descartes held back with publishing this hypothesis when he heard of Galileo's condemnation in Rome.

Dutch astronomer and mathematician Christiaan Huygens reasoned that if Sirius, the brightest star in the sky, was a twin of the Sun but so much farther away, its distance could be found by comparing its brightness to that of our star. The difficulty lay in comparing reliably the blinding brightness of the Sun with that of a pinprick of light. Scottish mathematician James Gregory proposed to use a planet such as Saturn as a stepping-stone toward the result, as its brightness was more readily comparable to that of Sirius. The question then was reduced to finding the amount of sunlight reflected by Saturn. With his method, Gregory arrived at a distance to Sirius of eighty-three thousand times that between the Earth and the Sun. Isaac Newton, using different data, placed Sirius a million times farther away than the Sun (the correct value is about half Newton's result—remarkably close given the many uncertainties involved in Newton's calculation!).[45] Enormous as this relative distance clearly was, nobody could reliably assign it an actual numerical value, for the Earth-Sun separation remained unknown. Its determination would have to wait for Cook's transit-of-Venus data from 1769, as we discussed.

Whereas Ptolemy had drawn the surface of a sphere, with a radius a paltry twenty thousand times that of the Earth, and populated it with an atlas of one thousand stars, a possible infinity of stars now lurked. Since gravity was universal, Newton worried that the stars should all fall toward each other, attracted by their mutual gravity. Yet stars truly appeared fixed in the sky—they aren't, but this wouldn't be recognized until 1718, when Halley discovered differences in the positions of contemporary stars with respect to measurements made by Ptolemy in antiquity. Newton concluded, "It must be that the stars are separated from each other as they are from the Sun."[46] Together with his estimate that Sirius is a million times farther than the Sun, he opened up an immense vista onto the trackless abysses of space.

By 1750, the kinds of speculations that had cost Giordano Bruno his life were being openly discussed. Astronomer Thomas Wright had jumped right into this cosmic breach through Ptolemy's Empyrean and posited that "there cannot possibly be less than 10,000,000 Suns, or Stars, within the Radius of the visible Creation; and admitting them all to have each but an equal Number of primary Planets moving round them, it follows that there must be within the whole celestial Area 60,000,000 planetary Worlds like ours."[47]

To pin a star with certitude in the uncharted depths of such a vast universe demanded the determination of its parallax, something that continued to escape bigger and more refined telescopes until 1838. At last, the self-taught director of the Königsberg observatory, Friedrich Bessel, caught it: the tiny shift in position on the sky of the star 61 Cygni (actually a binary system, with two stars orbiting each other) when observed from opposite ends of the Earth's orbit, six months and three hundred million kilometers apart. Bessel's measurement was described as "the greatest and most glorious triumph which practical astronomy has ever witnessed" by Royal Astronomical Society president John Herschel as he awarded to Bessel the society's Gold Medal in 1841.[48] A shift the width of a sideways

US cent seen from one kilometer away meant a distance to 61 Cygni of 657,700 times the Earth-Sun separation, or one hundred thousand billion kilometers. Nobody could picture such a length of space—not even Bessel himself, who helpfully translated it into the time light takes to cross it (10.3 years).[49] The light-year—a new unit of measurement commensurate to the daunting task of charting the depths of the universe—was born.

But the question of the size of the universe was merely displaced from the province of the stars, whose distances could now be routinely measured up to thousands of light-years away, to the mysterious realm of the nebulae, from a Latin word for "cloud." The Andromeda Nebula, a smudge of haze in the Andromeda constellation, had escaped the attention of the ancient sky watchers, until it was described for the first time in AD 965 by the Persian astronomer al-Ṣūfī, who included it among the fifteen "nebulous smears" he described in his *Book of the Fixed Stars*.[50] German astronomer Simon Marius didn't fare much better even with a telescope in 1612, describing what he saw as "like a candle seen at night through a horn."[51] As telescopes improved, astronomers dredged up from the depths of space many more such faint puffs of bluish haze, sometimes showing the hint of twirling lanes and arms. My own experience of contemplating one such nebula through a modern telescope on Monte Verità spoke to the impossibility of judging distance to such ethereal apparitions: Were nebulae nearby vaporous masses of gas or mighty galaxies in their own right, reduced to faintness by their unimaginable distance from us? No one could tell.

At the turn of the twentieth century, a new, powerful tool came to the rescue: the photographic plate, which could record fainter objects than the human eye and capture finer details. The number of stars grew into the hundreds of thousands, and hundreds of nebulae cropped up. The ever-growing collection of plates required an army of "computers," the heirs of Nevil Maskelyne's number crunchers, to measure position, brightness, color, and other characteristics of each nebula or star. The director of the Harvard observatory, Edward Pickering, who pioneered the systematic use of photography in astronomy, felt that hiring women for the task would be both expedient—at two-thirds the pay commanded by male assistants—and socially valuable. "Many ladies are interested in astronomy and own telescopes. . . . Many of them have the time and inclination for such work," he wrote when he put out a call for volunteer observers in 1882. "The criticism is often made by the opponents of the higher education of women that, while they are capable of following others as far as men can, they originate almost nothing, so that human knowledge is not advanced by their work."[52] Pickering's answer was to give women the opportunity to contribute to cutting-edge astronomical research—and many did so with extraordinary results.

The most prolific of the lady astronomers was Annie Jump Cannon, who, over forty years of work at the observatory, personally classified over two hundred thousand stars. But she was by no means the only woman whose unsung contributions transformed astronomy. In 1903, Henrietta Swan Leavitt was paid thirty cents an hour to first identify variable stars in plates of the Magellanic Clouds and then measure the changes in their brightness over time—a painstaking, meticulous job. By 1908 she had amassed 1,777 variable stars and noticed a strange pattern among sixteen of them in the Small Magellanic Cloud, a puff of white haze observable from the Southern Hemisphere even with the naked eye. "It is worthy of notice—she wrote—that the brighter variables have the longer periods."[53]

The brighter the star, the longer the period. Since the stars were all in the Small Magellanic Cloud, it was reasonable to assume that they shared a similar distance from Earth. Therefore, their different observed brightness reflected a difference in their true brightness, not their distance. This was a groundbreaking result: Leavitt's law gave astronomers the means to establish the absolute (i.e., intrinsic) brightness of a variable star by observing its period; from the observed brightness, they could then find the distance, as the observed brightness drops as the square of the distance. We today call Leavitt's variable stars *Cepheid variables*, after the prototype star of this kind in the constellation Cepheus. This kind of star, several times more massive than the Sun, traps radiation in its outer layers, thus expanding and eventually cooling. The cooling allows the radiation to escape: the star brightens and contracts, and the cycle begins anew. It is as if the star were breathing, letting light out at regular intervals.

Soon Leavitt's discovery was being put to use by Edwin Hubble, who had been captivated by astronomy since he experienced a total lunar eclipse as a nine-year-old in 1899 in Missouri, just as Tycho Brahe was over three centuries earlier. After hundreds of nights spent in freezing cold at the gargantuan one-hundred-inch Hooker telescope at Mount Wilson Observatory in California, he tracked down two of Leavitt's variable stars in the Andromeda Nebula. His historic photographic plate still bears his handwritten, exclamatory annotation next to the first one he found: *VAR!* Thanks to Leavitt's relationship, Hubble was able to estimate the distance to Andromeda from the period of the variable, obtaining a million light-years: less than half the actual distance, but still enough to deal a fatal blow to those who argued that Andromeda wasn't a galaxy in its own right but just a relatively nearby gas cloud.[54]

Hubble's discovery had stretched the plumb line into the millions of light-years, thus paving the way to extragalactic astronomy. The Copernican conception of the cosmos had been pushed to its ultimate conclusion: the Earth is but one of the planets orbiting the Sun, itself a nondescript middle-aged starlet like hundreds of billions of others, belonging to a fairly massive spiral galaxy swept along like another fifty billion cousins in an expanding universe.

THE CLOCKWORK UNIVERSE

The human mind, it seemed, could stretch beyond earthly confines and "pierce the lofty limits of the Universe," as Halley had written. Nature had yielded her most intimate secrets to mathematical analysis and the scientific method, but the "blessed Gods' Abode" wasn't to be found in the solar system or among the stars or even in the dark vastness between the nebulae. Where did that leave the Prime Mover?

As discovery succeeded discovery in the eighteenth century, the cogs of the "clockwork universe" metaphor introduced by medieval philosopher Nicole Oresme began spinning in earnest. If the universe was understood as a giant mechanism, it followed rationally that there had to be a watchmaker who had designed it all. Contrary to an enduring popular myth, Newton didn't believe that God no longer had an active role to play in the universe. Not only did he see the delicately balanced arrangement of the solar system as proof that its cause was "not blind & fortuitous, but very well skilled in Mechanicks & Geometry," but he also believed that God intervened in keeping that arrangement working by sourcing the gravitational force and tweaking it as necessary.[55] Newton brought a technical argument against God's inaction: since the mutual gravitational influence of Saturn and Jupiter, Newton observed, creates perturbations that over time would imperil the stability of the solar system, "blind Fate could never make all the Planets move one and the same way in Orbs concentrick."[56] Without God's continuing intervention in the movements of the heavens, Newton argued, the planetary motions would soon become chaotic. This provoked Gottfried Leibniz to retort, with blistering sarcasm perhaps colored by acrimony over the invention-of-calculus dispute,

> Sir Isaac Newton, and his Followers, have also a very odd Opinion concerning the Work of God. According to their Doctrine, God Almighty wants to wind up his Watch from Time to Time: Otherwise it would cease to move. He had not, it seems, sufficient Foresight to make it a perpetual Motion. Nay, the Machine of God's making, is so imperfect, according to these Gentlemen; that he is obliged to clean it now and then by an extraordinary Concourse, and even to mend it, as a Clockmaker mends his Work.[57]

The natural philosophers of the sixteenth and seventeenth centuries had seen the revelation of Nature's mysteries as proof of God's magnificent plan: Kepler was a Platonic mystic who considered the geometrical structure of the universe a mirror of God's mind, while Galileo, who despite his clashes with the church remained an ardent Catholic, never doubted that "God ordered the movements of the celestial spheres with proportions . . . totally imperceptible to our mind."[58] But by the nineteenth century, God would be exiled well beyond the orbit of Saturn. In 1802 Napoléon Bonaparte asked mathematician Pierre-Simon Laplace about the role of God in the heavens, to which Laplace is alleged to have retorted, "Sir, I had no need of that hypothesis."

Although Laplace's reply (probably never uttered quite in these terms) is often taken as a demonstration of the mathematician's atheism, its meaning is subtler, as illuminated by an eyewitness to the exchange: astronomer

William Herschel, famous for discovering the first new planet since antiquity, Uranus. On the hot August day when Napoléon met Laplace, Herschel, and his wife in the garden of the Château de Malmaison, outside Paris, the conversation veered genially from English breeding horses to opera, between servings of ices "of an excellent flavour, and very refreshing," as Herschel recorded. When the discourse inevitably fell on astronomy, Napoléon quizzed both men at length about the structure of the heavens and became much agitated by Laplace's replies, exclaiming, "And who is the author of all this!"

According to our informant, Laplace then argued that "a chain of natural causes would account for the construction and preservation of the wonderful system."[59] By this he meant that there was no need for God to intervene to fix the orbits of Jupiter and Saturn, like Newton had concluded: Laplace had discovered that the gravitational perturbations that had so preoccupied Newton reversed their influence over time, and therefore the planetary orbits remained stable indefinitely. The "Machine of God's making" didn't need any adjustment.[60]

This encounter took place shortly after Laplace had published the third volume of his monumental treatise in five volumes, *La méchanique céleste*, which he astutely dedicated to Napoléon, "Pacificator of Europe" (with the downfall of the emperor, the dedication disappeared from further editions). Once completed in 1825, *La méchanique* surpassed everything that had been written since the *Principia*. Laplace had taken Newton's ideas and developed them mathematically to a virtuoso extent, not only demonstrating the long-term stability of planetary orbits but introducing new mathematical methods to account for gravitational perturbations and improving the treatment of the lunar motion and the prediction of tides. It was a masterwork, crowning the program started by Ptolemy seventeen centuries earlier: to explain the motions of the heavenly bodies in mathematical terms. Laplace was feted as the Newton of France.

The English translation of Laplace's book, *The Mechanism of the Heavens*, won such acclaim that it became for decades the standard textbook on celestial mechanics in both Oxford and Cambridge. Much was due to the gifts of the translator, Mary Somerville, a self-taught Scottish mathematician and astronomer who explained in detail Laplace's fiendishly difficult calculations. She is today remembered on Scotland's £10 banknotes and by a nondescript crater on the Moon—one of only thirty-two women thus honored among the 1,546 named Moon craters.[61]

Tellingly, the English title describes the universe as a "mechanism," thus closing the circle started two millennia earlier: the heavens had inspired Archimedes and the unknown author of the Antikythera mechanism to build a mechanical model for it, which in turn instilled the notion that nature itself was best understood as a full-scale mechanism, a "clockwork universe" that, like John Harrison's marine chronometers, keeps running, indifferent to the intervention—and indeed even to the existence, Laplace might have said—of its clockmaker. The beauty of the night sky had been transformed into the order of mathematics.

Yet the exactitude of Newtonian physics wouldn't last. Brahe's high-precision data had enabled Kepler, the theorist, to prove the Copernican model; Galileo's telescopic observations had spurred the development of physical astronomy; Newton had synthetized it all in the grand construction of universal gravitation; Halley had used Newton's equations to predict the return of his comet and motivated Cook's expedition to measure the size of the solar system. After Laplace was done with *La mécanique*, the Newtonian model of the universe could explain and predict phenomena ranging from a falling apple to perturbations in the orbits of Jupiter's moons.

At the beginning of the nineteenth century, the pendulum swung back. As the theoretical description of the cosmos became more and more precise, its predictions grew so exacting that the small errors that crept into the astronomers' observations could no longer be overlooked. Telescopic data piled up, but a new kind of organizing principle was needed to make sense of everything, especially of the errors—indeed, a new kind of mathematics.

It was the Newton of France who stepped up to the task.

THE CALIGO TALES

Bison-Seeker's Tale

Even before Way-Finder had finished, Bison-Seeker jumped up—her voice gushing from her mouth like a spring from a hole in the ground. The Remembering had taken hold of her.

> I remember! I remember the story of one of the husbands brought back from the Big Swap, so many bat awakenings ago. His old name is forgotten, but our people called him "Stone-Thrower" after what happened.
>
> Long-Ago-Elder-Bison-Seeker was out one glow, looking at the length of grass blades to see when the Bison would return. Stone-Thrower was still finding his way among our people, and didn't know yet who he was to become, apart from being a strong husband to many women. He was no Cloud-Watcher, that much was certain, and only a mediocre Spear-Maker. Perhaps he could learn to Remember how to be Bison-Seeker. That's why he was out that glow with Long-Ago-Elder-Bison-Seeker.
>
> Stone-Thrower and Long-Ago-Elder-Bison-Seeker were lying on the ground, their noses almost touching

the grass, examining it for clues, when they heard a scream from behind the hill. They jumped to their feet and ran like the wind. When they got to the top of the slope, they saw a boy: he had seen perhaps as many bat awakenings as fit in one hand, and had no name yet. The boy was screaming and pointing at the Cloud, they couldn't tell whether in excitement or fear. When they got to him, he told them that the Cloud had ripped, showing a hole the color of the feathers on the kingfisher's wings. They searched the Cloud for the hole, in vain.

Long-Ago-Elder-Bison-Seeker had his back turned, his face still craning up, when he heard a soft thump and a whimper. Stone-Thrower was bending to pick up another stone, which he hurled like the first at the boy. After the second stone the boy moved no longer, but Stone-Thrower flung another, and another.

Long-Ago-Elder-Bison-Seeker looked at him from a distance, and asked, "Why?"

Without pausing, Stone-Thrower glanced up and replied, "He lied. There is no hole."

8

THE DEMON UNLEASHED

When all the stars were ready to be placed in the sky,
First Woman said: "I will write the laws that are
to govern mankind for all time. These laws cannot
be written on the water as that is always changing
its form, nor can they be written in the sand as the
wind would soon erase them, but if they are writ-
ten in the stars they can be read and remembered
forever."

—Navajo creation story[1]

THE MATHEMATICS OF UNCERTAINTY

When a new crop of freshmen shows up to their first university-level physics courses, they're often smitten with theoretical physics. Asked what attracted them to the field, they mention the elegance of general relativity, the weirdness of quantum mechanics, the majesty of cosmology, the lure of a Theory of Everything. What no one ever talks about is how those ideas are put to the test of reality, a fundamental yet often underappreciated component of how physics—indeed, science—works. The comparison of a theoretical model with observational data makes use of what is perhaps the most disparaged of all mathematical branches: statistics. Blame Mark Twain, who in 1895 spread the quip, incorrectly attributing it to Benjamin

Disraeli, "There are three kinds of lies: lies, damned lies, and statistics." Perhaps fittingly, the true origin of the phrase remains uncertain.[2]

What is certain is that the theoretical perfection of Johannes Kepler's and then Isaac Newton's models of the cosmos clashed with the imperfect reality of astronomical observations. Kepler had struggled with this when sweating over Tycho Brahe's Mars data; Brahe, in his obsessive pursuit of precision, had taken at least two measurements for every observation he had made, which left Kepler puzzling over which to choose whenever they disagreed. To prove the laws governing the universe, post-Enlightenment natural philosophers could no longer appeal to common sense, elegance, or beauty as Aristotle and Plato had done. The ultimate—indeed, the only—judge lay in data. But at the turn of the nineteenth century, no one had the math or the concepts to bring theory and data into quantitative contact.

With *La méchanique*, Pierre-Simon Laplace increased the explanatory power of Newton's theory of gravitation by applying it to fiendishly difficult astronomical problems. Inevitably, the problem arose of how to check his calculations against the available data. Faced with combining observations of the position of Saturn to verify his predictions, Laplace made a conceptual breakthrough that inaugurated the precise treatment of error, thus spawning a new science, statistics, which he felt was the underpinning of "the entire system of human knowledge."[3] The fruit of Laplace's star-powered insight would be a mathematical ability to learn from the past to predict the future—which even today underpins the artificial intelligence revolution.

Unlike the reclusive Newton, Laplace vied for high-powered academic and public positions: elected to the exclusive Académie Royale des Sciences at twenty-four, he was, at one time or another, professor at the École Militaire in Paris (where he examined a sixteen-year-old sublieutenant called Napoléon Bonaparte in 1785), professor at the École Normale, a member of the Bureau des Longitudes, and chancellor of the Senate. He rose to become minister of the interior under Napoléon, who tired of him after just six weeks, complaining, "He carried the spirit of the infinitesimally small into administration."[4] A revealing anecdote about Laplace the micromanager is told by American astronomer Joseph Lovering, whose admiration for Laplace was shattered when he overheard Laplace's wife ask him for permission to get the key to the sugar cupboard.[5] A shrewd opportunist, Laplace somehow managed to navigate the turbulent times in French politics, switching his allegiance from the republic to the emperor when he came to power in 1799, and then to the Bourbon monarchy during the restoration of 1814, somehow acquiring the title of "Marquis de Laplace" in the process.

Laplace considered the Swiss mathematician Leonhard Euler "our master in everything"; yet in his own 1812 book *Théorie analytique des probabilités*, he arguably surpassed his hero, achieving a synthesis that had eluded Euler fifty years earlier.[6] Laplace's insight was that the mathematical combination of many observations, each subject to a random error due to the limited precision of the measurement, leads to a more precise determination of the position of an object in the sky than any of the observations taken individually. This might seem counterintuitive, and it certainly appeared so even to Euler. But following the math, if two observers make a mistake of, on average, a sixth of a degree when measuring the position of Saturn, combining the two erroneous measurements gives a value that is closer to the truth—only about an eighth of a degree off—than either one on its own. The key is that errors must be *random*, with smaller errors more probable than large ones, but symmetrically distributed on either side of the truth. Some measurements will thus be larger than the true value and some smaller. Combining them gives a more precise estimate because errors, on average and in the long run, cancel each other out. Reasoning along these lines, Laplace was able to explain the highly successful "method of least squares," devised by French mathematician Adrien-Marie Legendre to combine measurements of comets, by marrying it with the burgeoning theory of errors developed by Prussian mathematics prodigy Carl Friedrich Gauss.

Gauss had become a sensation across Europe, following the astonishing discovery by Italian astronomer Giuseppe Piazzi on January 1, 1801, of what he suspected was a new planet between the orbits of Mars and Jupiter. Named Ceres after the patron goddess of Sicily, the new planet filled an empty place in the sequence of planetary distances from the Sun, whose proportions seemed to follow the arithmetic relationship 4 (Mercury), 4 + 3 (Venus), 4 + 6 (Earth), 4 + 12 (Mars), 4 + 48 (Jupiter), 4 + 96 (Saturn), where each term is given by four plus a number that is double that of the previous planet. But there was a gap at 4 + 24, which led German astronomer Johann Bode to conclude in 1772, "Can one believe that the Creator of the Universe has left this position empty? Certainly not!"[7] The search for the missing planet had been going on for thirty years when Piazzi made his discovery, which, however, nobody could confirm: Piazzi had sat for weeks on his observations, and in the meantime, in February 1801, the mysterious Ceres had disappeared behind the Sun. The astronomical community was incensed: by not sharing his discovery in a timely fashion, Piazzi had lost the new planet before its orbit could be properly measured, and now it was anybody's guess where Ceres would show up next. Nevil Maskelyne wrote scathingly in the summer of 1801, "There is great astronomical news: Mr. Piazzi, Astronomer to the King of the two Sicilies, at Palermo, discovered a new planet the beginning of this year, and was so covetous as to keep this delicious morsel to himself for six weeks; when he was punished for his illiberality by a fit of sickness, by which means he lost the track of it."[8]

Yet Gauss recognized an opportunity in "this crisis and urgent necessity": a chance to test and "[show] most strikingly the value" of his new theory of errors, at a moment "when all hope of discovering in the heavens this planetary atom . . . rested solely upon a sufficient approximate knowledge of its orbit to be based upon these very few observations."[9] Working off Piazzi's meager and uncertain observations covering a short three degrees of Ceres's orbit, Gauss predicted the time and location of its reappearance in the night sky, almost a year after it had been lost and in a completely different region of the heavens. The fugitive planet was located at the first attempt, precisely at the location forecast by Gauss.[10]

The glamour and excitement faded when, in February 1802, William Herschel proved Ceres to be merely a large rock, a new class of objects he called *asteroids* (meaning "star-like"). But Laplace was so impressed by Gauss's success as to describe him as "a super-terrestrial spirit in a human body."[11] Crucial to the recovery of Ceres was Gauss's treatment of measurement errors, which he described as the now famous bell-shaped curve that today bears his name, the *Gaussian distribution*. The synthesis of Gauss's method with Laplace's derivation of the least-squares method— "one of the major success stories in the history of science," according to statistician Stephen Stigler—would open the door to quantitative analysis of the kind that underpins all of contemporary science.[12]

THE BIRTH OF AVERAGE MAN

Modern statistics grew out of the need to compare a model of the heavens with imperfect observations. Being the first of the sciences to gather a large number of accurate data points, astronomy was also the first to need a method to deal with uncertainty. This new, powerful mathematical technology was immediately adopted across astronomy and geodesy (i.e., the measurement of the shape and area of the Earth). But a conceptual leap was needed if it was to be applied to the sphere of human concerns, where, Laplace argued, his method had the potential to revolutionize policymaking, economics, and public health.

During the 2020–2021 Covid-19 pandemic, we became accustomed to almost real-time projections of the contagion's spread, as policymakers turned to epidemiologists for advice on containment measures, from mandatory face coverings to full-blown lockdowns. But in the early nineteenth century, nobody knew how to interrogate even simple census data to better understand the factors driving, for example, infant mortality, birth, suicide, and divorce rates, or crime and marriage statistics. The human sphere remained opaque.

It had taken almost two millennia for the Ptolemaic crystalline spheres to become amenable to mathematical dissection. But in just a few decades, the concepts and methods of statistical analysis that had dispelled the fog of measurement errors in astronomy would spawn entirely new fields in the social sciences: the quantitative study of society (which we today call sociology), experimental social psychology, criminology and econometrics, and others. The same root also sprouted the twisted branch of eugenics, the selection of "desirable" physical or intellectual traits by enforced birth control practices, which—in its worst but not only legacy—prepared the ground for the horrifying exterminations in Nazi Germany.

One astronomer did more than anybody to bring his field's powerful tools to bear on the human condition: Adolphe Quetelet, born in the city of Ghent (in the Flanders region of today's Belgium) in 1796. His name is largely unrecognized today, except by social scientists with a penchant for the history of their subject, so much so that he has been described as having sunken "from the pedestal to oblivion."[13] But chances are you have heard of one of his brainchildren: the body mass index (or BMI), which measures an individual's body weight relative to his or her height. Until 1972, it was called the "Quetelet Index" after its inventor.

Quetelet trained as a mathematician and an astronomer and successfully petitioned the government to build an observatory in Brussels, to which he was appointed director. In preparation for this role he traveled to Paris in 1822 to investigate the workings of the Parisian observatory. During that fateful trip he met with the leading French physicists and mathematicians of the time: Jean-Baptiste Fourier, Siméon Poisson, Sylvestre Lacroix, and, most influentially of all, Laplace. Quetelet learned at the foot of the master the burgeoning field of probability and, in particular, the newly developed theory of errors.

When he returned to Brussels, he looked for regularities where nobody expected to find them. He discovered that the distribution of chest measurements of Scottish soldiers followed the very same bell-shaped curve of Gauss and Laplace; the same was true of the height of Italian conscripts, the weight of German servicemen, the proportions of French men and women. Quetelet's reaction was incredulity at first, followed by awe. The chaos of human variations and irregularities abided by the iron discipline of the bell-shaped curve, just like astronomical errors did.

It was a revelation. He began to suspect that the same might hold true of moral and intellectual qualities: that the propensities to commit murder, to give in to drunkenness, to marry, or to suffer from mental health problems were subject to regular and predictable statistical patterns rather than products of choice and chance. Writing in 1850, Charles Dickens wryly commented on how Quetelet's ideas had spread to domains previously thought untouchable: "Not content with making lightning run messages, chemistry polish boots, and steam deliver parcels and passengers, the savants are superseding the astrologers of old days, and the gipsies and wise women of modern ones, by finding out and revealing the hidden laws which rule that charming mystery of mysteries—that lode store of young maidens and gay bachelors—matrimony."[14]

The average of any human quality, physical or intellectual, could be measured from its relative rate of occurrence in a given society, and from this collection of data a portrait of what Quetelet called "the average man" could be constructed—a prototypical, imaginary ideal individual. In Quetelet's interpretation (since discredited), the average man was not merely the most common type in a given population at a given epoch. Quetelet presented him as the "perfect" type, embodying "all that is great, good or beautiful," the one exemplar toward which Nature attempted to converge.[15]

By applying to human beings the same statistical laws that govern the observational errors of the stars, Quetelet inaugurated a new discipline: that of "social physics" (which he initially called "social mechanics," on the blueprint of Laplace's celestial mechanics), whose purpose was to reveal the laws, existing "independently of time and of the caprices of man," that regulate humanity as a whole.[16] Free will itself came under scrutiny, for how could individual choice be reconciled with the inevitable statistical distribution of a population? Whereas Newtonian dynamics had removed the necessity for God to intervene in the orbits of the planets, so statistical regularities in the human domain appeared to unburden the individual from the necessity of choice. Quetelet explicitly linked his laws of social physics with the regularities of the heavens, comparing the blind adherence of stars to Newton's laws with man's unwitting following of inscrutable, but equally unavoidable, divine patterns: "We find laws as fixed as those which govern the heavenly bodies: we return to the phenomena of physics, where the freewill of man is entirely effaced, so that the work of the Creator may predominate without hindrance."[17]

LAPLACE'S DEMON GOES PLACES

Quetelet had been seduced by what is now known as "Laplace's demon"—not an evil spirit or any other supernatural creature but an imaginary intelligence capable of collecting sufficient data and subjecting them to mathematical analysis to uncover the laws governing them. To such an intelligence, Laplace observed, the universe would become eternally transparent, the power of mathematics allowing it to project indefinitely the present state of the universe in both directions of time. It is in astronomy that Laplace identifies the leading light, the paragon of such analysis:

> We ought then to regard the present state of the universe as the effect of its past state and as the cause of the one which is to follow. Given for one instant an intelligence which could comprehend all the forces by which nature is animated and the respective situation of the beings who compose it—an intelligence sufficiently vast to submit these data to analysis—it would embrace in the same formula the movements of the greatest bodies of the universe and those of the lightest atom; for it, nothing would be uncertain and the future, as the past, would be present to its eyes. The human mind offers, in the perfection which it has been able to give to Astronomy, a feeble idea of this intelligence.[18]

Laplace's articulation of scientific determinism has been considered the definitive manifesto of a positivistic view of the world (a later name for the "intelligence" he spoke about). From the second half of the nineteenth century onward, the influence of Laplace's demon, born among the stars, spread like a mighty river fanning out in innumerable rivulets as it reaches its estuary. At times, like the Timavo—a famously secretive river that dives into the depths of the Karst for forty kilometers before reappearing, cold and self-possessed, on the shores of the Adriatic Sea, where it was venerated as a god from prehistoric times—a stream of star-inspired ideas disappears from view, only to resurface at a different time and in a different field, transformed. We shall follow some of these divergent streams before they dissolve into the ocean of the present, where their original, celestial headspring is all but forgotten.

The quintessential ability of Laplace's demon is predicting the future from observations of the past, be it the location of Ceres in the sky a year after the last observation, the average life expectancy of an overweight smoker, or the propagation of heat in the Earth's atmosphere and oceans.

This it accomplishes by mathematical analysis, by establishing the laws, expressed in mathematical language, followed by the phenomenon at hand, then projecting its behavior into the future, when new observations are made that can be used, if necessary, to perfect the theoretical model.

We recognize in this approach the blueprint of all modern science and technology. The "vast intelligence" of Laplace's demon led to the discovery of the structure of the atom and of the alphabet of life; it bent invisible electromagnetic rays to our service and exterminated impalpable life-threatening viruses, multiplied the yield of wheat, and filled our skies with planes. To the demon we also owe artificial suns precariously snoozing in their underground launching silos and the sinister crashes that signal, season after hot season, ice shelves the size of small cities disappearing into the sea.

The demon's conquest of natural and human phenomena became unstoppable, as natural and social scientists alike bent its omniscient eye to their purposes. In his 1822 masterpiece *Théorie analytique de la chaleur*, mathematical physicist Jean-Baptiste Fourier observed that mathematical analysis, by describing phenomena inaccessible to direct human experience, seems "meant to compensate for the shortness of [human] life and the imperfection of our senses."[19] Laplace's demon supplied scientists with capabilities that only a century earlier the philosopher John Locke had attributed to angels, some of whom he said enjoyed "perfect and exact views" of the world and could thus achieve "certainty of conclusions," as opposed to the "lame and defective" decisions of men of study and thought.[20] The demon would banish angels and gods farther than ever, until they had fallen back all the way to the first instants after the Big Bang and curled up in the inaccessible extra dimensions imagined by string theory.

FEEDING THE DEMON

The old mathematical tools were no longer sufficient for handling the data the demon craved. The invention of logarithms in 1614 by John Napier had simplified trigonometric calculations, the bread and butter of computational astronomy, to the delight of Laplace, who maintained that this method "by shortening the labors . . . doubled the life of astronomers."[21] Nevil Maskelyne had set up a network of human computers to crunch the numbers needed to fill the tables of his *Nautical Almanac*. But mistakes were inevitable in the long and repetitive manual calculations required to create the numerical tables on which Victorian society increasingly depended. Tables of logarithms, tables of trigonometric functions and their logarithms, tables of interest, tables of annuities, and tables of the hourly locations of the Moon (and its distance from a select number of stars), the Sun, and the planets: these hefty, well-thumbed volumes were found in the captain's cabin as in the astronomer's observatory, on an insurance broker's desk as on an architect's shelf, in a mathematician's library just as in a civil engineer's. The problem was, they couldn't be relied upon.

A survey in 1834 by Dionysius Lardner, professor of astronomy at London University, found no fewer than thirty-seven hundred acknowledged errors in forty random volumes of tables; corrections of corrections of corrections piled on top of each other, including for the *Nautical Almanac* used daily by mariners the world over, for whom an error in the second significant digit could (and sometimes did) make the difference between life and death.[22] Even worse were the errors nobody had found yet. John Herschel, son of astronomer William Herschel (whom we encountered earlier in the company of Napoléon and Laplace), writing to the chancellor of the exchequer in 1842, lamented, "An undetected error in a logarithmic table is like a sunken rock at sea yet undiscovered, upon which it is impossible to say what wrecks may have taken place."[23] A mechanical replacement for imperfect human calculation was called for.

No one was more frustrated by the mistakes found in human-produced numerical tables than Charles Babbage, whose private collection grew to over 140 tomes, from which Lardner had culled his 40 error-ridden volumes. As a well-to-do mathematics undergraduate at Cambridge in 1812, Babbage had already daydreamed about computing logarithmic tables by machinery.[24] While at Cambridge he befriended a young John Herschel, an encounter that would have a profound impact on his life. Forged over discussions on "all knowable and many unknowable things" during Sunday breakfasts after chapel, at meetings of the Analytical Society they cofounded, or as they worked shoulder to shoulder in a makeshift chemistry laboratory set up in one of Babbage's spare rooms, their friendship would last a lifetime—only fraying at the very end.[25]

After their Cambridge days, Babbage and Herschel started to make a name for themselves in the upper echelons of Britain's "scientific people," as they were then called: in 1820, they were among the fourteen founders of what would later become the Royal Astronomical Society. Astronomy was a natural interest for Herschel, son of the discoverer of Uranus, but not so obvious for Babbage, scion of a wealthy banker. But the stars were to prove a fateful attraction for Babbage, who, as a young man of means, education, and ambition in 1820, could not have imagined that fifty-one years later his obituary would state, "Unhappily for himself, he chose a path which . . . led only to loss of fortune and embitterment of mind."[26]

Perhaps Babbage would have hit on his "analytical engine" scheme anyhow. But it was Herschel and his astronomical tables that tipped him over, one day in 1821. One of the stated aims of the newly established society was to improve lunar tables and to map out the whole universe of stars through "regular systematic examination of the heavens."[27] Its first president, John Herschel, called upon Babbage's help to improve the *Nautical Almanac*, and that meant checking over the calculations of the human computers.

Babbage later recalled the momentous visit thus: "My friend Herschel, calling upon me, brought with him the calculations of the computers, and we commenced the tedious process of verification. After a time many discrepancies occurred, and at one point these discordances were so numerous that I exclaimed 'I wish to God these calculations had been executed by steam.'"[28]

Babbage dedicated the rest of his life to building a machine capable of achieving that vision. In 1823, he felt sure that in just a few years, his machine would churn out "logarithmic tables as cheap as potatoes."[29] In 1824, he was the first recipient of the Gold Medal of the Royal Astronomical Society for his "calculating machine" proposal—an accolade that proved premature. Babbage's vision was revealed first as a mirage and later as a nightmare: by 1829, his friend William Whewell reported to a correspondent, "[Babbage's] anxiety about the success and fame of his machine is quite devouring and unhappy."[30] After a decade of efforts, the project to realize his first design, the "difference engine," foundered when, due to a money dispute, Babbage fired the engineer who had been handcrafting the thousands of gears required, even though the completion of a working prototype was within reach. By then, Babbage had realized that logarithmic and astronomical tables were only a tiny fraction of what a general-purpose, fully programmable computing machine could achieve. He abandoned the difference engine and set about designing an even more ambitious calculating machine.

Nobody had conceived of anything like it before: Babbage aimed at nothing less than "substituting mechanical performance for an intellectual process."[31] In his quest, he was pursuing a mechanical embodiment of Laplace's demon, with whom he was well acquainted. He had met Laplace during a visit to Paris in 1819 and earned the great man's esteem, so much so that Laplace wrote him a recommendation letter for a university professorship. Babbage had been greatly impressed by Laplace's *Theory of Probabilities* and reprinted the excerpt on the "intelligence" quoted in the previous section in his *The Ninth Bridgewater Treatise*, a rambling polemic on topics ranging from free will to the properties of granite, from the likelihood of miracles to the calculating prowess of his engine.

In pursuit of his dream of a mechanical computer that could perform any calculation whatsoever, Babbage resigned the hard-won Lucasian Chair that had been Newton's and squandered vast government grants and the majority of his own considerable fortune. He never succeeded—partially because of the immense complexity of building a machine the size of a locomotive with twenty-five thousand finely chiseled moving parts with Victorian-era mechanical technology, but largely due to his own character flaws. Abrasive, haughty, and self-important, Babbage often alienated the supporters he should have been charming. Yet his vision was truly ahead of the times. The first general-purpose computer would be built in the 1940s, using electronic components, whereas Babbage's design called for rotating drums with pins and geared wheels—all to be operated by steam. To commemorate the two hundredth anniversary of Babbage's birth in 1991, a team at London's Science Museum set out to build the full difference engine from his original blueprints, endeavoring to use only technology and processes available during Babbage's time. The engine worked as intended.[32]

Although never completed, Babbage's mechanical realization of Laplace's demon did have lasting impact, if not in the direction its maker intended. During the 1830s, Babbage put on display in his drawing room a prototype of the full difference engine, a seventh of the machine he had in mind. To the Victorians, it was a marvel: three columns of burnished brass, each carrying six rotating drums, each of those elegantly engraved with the numbers from zero to nine and driven by a precisely interlocking gear. More gears sprouted from the top and disappeared from view into a stand of polished rosewood. Babbage used it to demonstrate the profound implications of his ideas to his guests: the engine was programmed to display a sequence of increasing numbers—zero, one, two, three, and so on—until at a certain point, and without intervention from Babbage, the sequence would change, apparently miraculously, and follow a new progression. Such apparently disparate laws, Babbage explained, were in fact the product of a higher law that contained not just the potential but the necessity of change.

His parties attracted the most fashionable London crowd. Young aristocratic ladies in evening gowns mixed with "scientific people" in swallowtail jackets; bishops and statesmen conversed with artists and writers. All of them were treated to the finest oysters, fowl, and salmon. Among the regulars were Mary Somerville, Laplace's English translator, who asked Babbage to explain his analytical engines to her and judged him a "first-rate mathematician"; Lady Ada Lovelace, Lord Byron's estranged daughter, who would become the first coder in history by describing a program intended for Babbage's engine in 1843; philosopher William Whewell, a close friend of Babbage's from his Cambridge days, who in 1834 had come up with the word *scientist* to describe "a cultivator of science in general"; and a young but already well-traveled naturalist, a Mr. Charles Darwin, recently returned from his round-the-world trip aboard the *Beagle*.[33]

The *Beagle*'s mission may have been to improve longitude measurements in South America and the Pacific, but the voyage's historical significance lies in instilling in Darwin the insights on natural selection that would ground his theory of evolution. Back in London, Darwin was mystified by how precisely natural selection came about. How could the incredible adaptations of organisms he had witnessed most spectacularly in the finches of the

Galapagos Islands be brought about by the blind rule of a natural law, without resorting to the intervention of an intelligence? And how long would it take for such changes to appear? Fresh from his impressions of remote nature, in February 1837 Darwin was introduced to London's most glamorous parties by his friend and mentor, geologist Charles Lyell. "Lyell says Babbage's parties are the best in the way of literary people in London—and that there is a good mixture of pretty women," Darwin enthused.[34]

In early March 1837 Charles Darwin attended his first Babbage bash, where the mathematician, perhaps sporting one of his trademark colorful waistcoats, no doubt cornered the new arrival and, as was his custom, lectured him about his "thinking machine," as Mary Somerville called it. Babbage published that same year in his *The Ninth Bridgewater Treatise* a summary of how the difference engine offered a point of comparison for the workings of evolution:

> In turning our views from these simple results of the juxtaposition of a few wheels [a reference to the difference engine], it is impossible not to perceive the parallel reasoning, which may be applied to the mighty and far more complex phenomena of nature. To call into existence all the variety of vegetable forms, as they become fitted to exist, by the successive adaptations of their parent earth, is undoubtedly a high exertion of creative power. . . . To change, from time to time, after lengthened periods, the races which exist, as altered physical circumstances may render their abode more or less congenial to their habits, by allowing the natural extinction of some races, and supplying by a new creation others more fitted to occupy the place previously abandoned, is still but the exercise of the same benevolent power . . . to have foreseen all these changes, and to have provided, by one comprehensive law, for all that should ever occur, either to the races themselves, to the individuals of which they are composed, or to the globe which they inhabit, manifests a degree of power and of knowledge of a far higher order.[35]

Just as the difference engine changed its output following a program with which he, Babbage, had furnished it, so surely God could have had the foresight to build into the laws of nature rules that would, at a later point in time, bring about a change that would appear to us "a miracle"—like the emergence of new forms of life or the disappearance of others of whose existence the large troves of unexplained fossils spoke.

As the cogs of the difference engine spun on their well-oiled shafts, so must have the wheels inside Darwin's head rearranged themselves into never-before-thought ciphers. If Babbage could build a machine with instructions to modify its behavior under the appropriate circumstances, one can imagine Darwin realizing, could God not write into the laws of Nature, at the dawn of time, the appearance of humankind? And if the trajectory of every molecule in the atmosphere could, in principle, be predicted infinitely into the future, could the direction of evolution not be subject to similar forces, with no need for the hand of God to guide its course? A couple of weeks after Babbage's party, Darwin confided to his notebook, for the first time, his fledgling ideas around the transmutation of species: "If one species does change into another it must be per saltum [by sudden transitions]"—precisely as Babbage's engine suddenly, seemingly miraculously, changed its output and spit out a new sequence of numbers.[36]

Babbage himself remained resistant to change. His lifelong friendship with John Herschel eventually succumbed to the same forces that doomed his difference engine; while Herschel more than filled his father's astronomical shoes and became one of Britain's most eminent scientists of his time, Babbage, by contrast, was fated to be remembered as "essentially one who began and did not complete."[37] The last letter he sent was to Herschel's widow, in which he morosely complained that his late friend's illustrious name had opened to him avenues "inaccessible to others." Five months after Herschel was laid to rest with all pomp next to Newton in the nave of Westminster Abbey, Babbage was buried without fuss in the public cemetery of Kensal Green.[38]

UNEXPECTED CONNECTIONS

Both Babbage and Darwin were familiar with the work of Quetelet, and they weren't the only ones to be influenced by it. The meandering course of the river that started with Gauss's and Laplace's efforts to understand observations of heavenly bodies veers off, thanks to Quetelet, in a myriad of directions. Starting with astronomy, it doubles back onto itself and, via Quetelet's social physics, sources a new branch of physics, today known as statistical mechanics.

Scottish physicist James Clerk Maxwell discovered social physics from a review by Herschel of Quetelet's work. Deeply impressed, he set about applying the same statistical reasoning to the equally inscrutable realm of atoms. Maxwell's insight was that just as one didn't need to comprehend each individual's heart to predict the homicide rate in a population, so the physicist didn't have to measure every particle in a room to describe the laws regulating their average properties, such as temperature or pressure. A statistical description would suffice—precisely the same idea that Ludwig Boltzmann, considered with Maxwell the cofounder of thermodynamics, arrived at in 1872, having been similarly struck by social statistics. Statistical mechanics has since blossomed to describe the beginning of the universe and the behavior of financial markets, the growth of forests and the learning patterns of the brain.

Another Victorian polymath, Francis Galton (Darwin's half cousin), took Quetelet's work one step further: instead of focusing on the average, Galton considered the whole distribution of qualities, in particular intelligence. He looked closely at deviations from the norm, especially at genius as an extreme form of intelligence. He coined the term *normal distribution* to describe the Laplace-Gauss bell-shaped curve describing the distribution of "mental capacity," as he called it, with the implication that any deviation from the average (corresponding to the peak of the curve) was "abnormal." Modern-day intelligence tests are a rivulet flowing from Galton's ideas; so was eugenics, another term invented by Galton. He championed the notion that the human race should be "improved" by socially engineering the selection of qualities deemed desirable, like class, intelligence, or ethnicity. Just as farmers take over and accelerate the job of natural selection by replanting only the better, stronger varieties of crops each season, or breeders create the ideal racehorse by choosing the fastest stallion for the mount, so eugenics aimed at stopping the perceived uncontrolled breeding of the "feebleminded," which threatened to overrun the "good stock." One of the foremost eugenics experts in the United

Kingdom, neurologist and psychiatrist Alfred Tredgold, maintained that "the problem of the feeble-minded is no isolated one, but . . . it is intimately connected with those of insanity, epilepsy, alcoholism, consumption, and many other conditions of diminished mental and bodily vigour. And when we remember that these are the conditions which connote social failure and which give rise to such a large proportion of our criminals, paupers and unemployables, we begin to see how far-reaching this question is."[39] The eugenicists' solution was to segregate people deemed unfit, control their fertility, and even, in Nazi Germany, exterminate them.

Eugenics rapidly became a swamp from which all sorts of tentacles of inhumanity emerged. In the United Kingdom, eugenics motivated and justified the Mental Deficiency Act of 1913, which allowed the government to forcibly lock up in asylums people deemed to be "feebleminded" or "morally deficient"—destroying the lives of sixty thousand people by 1957. In the United States it resulted in forced sterilizations of people with disabilities—twenty thousand in California alone before the practice was forbidden in 1979. Eugenics took an even more horrific turn in Nazi Germany, where over two hundred thousand people with disabilities were murdered in "euthanasia" centers. The Nazi "Euthanasia Program" of 1940–1941 became the model for the Holocaust, eugenics' ultimate, horrifying apogee: Adolf Hitler's obsession with the superiority of the Aryan race led to the genocide of six million Jews and between 250,000 and 500,000 Roma and the systematic murder of thousands of homosexuals.[40] The stench of this degenerate branch of the river must never be forgotten.

THE HUMAN MACHINE

Another astronomer, working by his own admission in a third-rate observatory, faced a problem that could only be solved by "measuring thought"—and in doing just that, he became a pioneer of experimental psychology.

Swiss watches are today synonymous with precision and, often, exclusivity. I remember gaping as a doctoral student in Geneva at the diamond-studded dials of high-end chronometers sparkling like stars under their bulletproof glass. To achieve their exquisite reputation, the key ingredient from the start of the Swiss clock and watch industry was an equally exact reading of the time of the stars—the raison d'être of the observatories of Geneva and Neuchâtel, not coincidentally the two main centers of Swiss clockwork manufacturing. Every aspect of their functioning was "calculated to fulfil that goal to the highest possible degree," wrote Adolph Hirsch, director and lone astronomer of the Neuchâtel observatory, in his biannual report in 1861.[41]

But Hirsch soon realized that a hurdle stood in the way of his quest for ultimate chronometric precision: himself. The observer's reaction time to the passage of a clock star across the spider threads of the telescope was an unknown quantity, which astronomers since Friedrich Bessel had termed the *personal equation* and we would today call *reaction time*. In 1816, Bessel had been spurred to investigate the phenomenon upon learning that Nevil Maskelyne had dismissed his young assistant, David Kinnebrook, because of his "vitious way of observing the times of the Transits too late" by a full eight hundred milliseconds. That Kinnebrook had rejected Maskelyne's recommendation to marry the niece of a colleague only a fortnight before his dismissal may or may not have swayed the royal astronomer's hand.[42]

The personal equation varied from individual to individual, even during the same night for the same person, introducing error into the measurement—error that wasn't amenable to treatment by Laplace's method, as it occurred not at random but always in the same way (albeit by a different amount in different observers). Hirsch therefore cast aside the timing of stars and proceeded to systematically study his friends' and his own reaction time to auditory, tactile, and visual stimuli, using a "chronoscope" capable of measuring the speed of bullets, later adopted by experimental psychologists everywhere. He had an artificial star apparatus built that created controlled experimental conditions, with which he measured the personal equation of each observer. Perhaps inevitably for someone in the service of the clockwork industry, Hirsch concluded that man was "exactly like a precision machine," and like any other temperamental instrument, the human observer needed careful calibration in order to operate accurately.[43]

The clockwork universe had cycled back onto us, as the surveillance of the stars led to the scrutiny of human senses as part of the measuring apparatus. Soon the human observer would be written out of the picture altogether, replaced by the then new electrochronograph for the timing of transits, the photographic plate, and finally digital imaging. The Romantic poets raged against the mechanization of man and nature, often appealing to the stars to bear witness to their anguish—ironic given the stars' ongoing role as paragons of clockwork. As Walt Whitman put it in 1865,

> *When I heard the learn'd astronomer,*
> *When the proofs, the figures, were ranged in columns*
>   *before me,*
> *When I was shown the charts and diagrams, to add,*
>   *divide, and measure them,*
> *When I sitting heard the astronomer where he*
>   *lectured with much applause in the lecture-room,*
> *How soon unaccountable I became tired and sick,*
> *Till rising and gliding out I wander'd off by myself,*
> *In the mystical moist night-air, and from time to*
>   *time,*
> *Look'd up in perfect silence at the stars.*

From Victorian times onward, society itself became mechanized. After all, were we not complex mechanisms ourselves, the result not of loving design but of the survival of the fittest? Unpredictable and haphazard as individual choices appeared, had the cumulative effect of each person's free will not been shown to follow the iron laws of probability? And had Karl Marx not intimated that the whole of society followed natural laws of economic development?

The standardization of manufacturing processes, the invention of the assembly line, the skyrocketing demand for wood, coal, iron, and later oil for travel (by steam trains first, the motorcar later, then the double-decker planes of today), and the new means of production (the power loom, the Bessemer process for making cheap steel, the combine harvester) and of destruction (the machine gun, the iron-plated gunboat, the bomber plane, the intercontinental ballistic missile) required that men and women be put at the service of the machine. The repetitive, sequential processes that the new world of manufacturing required transformed the individual, according to Marx, "into the automatic motor of some partial operation."[44]

The world was reshaped in a new cosmic order. Just as stars rose and set at predictable times, so steamships crisscrossed oceans indifferent to the caprices of trade winds, passenger trains traversed continents on gleaming steel tracks, and electric news rode telegraph wires across the globe at the rhythm of Morse code. The revolutions of the heavenly bodies had been imperfectly mimicked by the Antikythera mechanism, imprinted onto the face of Jacopo de' Dondi's magnificent clock, transmuted into mathematical laws by Newton, breathed into brass and steel by Babbage, and at last broadcast far and wide by the pervasive influence of Laplace's demon.

THE IMMENSITY OF TIME

Laplace's demon likely influenced Darwin's early thinking about evolution through his brush with Babbage and his reading of Quetelet. But Darwin's theory needed a vast supply of a key ingredient that Victorian science was unable to provide in the required quantity: time.

Here, too, the stars obliged. First, the astronomical distance scale jumped into the previously almost inconceivably large distance of the light-year, which Friedrich Bessel introduced in 1838, as we saw, while Darwin was musing on evolution. This is, I think, the meaning of Henri Poincaré's remark that "astronomy taught us not to be afraid of big numbers": thanks to the example of astronomy, geologists and evolutionary biologists were ready to entertain an almost bottomless abyss of time, the counterpart to the indefinite extension of space that the telescope had uncovered.[45] But just as importantly, the stars also provided positive evidence that the cosmological timescale was far more expansive than the biblical account of creation allowed. To uncover this subterranean stream connecting astronomy and the theory of evolution, we must follow its meandering through geology.

James Hutton, considered today the founder of modern geology, intuited the immensity of geological time not by looking at the heavens but by considering the humble, slow work of a drop of water carrying away fertile soil on his Berwickshire farm, one grain at a time. Given enough time, he reasoned, all the soil would eventually end up in the sea, unless new soil was formed somewhere else—by frost, wind, and rain wearing down mountains. Hutton roamed the Scottish countryside, inspecting riverbeds and cutbanks and puzzling over fossil clams in the Cheviot Hills, and concluded that they hadn't gotten there in the biblical flood in 2350 BC, as the literal interpretation of Genesis had it. "Old continents are wearing away and new continents forming at the bottom of the sea," he claimed in 1785.[46] But even Hadrian's Wall hadn't changed much in the sixteen centuries since it had been built, as anyone could see. How much time was necessary for Ben Nevis, the highest mountain in Scotland, to be ground to dust? And how much time had been required for it to raise from the sea, layer after deposition layer? None of this was possible in the mere six thousand years since the Creation allowed for in the Bible.

Just as Nicolaus Copernicus was largely ignored until Galileo Galilei wrenched from the heavens hard-biting evidence for heliocentrism, Hutton's assertion of the Earth's fantastically great age was overlooked until Charles Lyell freed geology from Moses in his masterwork, *Principles of Geology*. Gifted with an eloquence Hutton never could muster, Lyell made the case that continents and mountains had not been shaped by sudden catastrophes or emerged from the biblical flood but were rather slowly molded by the prosaic processes of erosion due to the wind, the water, and the Sun over millions of years. Lyell's gradual "uniformalism" was ridiculed at first and only received wide acceptance with the next generation of geologists.[47] But John Herschel, as an astronomer familiar with the immensity of space, was immediately onboard with similar views about time.

Herschel wrote to Lyell in 1836, "Time! Time! Time!—we must not impugn the Scripture Chronology, but we *must* interpret it in accordance with *whatever* shall appear on fair enquiry to be the *truth* for there cannot be two truths."[48] In February 1837, in a letter to his sister, Darwin found Herschel's a refreshingly novel idea when applied to the time elapsed since the appearance of man:

> You tell me you do not see what is new in Sir J. Herschell's [*sic*] idea
> about the chronology of the old Testament being wrong.—I have

used the word Chronology in dubious manner, it is not to the days of Creation which he refers, but to the lapse of years since the first man made his wonderful appearance on this world—As far as I know everyone has yet thought that the six thousand odd years has been the right period but Sir J. thinks that a far greater number must have passed.[49]

The first volume of *Principles of Geology*, fresh from the printer, was in the small library that Charles Darwin brought with him onboard the *Beagle*, a gift of Captain Robert FitzRoy, and the young naturalist soon became "a zealous disciple of Mr Lyell's views"—with regard to not only geology but also the giddy stretches of past time that Lyell had opened up.[50] Time was to be one of the essential ingredients of Darwin's theory of evolution: heaps and heaps of it, for in the biblical six thousand years since the creation, the peacocks' feathers could never have evolved from a T. rex, but given six hundred million years, opposable thumbs and self-consciousness might have a chance.

"The geological age plays the same part in our views of the duration of the universe as the earth's orbital radius does in our views of the immensity of space," wrote Irish geologist John Joly in 1915.[51] In other words, the age of the Earth was the plumb line with which to probe the immensity of time. Would it touch bottom before giving mountains a chance to raise from the depths of the oceans, like Lyell posited, and humans to sprout from the rib of a mouse? Joly estimated the age of the Earth to be about one hundred million years by looking at the time required for the salinity of the oceans to build up to current levels—an idea first proposed in 1715 by Edmond Halley, who presciently speculated that "perhaps by it the world be found much older than many have hitherto imagined."[52]

Late-nineteenth-century physicists had a hard time granting the geologist and the evolutionist the unfathomable stretches of time they clamored for. Lord Kelvin was adamant—and correct—that if the Sun's energy was provided by gravity, "unless sources now unknown to us are prepared in the great storehouse of creation," our star could not be older than twenty million years.[53] The physicists' plumb line had hit the rock bottom of their ignorance of the true energy source of the Sun. But the discovery of radioactivity in the early twentieth century allowed geologists to determine the age of rocks by measuring the ratio of lead and uranium, resulting in estimates in excess of a billion years. George Darwin, son of the father of evolution and a noted astronomer, shored up the geologists' position by showing that the heat from even a small sprinkling of radioactive elements in the makeup of the Sun could keep it going for hundreds of millions of years.[54] A complete resolution wouldn't be reached until the 1930s, with the discovery of thermonuclear fusion as the origin of the heat of the Sun, which we today know is about five billion years old.

The universe itself is far older. Just as the distance to the stars was almost inconceivable before Bessel's discovery of the stellar parallax, so the age of the cosmos wasn't determined until high-precision data were fed to Laplace's demon. Observations of the leftover light from the Big Bang, first obtained in 1992, supply the demon with a snapshot of the baby universe as it was a mere 380,000 years after it all began. Running on the Laplace-Gauss bell-shaped curve, an excellent description of the distribution of energy in the aftermath of the Big Bang, the demon moves forward in time, following the expansion of the universe until the present day, predicting the ignition of stars, the formation of galaxies, and their statistical distribution in the sky. The result is an extraordinarily precise estimate for the age of the universe: 13.80 billion years since the beginning of all there is, give or take 3 million years, which is ten times more precise than the margin of error on the age of the Earth. Demon-powered cosmologists thus gave us a universe vastly older than anybody had dared imagine before.

THE DEMON REBORN

With the convergence of data, statistical analysis, and computing power, the demon became capable of surveying and predicting phenomena ranging from the subatomic to the extragalactic, over timescales encompassing the fleeting collisions of particles in accelerators and the formation of galaxies.

The demon outgrew the juvenile forecast of the position of heavenly bodies based on past observations, as in the case of Halley's Comet or Ceres. It had moved on to the harder problem of foretelling what had never been seen before: a new planet, like Neptune, which French astronomer Urbain Le Verrier had discovered in 1846 "with the tip of his pen," as his colleague François Arago memorably put it, by analyzing the perturbations in the orbit of Uranus; or the change in the apparent position of stars around the Sun during a solar eclipse, calculated by Albert Einstein in 1915 on the basis of his new theory of general relativity and verified by Arthur Eddington in 1919. Another prediction of Einstein's—the existence of gravitational waves, perturbations in the fabric of space time that travel at the speed of light—took a century to verify.

Emboldened by the demon's successes in astronomy, early-nineteenth-century physicists had become comfortable in running the present state of the world backward and forward in time by solving so-called differential equations, which describe the dynamical evolution of a system—be it a planet in orbit around the Sun, the plucking of a violin string, or a parcel of heat diffusing along a metal bar leaned on a hot stove. Boltzmann and Maxwell taught the demon to work with average properties of large ensembles of identical particles: it didn't need to follow the detailed trajectory of each single atom to compute the temperature and pressure of air in a piston being compressed, for example.

At the start of the twentieth century, quantum mechanics forced the demon to pick up new tricks: at the atomic level, probabilities replaced certainty, and the demon was prevented by Werner Heisenberg's uncertainty principle from acquiring all the data it craved. The uncertainty in the data that had evoked the demon's first incarnation in Laplace's method was unavoidable at the atomic level. In the quantum world, knowledge of a particle's position translates into ignorance about its velocity (or more precisely, momentum), and vice versa, so the demon could never be entirely sure of both position and velocity. Still, the demon learned to predict the probability of any possible measurement outcome, and in a few decades it built up a detailed understanding of the structure of a subatomic world that no human being had ever experienced. The method was working beyond Laplace's wildest dreams.

At the foot of the theoretical physicist, the demon concentrated its powers on the one language it knew: mathematics. Galileo was the first to describe mathematics as the language of Nature, whose characters, he said, are "triangles, circles, and other geometric figures without which it is humanly impossible to understand a single word."[55] To Galileo, mathematics was simply geometry, but as the field grew more and more abstract, mathematicians created a web of concepts and rules to derive ever more general, elegant results after a long chain of logical arguments. With the advent of quantum mechanics at the beginning of the twentieth century, Galileo's words assumed a new meaning: the behavior of subatomic particles—insofar as it could be glimpsed from a human scale as immense compared to the atom as the galaxy is compared to us—was incomprehensible in the absence of a mathematical model. The time-honored process of going from data to theory that had guided Kepler and, to a point, Newton appeared to have run its course. Theoretical physicists appropriated pure mathematics with a view to building around it new models for the working of the world and equipped the demon with the predictive power of mathematical necessity. When Albert Einstein was asked what he would have done had Eddington's astronomical observations proven his theory of general relativity wrong, he retorted, "I would have felt sorry for the dear Lord. The theory is correct."[56]

Stunningly, the physicists' faith in pure math paid off. The structure of the natural physical world appeared to be intimately connected with abstract products of human imagination. Even theoretical physicist and Nobel Prize winner Paul Dirac, whose faith in math led him to conjecture the existence of antimatter, was surprised by this turn of events: "Non-euclidean geometry and non-commutative algebra, which were at one time considered to be purely fictions of the mind and pastimes for logical thinkers, have now been found to be very necessary for the description of general facts of the physical world," he wrote in 1931.[57]

The language Galileo glimpsed in the stars has been found to apply everywhere. Why do constructions hatched in the insular world of a mathematician's mind describe, with uncanny accuracy, natural phenomena occurring in the real world? How does an apple know to fall precisely to the beat meted out by a second-order differential equation? These questions remain unanswered to this day, even as the "unreasonable effectiveness of mathematics in the natural sciences," as theoretical physicist Eugene Wigner put it, has been proven over and over again, an everyday miracle.[58]

As the second decade of the new millennium rolled in, Laplace's demon underwent yet another remarkable metamorphosis. The capillary diffusion of the internet and the ubiquity of mobile devices, together with simultaneous breakthroughs in machine learning algorithms and computational power, supercharged the demon's abilities and its realm of application. "The new availability of huge amounts of data, along with the statistical tools to crunch these numbers, offers a whole new way of understanding the world," announced tech journalist and entrepreneur Chris Anderson in a much-debated article in 2008—one in which, Anderson argued, theoretical models were passé and the scientific method itself needed overhauling: "this approach to science—hypothesize, model, test—is becoming obsolete."[59]

Today, artificial intelligence (AI) is everywhere, exhibiting capabilities that were unthinkable just a decade ago. From fluent conversation with a human-sounding chatbot like ChatGPT to the on-demand production of artwork from a natural language description, from speech recognition to autonomous driving, AI is rapidly changing all aspects of our lives, even redefining what it means to be human in our technological age. Some fear the moment when this latest incarnation of the demon might tip over to becoming superhuman in its powers—a demon-driven technology descended from the stars, once our gods.

"Astronomy has given us a soul capable of comprehending nature," wrote Poincaré, and Laplace's demon translated that understanding into prevision and, ultimately, action.[60] But the demon's powers haven't acted only on the material, external world. Between Galileo's first rudimentary *cannocchiale* and the James Webb Space Telescope's spaceborne infrared eye, humanity has been changed from within by the new outlook on reality, and on ourselves, afforded by Laplace's demon, trained among the stars. As we leave the world of quantitative science behind, the light of a full Moon beckons us to reconnect with another side of ourselves.

THE CALIGO TALES

Fire-Keeper's Tale

When Bison-Seeker slumped back onto her sitting stone, exhausted, Shepherd moved to where Fire-Keeper was crouching and said, "Fire-Keeper, thanks to you our cave is warm and the Dark stops at its entrance. Remember how we care for Fire!"

Fire-Keeper put down his stoking branch, and thus began his tale.

Fire and Lightning are one—only Lightning is purer. Fire is a gift from Lightning, and it is our friend when we honor it well. It cooks our meat and stays the cold; it scares the wolf, smokes our hides, and dries our furs. It shows us the way in the Dark, burns fiercely when fed with sap. It travels with us, sleeping in embers, ready to rise again with a gentle blowing onto dry leaves. It summons the Cloud among us during Remembering. I Remember how to care for its bounty and soothe its wrath.

Like Lightning, which comes out of darkness, so does Fire bring forth the Dark inside: the brighter it burns, the deeper the Dark. That darkness is inside each of us, unseen during the glow: we call it "the Shadow."

The Shadow rises to dance behind each of us now, on the walls of our home. But we fear not: we know it is a piece of darkness cleaved from the Great Blackness outside. I Remember how to harness the Shadow! Watch the Shadow teach us!

With this, Fire-Keeper gathered a burning stick and lit it from the big fire. He stuck it among two large rocks, and, moving around it, coaxed the Shadow out of his hands. Onto the cave wall, the Shadow wobbled, then took the shape of a mammoth, slow and powerful as it munched a tree. We watched it for a long while, and I sensed in my blood the thrill of the hunt. At last, from the Shadow the shape of one of us sprang up—it must have been Spear-Thrower, for he was carrying a spear, shaking it with force, unafraid. Spear-Thrower ran up to the mammoth from its backside, and hit it over and over, until the giant fell on its side, vanquished. Spear-Thrower jumped on it and danced, until the Shadow trembled, then faded away. A plume of smoke rose from the burning stick. The Fire had gone.

"Let us learn from the Shadow!" said Fire-Keeper. "Fire has wrenched a piece of the Dark and put it to our service, so it may tell us what is to come! The mammoth will be conquered!"

A roar followed Fire-Keeper's last words—Fire did not lie and we felt ready for the hunt that was to come. Shepherd rose again, and said, "Mist-Catcher, after the Fire, let us Remember the ice. Tell us about Name-Giving!"

Mist-Catcher stood, and thus began her story.

9

A MIRROR TO OURSELVES

But if one once takes it for granted that these things are true, one is confronted with the terribly serious question, what have we to do with the stars?

—CARL GUSTAV JUNG, *Dream Analysis*

THE SUN WORSHIPPERS

I was bewitched the moment I started down the incline. Almost two decades later, I still remember my feet taking over and marching me toward the far end of the cavernous hall. The orange haze transformed the space into another planet, where I imagined my light-headedness would be explained by the lower gravity, or a different atmosphere, or the sheer alien nature of the place. The silhouette of people crowding the bridge that spanned the shorter side of the hall reminded me of passengers on an ocean liner, leaning over the railing for a last good-bye. Or perhaps the black figures bathed in apocalyptic light were saluting for the last time their home planet, the cradle of their space-faring species, after their once benign star had turned into a planet-eating red giant.

As I emerged on the other side of the bridge, the huge, glowing disc towered in front of me. "Daddy, look, it's the Sun!" cried a little girl nearby, pointing. Around me, people walked like automatons toward the Sun, as if in a dream, staring at it, transfixed. Others were scattered on the floor, their clasped hands cushioning their heads, looking up. When I craned my neck to follow their gaze, I saw myself looking down from the mirrored ceiling five stories above, floating in a depthless orange mist.

The Sun didn't budge; its glow didn't change; time didn't tick. People hung around, unhurried, unbothered by the urgency of taking selfies—phones had no camera then. I was now only a few meters away from the disc, feeling like a modern-day Icarus who had failed to be incinerated. I could feel no temperature change. Swirls of aromatic mist, like sage grown in space, wafted every now and then from concealed fog machines. I sat on the floor and basked in the heatless, permanent sunset.

The site-specific installation by Icelandic-Danish artist Olafur Eliasson more than met the challenge offered by the sepulchral space of the Tate Modern's Turbine Hall in London, five hundred feet long and seventy-five feet wide. Eliasson filled the huge, empty volume delimited by brick, industrial concrete, and steel with the otherworldly light of an artificial sun, taking the form of a fifteen-meter half disc backlit by two hundred monochromatic lamps. A fully mirrored ceiling created the illusion of a complete disc for the Sun and gave visitors the opportunity to look at themselves when lying on the floor, encouraging an element of choral participation. By the evidence of my visit, it was working.

Titled *The Weather Project*, Eliasson's work ostensibly revolved around simulating an indoor, artificially controlled weather phenomenon. But its power, experienced by more than two million people over its five-month life, came undoubtedly from the Sun itself. Eliasson resuscitated in spectacular fashion the symbolic potency of the Sun. For all of the installation's clever technology, visitors to Tate Modern were confronted with a primal, numinous feeling, one that had infused humankind's psyche for millennia: the sense of being at the mercy of a higher power, a deity that could turn from benevolent life giver to deadly scorcher on a whim.

Eliasson's installation speaks also of the role of the Sun in humankind's search for the meaning of existence. It invites us to confront the question of how the stars have shaped not just our technology, the way we explore, communicate, measure time, and organize our lives, but also our collective psyche, the scaffolding underpinning what it is to be human. All of this symbolism, which we already encountered in the meaning attributed by our forebears to eclipses and comets, the stories told about constellations and stars, the godlike powers ascribed to the planets, and the death-and-rebirth cycle seen mirrored in the Moon, didn't disappear with the Scientific Revolution. If anything, it has been pushed into deeper wells, less-examined interior spaces. This chapter explores the firmament's influence on our very souls.

SOL INVICTUS

To trace the power behind Eliasson's sun, like Gilgamesh I had to descend underground. A few hundred yards from where the Timavo river resurfaces to flow into the Mediterranean in northeastern Italy, a natural cave bears witness to the ancient cult of Mithra, the Sun god whose worship spread in the death throes of the Roman Empire, carried by soldiers, merchants, and slaves from its obscure origins in the Middle East to the four corners of the Roman dominions.

The site is one of hundreds scattered from the edge of the Sahara to the Scottish mountains, from the banks of the Danube to the valleys of the Asturias, where pontiffs of the Sun garbed in robes encrusted with jewels sacrificed bulls resplendent with golden leaf in honor of Mithra, the Unconquered Sun. But the Duino Mithraeum is the sole example known in Italy of a temple to Mithra located inside a natural cave rather than purpose-built. To reach it, I hiked among the blackened remnants of what used to be indigenous woodland, until megafires devoured it the previous summer after weeks of droughts—the Sun god flexing his muscle, one could say, or perhaps the weather eschewing human control. Access to the cave used to be via a vertical shaft, leading to a vaulted chamber illuminated by hundreds of lanterns, whose remnants were discovered with the cave in 1963. The simple limestone altar, still intact, was probably at the center of the ceremonial slaughtering, depicted in a nearby bas-relief in which a priest cuts a bull's throat with a knife, while a scorpion gnaws at the animal's testicles. The Sun and the Moon bear witness to the offering. The cult was so popular in the second and third centuries AD that it nearly smothered another up-and-coming religious faith: Christianity.

The ascendancy of Mithra among the Roman legions is witnessed by the account of a crucial victory of Vespasian during the first Roman civil war in AD 69, in which one could say the Sun god himself intervened to clear the way for Vespasian to become emperor. Vespasian's forces were locked in a finely balanced nighttime battle against the legions of his adversary, Vitellius, near Cremona, in northern Italy, when dawn broke. Vespasian's legions, who had served in Syria and there become worshippers of Mithra, turned their backs to the battle (thus putting themselves at the mercy of their enemy) to face east and salute the rising orb with mighty cheers. The impact was immediate and decisive: Vitellius's forces mistook their gesture for a celebration of the arrival of reinforcements and fled in despair. By the day of the winter solstice, Vitellius was dead and Vespasian proclaimed emperor by the Roman Senate, while he was in Alexandria being hailed as the new Egyptian pharaoh and king—and thus proclaimed the son of Ra, the Egyptian Sun god.[1]

In AD 274, the worship of Mithra was given the official seal of approval by none less than Emperor Aurelian, whose mother was said to be a priestess in the Sun temple of her village and who was himself the scion of an old noble family who had honored the Sun (*ausel*) already in pre-Roman times, when they were called *Auselii* (which later morphed into *Aurelii*). From the same root come the words *aurora* (the dawn) and, in Romance languages like Italian, *aureo* (golden)—for this reason, the chemical symbol for gold is Au. The resplendent, unblemished appearance of gold is reminiscent of the bright shine of the Sun, which is one reason why the metal has been so prized since antiquity. Its very name in modern Latin-based languages (*oro* in Italian) remains tied to the solar god of antiquity. "Sol gold is and Luna silver we threpe [The Sun is gold and the Moon silver, we say]," explains the canon's yeoman when describing his master's alchemical studies in Geoffrey Chaucer's *The Canterbury Tales*. When a golden bracelet catches a sunbeam, we should spare a thought for the long lineage of their connection.[2]

With the imperial favor of Aurelian and his successors, the cult of Sol Invictus (the Unconquered Sun, now entirely identified with Mithra) took over the Romans. By the third century AD, Mithraism and Christianity shared many beliefs and held uncannily similar rituals, enough so to make the church fathers uncomfortable. Mithra was sometimes depicted as a trinity (with rising and setting versions of the Sun as two other personifications of the deity); its followers were baptized by being marked on the forehead, received a form of eucharist as an offering of bread and water, and were promised resurrection, which was ceremonially enacted by pretending to kill a human victim, whom the priest would then bring back to life. In a ceremony of purification that early Christians found abhorrent, the high priests of Mithra washed themselves of sin in a pit where they were showered with the blood and steaming entrails of a freshly slaughtered bull.

By the mid-fourth century, Sol Invictus was winning the battle for the hearts and immortal souls of the citizens of the Roman Empire. Its major festival was celebrated with magnificent games in the circus on December 25, the winter solstice—the day when the Unconquered Sun started to regain its strength and bestow again its life force on the world. In actual fact, winter solstice had drifted by then to December 21, having fallen on December 25 when Julius Caesar reformed the calendar in 46 BC but lost four days in the intervening centuries, since the Julian year was eleven minutes shorter than the actual solar year. December 25 was nevertheless chosen as the traditional date of the solstice, despite no longer coinciding with the astronomical event. The Roman church fathers saw it as a case of "if you can't beat 'em, join 'em" and cunningly shifted the day of Jesus's birth, until then associated with Epiphany on January 6, to December 25.

The tide was turning against Mithraism, particularly after Emperor Constantine, a member of the Sol Invictus cult, converted to Christianity following a religious vision on the eve of battle and legalized the Christian religion with the Edict of Milan in AD 313. By the late fourth century, the Christian sublimation of the Sun into pure symbol was complete. In one of his nativity sermons, St. Augustine draws the parallel between Christ and the returning light while admonishing the faithful to avoid the mistake of the heathen, who adore the actual Sun as opposed to the symbol of the Sun:

> Since that infidelity which had covered the whole world with the darkness of night had to be lessened by an increase of faith, therefore,

on the birthday of our Lord Jesus Christ, night began to suffer dim-
inution and day began to increase. And so, my brethren, let us hold
this day as sacred, not as unbelievers do because of the material sun,
but because of Him who made the sun. . . . Now, in truth, even in the
flesh He is above the sun which is worshiped as a god by those who,
blinded in mind, do not see the true Sun of Justice.[3]

The Sun's theological centrality was of little help to Nicolaus Copernicus and Galileo Galilei when they argued for its physical centrality in
the solar system. But next time you decorate for Christmas with candles,
tinsel, or strings of colored lights, remember how this day used to be the
celebration of a different kind of light. A persistent story about the origin
of the candle-lit evergreen holds that Martin Luther was walking by night
in a snow-covered forest of firs when he was struck by the sparkling beauty
of the firmament, which he then reproduced for his family as a Christmas
tree—a legend, infused with the enduring power of myth.[4]

The religious symbolism tied to the Sun spilled over to confer secular
power as well. Many cultures considered their rulers to be direct descendants of their supreme solar god, infused with power and authority by
this divine ancestry: the Japanese emperor was "the Son of the Sun," the
king of the Incas invoked Inti as "the Sun, my Father," and the Aztecs
saw themselves as "the People of the Sun." In eighteenth-century France,
Louis XIV, the Sun King, surrounded himself with countless versions of
his emblem, a head resplendent with Sun rays (often made of solid gold),
the seat of illumination and intelligence: "L'état, c'est moi [The state,
that's me]," he quipped. Even the imposing row of colonnades adorning
the majestic eastern facade of the Louvre Museum, built by the Sun King's
order, was said to have been inspired by Ovid's description of the palace of
Apollo, the Roman Sun god.[5]

The radiance of the Sun can be obliterated, as we've seen, by cover
of the Moon. The world of symbols saw a reverse eclipse, however, as it
was the Sun that usurped the early primacy of the Moon. The historian's
equipment is no longer sufficient for this investigation; nor is the archaeologist's toolbox of any use. It's time we seek the help of a psychoanalyst.

SWEAR NOT BY THE MOON

Among the decorations that brighten up my office walls, a colorful drawing that my daughter Emma did for me when she was about nine often attracts my visitors' attention. On the left, a crescent Moon is portrayed in profile with shades of azure. Her feminine visage looks to the right, where it merges with the rotund face of a masculine Sun, his crown of resplendent rays ablaze with yellows, oranges, and reds. All around the pair, the sky is peppered with pointy stars, ringed planets, and clearly recognizable spiral galaxies.

My visitors never comment on the drawing, but I can see from their eyes that the scene tugs at something deep. The only symbols my visitors and I, people of science, are really fluent with are those indicating integration in calculus, the direct sum in algebra, and the like—although we remain ignorant that the symbol we use for the direct sum, an encircled cross, is an ancient glyph that used to denote the Sun, borrowed from the shape of the wheels of the chariot it was supposed to ride on.

Swiss psychologist Carl Gustav Jung, who had been Sigmund Freud's disciple before inaugurating a new branch of psychoanalysis, introduced the notion of the "collective unconscious," a "second psychic system of a collective, universal, and impersonal nature which is identical in all individuals," consisting of ancient primal symbols, the archetypes, which are inherited rather than developed individually.[6] Jung did think that children are "rooted in the collective unconscious," what the poet William Yeats called "a great Memory passing on from generation to generation."[7] Perhaps Emma's drawing dredged up associations from our shared murky depths.

The Moon's cycle of waning and waxing extended time reckoning past the simple alternation of day and night, as we saw in Chapter 4. But just as importantly, the lunar phases came to be associated with the ever-recurring cycle of life, the universal law of becoming, the inescapable wheel of human birth, growth, decline, and death, all fundamental Jungian archetypes. Its monthly three-day disappearance, when the Moon is new, was often considered a real death, from which the Moon could be resurrected only thanks to magic rites, dances, prayers, and sacrifices, on whose success the life of all creatures on Earth depended. "Mama Quilla, Mother Moon, do not die, lest we all perish," prayed the Incas of pre-Colombian Peru.[8] And when the first silvery sliver of the crescent Moon was sighted, a huge wave of relief swept the tribe: people would fall to their knees in Congo, crying, clapping their hands and singing; Native Americans would dance in a circle and chant, play celebratory games, and stretch their hands toward the Moon to help it back into the sky.

With its periodic resurrection from darkness, the Moon promised a similar destiny for humans too: "As the Moon dies and comes back to life, so we also, having to die, will live again," prayed Californian Native American tribes.[9] The three days during which the Moon is thought to have died coincide with the time that elapses between Jesus's death and his resurrection, which was foreshadowed by the prophet Jonah's being trapped "for three days and three nights in the whale's belly."[10] Three days and three nights is also how long the Sumerian goddess Inanna remains prisoner in the underworld, as described in fragments of an epic poem composed even earlier than the Epic of Gilgamesh.

At some point in history, however, Moon-centered religious worldviews shifted toward the Sun. With its fierce gaze at zenith, especially in hot and tropical countries, the Sun had often been considered a threat, but this began to change. Perhaps the tide turned with the Egyptian obsession with Ra, who commanded an army of thirteen thousand priests at his great temple in a now vanished city that the Egyptians called Pa Ra (House of Ra) and the Greeks, Heliopolis (City of the Sun).[11] Perhaps what demoted the Moon was the realization that, as the Greek philosopher and astronomer Anaxagoras wrote in the fifth century BC, "it is the sun that puts brightness into the moon."[12] Perhaps it was Plato's association of the Sun, as the source of light, with the good, the fountain of truth and knowledge in the *Republic*, together with his famous allegory of the cave. Perhaps the Moon's reduction in status was sealed by Aristotle's definitive dividing line between the corruptible world of nature and the ethereal divine province of the sky, which he drew at the orbit of the Moon, including our satellite as part of the lesser, earthly, perishable world. Cicero summarizes it thus: "Below the Moon there is nothing but what is mortal and doomed to decay, except the souls given to men by the bounty of the gods, whereas above the Moon all is eternal."[13] Perhaps it was the Christian association of Jesus as the Invincible Sun, metaphorically described as "the Sun of Justice" or "the Sun of Righteousness," while the Virgin Mary took over what remained of the lunar myth. In Christian iconography, she is often portrayed as the woman who appears triumphantly in the book of Revelation, "clothed with the sun, and the moon under her feet, and upon her head a crown of twelve stars."[14] From the fifth century, she also took on the moniker *stella maris* (star of the sea), guardian of childbirth and protector of mothers, just like pagan lunar goddesses before her.

While the Moon became associated in the Western world with the feminine principle—having always held sway over the spirit of growth, women's fertility, and the health of babies, as part of the larger cycle of life—the Sun was internalized as a masculine hero. According to Jung, an external observation of the Sun rising and setting "must at the same time be a psychic happening: the sun in its course must represent the fate of a god or hero who, in the last analysis, dwells nowhere except in the soul of man."[15] Our star's sheer, dazzling power became firmly the precinct of male kings like Louis XIV, while, under the assault of patriarchy, the Moon became a byword for weakness. Our language bears witness to the debasement of the Moon's cycles, once considered a heavenly manifestation of the very essence of life, reduced to an expression of (feminine!) inconsistency and mutability:

> *O, swear not by the moon, th'inconstant moon,*
> *That monthly changes in her circled orb,*
> *Lest that thy love prove likewise variable,*

begs Juliet of her ill-fated lover Romeo in William Shakespeare's tragedy. Even today, *lunacy* refers to a kind of intermittent, inconstant folly, as does *being moonstruck*, particularly in matters of the heart. *Moonshine* describes a silly, foolish idea, while *to moon about* is to act dreamily and listlessly. In Italian, somebody who has *la luna storta* (a tilted Moon) is in an irascible and uncongenial mood. What a fall in esteem from the

Gaelic *rath*, meaning "bounty, success, good fortune" and derived from the word for "full Moon"![16]

The patriarchal dichotomy of Sun/male and Moon/female papers over a huge variety of myths throughout history and across cultures. There are traditions of a Sun goddess and a Moon god (in Oceania and Japan or among the Māori, for whom the Moon is "the husband of all women," for example), who are sometimes married; in some myths the Moon is the Sun's mother (for the Hopi Native Americans), or the Sun is the sister of a male Moon (among the Norse). Sometimes the Moon is male during its waxing and becomes female in its waning (among tribes in the Andaman Islands in the northeastern Indian Ocean) or is the mother of two boys (representing the waning and waxing phases) when full (for the Navajo tribe); sometimes the Moon "partakes of either sex" and creates hermaphrodites, as in Plato's *Symposium*. Even in the heart of the Western canon, there are sparks of rebellion: both Chaucer (in the fourteenth century) and John Milton (in the seventeenth century) referred to the Sun as "she." As late as the twentieth century, folklore experts concede, the Moon remained "second to nothing in its influence on worldwide folk belief and practice."

THE GREAT MEMORY

Today, when presented with a picture of the Sun, the subconscious of the guest in my office is likely to resonate with archetypal images such as the eye of the world, the harbinger of justice, the divine light, the bridegroom, the sky father, the male power, the first cause of all that exists, the active principle, the yang, intuitive knowledge, the spirit, matters of the heart, the day, the higher, the right side. A guest who perceives a Sun standing at its zenith might think of immortality, the eternal now, the war of the light against darkness and evil, the safe passage through the netherworld that brings the Sun back to life in the east. A guest who imagines it rising and setting will ponder the cycle of life and death, rebirth, resurrection, salvation, escape from darkness, victory over death and chaos, the order principle, the divine essence of man.

The crescent of the Moon might instead stir up images of the night, the lower, the left, the female, the unseen aspects of nature, inner knowledge, the eye of the night, the bringer of change, the realm of becoming, the weaver of fate. The guest will be put in a mood for reflection, perhaps thinking of water, fertility, the passive, the receptive, the nurturing. A full Moon will come with associations of completeness, wholeness, strength, and spiritual power, the ship of light on the sea of the night, the abode of Archangel Gabriel in Christianity, the cup of the elixir of immortality in Hindu tradition, the passage from life to death, the yin to the Sun's yang.

Just like the Earth itself, the sky, and various aspects of nature, the Moon and the Sun transcended their astronomical natures to become receptacles of our consciousness as a species, a phenomenon that psychotherapists call "projection" and that Jung describes thus: "All the mythological processes of nature, such as summer and winter, the phases of the moon, the rainy seasons, and so forth, are in no sense allegories of these objective occurrences; rather they are symbolic expressions of the inner, unconscious drama of the psyche which becomes accessible to man's consciousness by way of projection—that is, mirrored in the events of nature."[17]

We see in the heavens a reflection of our souls, projected as symbols onto a cinema screen the size of the sky. As at the movies, the images flicker and change over time, starting vibrant and colorful at the dawn of consciousness and (reversing the history of real moving pictures) fading over the millennia into black and white and eventually into dim shapes, the fleeting ghosts of gods. Once glorious sound is reduced to hissing and scratching, and the contours of the characters all but vanish in shadow, but the power is still there. How many times did I praise my children by saying, "You are my sunshine!" without realizing I was leaning on an association (Sun equals good) inherited from Plato?

How different our collective psyche would have been in a world without a visible Sun and Moon! Would the Sun have commanded the same power had it been a giant red disc permanently hovering on the horizon, had Earth orbited at close quarters a red dwarf, where gravity forces the planet to spin in exact sync with its revolution around the star? Would our beliefs about death and resurrection have been deformed beyond recognition on Mars, where two misshapen, potato-shaped moons share the night? Ralph Waldo Emerson felt that the stars are essential to cultivate a sense of the divine: "If the stars should appear one night in a thousand years, how would men believe and adore, and preserve for many generations the remembrance of the city of God which had been shown! But every night come out these envoys of beauty, and light the universe with their admonishing smile."[18]

We can't easily imagine the psychological makeup of an alien race living in a world without stars, just as we can't predict their bodily shape on a planet subject to radically different evolutionary pressures. But if Jung was right and the psychological workings of a mind that emerges shaped by evolution are universal (a big if!), then the collective unconscious of Caligoans anywhere would lack some of the defining elements that make us who we are.

LED BY ONE'S STARS

For thousands of years most people believed the stars were more than symbolic actors in human affairs. Astrology—from the Greek for "knowledge of the stars," as opposed to astronomy, "the arranging of the stars"— is the belief that the positions, interrelations, and cycles of heavenly bodies determine our character and influence our fate. Often thought to be concerned only with horoscopes (fortune-telling based on the position of the stars and planets at one's birth), astrology is in fact a complex belief system with millennia of different traditions. Modern science considers astrology an empty practice, for (beyond the obvious physical and biological importance of the Sun's light and warmth and the gravitational connection of Moon and Sun to the tides) no known physical mechanism emanating from the stars and our solar system's planets could influence our affairs and free will. But if the stars have not actively steered human behavior, human belief in their influence certainly has.

Of special significance for astrology are the twelve constellations occupying a band on either side of the ecliptic, the path traced by the Sun in the sky: Aries (the Ram), Taurus (the Bull), Gemini (the Twins), Cancer (the Crab), Leo (the Lion), Virgo (the Maiden), Libra (the Scales), Scorpius (the Scorpion), Sagittarius (the Archer), Capricornus (the Mountain Goat),

Aquarius (the Water Bearer), and Pisces (the Fishes). Together, they form the familiar twelve zodiac signs and are grouped into four elements (water, air, fire, and earth), of which they inherit the character. Each of the planets in turn possesses its own temperament, and as they move around the zodiac, their astrological meanings are influenced by the qualities of the signs. Seen from the point of view of the Earth (astrology taking a strictly geocentric perspective), as the planets race around the zodiac they find themselves in geometrical relationships with the zodiac signs, and with each other, creating patterns of meaning. The astrologer blends the planetary aspects and many other elements (including a system of twelve houses, lunar nodes, the times of planetary transits, and others) into a reading of a person's fate, a prediction of future events, a divination on the most opportune moment for a certain activity, a characterization of a historical period, and so on.

That astrology has little to do with the physical universe is obvious when one considers that the twelve signs, each covering exactly thirty degrees of the ecliptic, do not match their namesake constellations, which are all of different sizes, ranging from seven degrees for Scorpio to forty-five for Virgo. Western astrology also ignores the precession of the equinoxes: the so-called first point of Aries, marking the location of the Sun on the spring equinox, used to be in Aries two thousand years ago, when Ptolemy codified the zodiac, but in the meantime precession has shifted it by about thirty degrees westward, well into Pisces. But the zodiac did not move, so March 21 remains the first day of Aries, even though the Sun is actually in Pisces at that time (and will be in Aquarius in another few hundred years). This means that your star sign, supposedly the sign in which the Sun was located at the moment of your birth, is an astrological convention, not an astronomical event.

Long seen as a sophisticated divinatory tool, astrology required knowledge fiercely guarded by an intellectual elite, who alone carried out the computation of astrological charts and the prediction of astronomical phenomena and planetary positions. Jung called it "the first form of psychology" and explained, "Instead of saying that a man was led by psychological motives, they formerly said he was led by his stars."[19]

Astrologers were hot currency in late imperial Rome, when the traditional divinatory practices of observing the flight of birds or the innards of slayed animals were supplanted by what was considered a more objective practice, as it relied on observational data and complex calculations. Roman emperors, among them Domitian and Caracalla, routinely had the horoscopes of potential rivals cast and eliminated those that the stars singled out as threatening. Possessing an "imperial horoscope" (i.e., a chart that predicted a raise to power) was a sure way to a short life in Rome.[20] Astrologers had an even stronger influence in China, where for millennia a deep belief in the correspondence between microcosm and macrocosm translated into the attempt to align society in harmonious resonance with the heavens. Astrology was tightly regulated by the state as a tool to record the future and hence control the present.[21]

Meddling behind the scenes of power, astrologers themselves could and did land in hot water. No fewer than eight mass expulsions of astrologers from Rome were ordered in the three centuries around the beginning of the common era, with a view to depriving the emperor's enemies of vital intelligence. Tacitus dryly describes them as "a tribe of men most untrustworthy for the powerful, deceitful towards the ambitious, a tribe which in our state will always be both forbidden and retained."[22]

But, as Tacitus implicitly acknowledges, in politics the efficacy of astrology was indisputable: an unfavorable augury weakened the status of a prince, pontiff, or ruler in the eyes of the people and emboldened his enemies. The divination became thus a self-fulfilling prophecy. Recall also the diametrically opposed effects of the sight of Halley's Comet on Harold Godwinson and William of Normandy (Chapter 2): the first saw it as a bad omen, the latter as a sign of future glory. Jung would say that their respective psychological states, and those of their commanders and troops, disadvantaged the English king going into battle. He didn't return alive.

Despite astrologers being officially condemned by the church already in the fourth century, even popes didn't shy away from employing their services under the grave apparition of a comet or eclipse. Urban VIII, who four years later would squash Galileo's Copernican views out of hurt pride, in 1628 asked the astrologer and convicted heretic Tommaso Campanella to help him fend off his premature death, which future-tellers had predicted would be brought about by a solar eclipse on December 25. Campanella sealed himself with the pontiff in a room hung with white silken linen and decorated with branches, where he recreated a version of the solar system shielded from the impending eclipse. He used candles to stand in for the Sun and the Moon and torches for the planets. While Urban was burning rosemary, cypress, laurel, and myrtle and drinking arcane liquors concocted by Campanella for the occasion, the partial eclipse came and went over the Eternal City. The astral menace had been seen off, and Campanella earned his freedom and the title of master of theology in return.[23]

After one of Urban's grandnephews suffered a similarly narrow escape during the eclipse of 1630 (also abetted by Campanella), the pope resolved to put an end to all astrologers' predictions, especially those concerning his own demise. In 1631, Urban VIII issued a bull (possibly drafted by Campanella himself) reiterating the condemnation of the practice of astrology for all members of the church—a prohibition still in force today. For good measure, the prediction of the deaths of popes and their family members up to the third degree was an offense to be repaid with death.

While it is tempting to sneer at superstition, Pope Urban's fear of unfavorable auguries and the magical countermeasures he took demonstrate how deeply the relationship between sky and Earth was embedded in the human psyche at the time. From cutting one's nails to choosing the date of one's wedding, from conceiving a child to laying the foundations of a new building or city, from buying or selling property to ploughing a field, so many decisions, big and small, hinged on astrology before its precipitous fall from esteem in the seventeenth century. Who knows how many battles were started (or averted), how many men murdered (or saved), and how many marriages arranged (or broken) by the advice of an unnamed astrologer. Even in recent history, astrology occasionally entered the nexus of power: after the failed assassination attempt on President Ronald Reagan in March 1981, First Lady Nancy Reagan regularly consulted an astrologer to vet her husband's itineraries and to identify auspicious dates for public events. When the story came out, the president declared, "No policy or decision in my mind has ever been influenced by astrology."

THE ESTRANGED MOTHER

In the Western world, a roll call of astrologers reads like a who's who of the history of astronomy: Hipparchus, Ptolemy, Regiomontanus, Galileo, Tycho Brahe, and Johannes Kepler were all practitioners, because it was simply part of their job description. Until late into the seventeenth century, astronomer and astrologer (and mathematician) were often the same person, described as a *mathematicus*. Often forgotten is how much the Scientific Revolution owes to astrology, whose pursuit motivated or even directly inspired key breakthroughs. We saw in Chapter 7 how Kepler's first geometrical insight into the structure of the solar system was sparked by the contemplation of astrological conjunctions. Before him, while studying law at Bologna between 1496 and 1500, Copernicus lived for a time with astronomy professor Domenico Maria Novara, a noted astrologer whose university job required him to issue yearly astrological prognostications. Copernicus became his assistant and probably helped him with astronomical observations and astrological charts.

Galileo, in his job as lecturer at the University of Padua, instructed medical students on the casting of horoscopes—a central skill for physicians, who relied on astrological charts to identify the most opportune time for letting blood, taking potions, applying ointments, taking medical baths, and so on. He also prognosticated for his family members, friends, patrons, and himself—and there are indications that he enjoyed a certain renown as an astrologer. Unfortunately, while he left behind detailed horoscopes of his daughters (of Livia, for example, he predicted, "Mercury rising is very strong for all things, and Jupiter which is conjunct gives knowledge and bounty, simplicity, humanity, erudition and prudence"), he didn't write down any of the interpretations of his own natal chart.[24] The Catholic Church saw fatalistic astrology—the idea that a horoscope determined destiny with absolute certainty—as heresy, for according to St. Augustine it both restricted God's ability to intervene directly if He so chose and made Him responsible, via the stars, for human sin.[25] In a shot across the bow for what was to come, in 1604 the Inquisition investigated Galileo for propounding astral determinism—a serious accusation, which was later dropped.

By the time Isaac Newton came onto the scene, astrology had largely fallen into disrepute among the natural philosophers. There is not a single horoscope among Newton's vast trove of alchemical and theological writings, and he had only four volumes on astrology in a library of over seventeen hundred books (over a third of which consisted of alchemical and occult texts, many well thumbed and covered with annotations).

However, astrology did play a tiny but potentially decisive role early in Newton's life. In the summer of 1663, Newton had been a subsizar at Trinity College, Cambridge, for two years—the lowest rank in college society, a student who earned his keep by polishing the boots and emptying the chamber pots of more affluent scholars. At the Sturbridge midsummer fair, Newton was tempted to buy a book about astrology "out of a curiosity to see what there was in it."[26] Entirely ignorant of trigonometry, Newton was frustrated by its undecipherable charts; so he turned to Euclid's *Elements* and from there to René Descartes's *Geometry*, until before long the self-taught Newton got to the outskirts of seventeenth-century mathematical knowledge. From there he was in a league of his own, inventing whatever new math he felt he needed. After that initial, perhaps triggering brush with astrology, Newton never looked back. His judgment later in life was scathing: he was "convinced of the vanity & emptiness of the pretended science of Judicial astrology."[27]

Astrology remains popular today, in an age when artificial intelligence, crunching enormous heaps of granular details about our searches, purchases, metabolic data, and intimate conversations, claims to divine our desires before we are even aware of them, to predict a movie or a romantic partner we might enjoy, and to foretell the probability of our death. Some see astrology as an evolutionary relic, as useless—and occasionally dangerous—to the psyche as the appendix is to the colon. Others still consider it a rich system by which to explore meaning in a world where the divine has fallen back beyond the stars. It is for sure a means of arousing the ire of contemporary astronomers when one incorrectly describes them as "astrologers." Perhaps Tacitus captured its enduring fascination when he concluded, "Human nature is especially eager to believe the mysterious."[28]

THE LAST RIPPLE

"A symbol," writes American poet John Ciardi, "is like a rock dropped into a pool: it sends out ripples in all directions, and the ripples are in motion. Who can say where the last ripple disappears?"[29] Astrology's ripples still surround us in small but significant ways, if only one knows how to look. A gloomy and melancholic character can be described as *saturnine*, from the slowness and astrological heaviness of Saturn and the element of lead that was associated with it; *jovial* (cheerful, merry, and convivial) and *mercurial* (subject to unpredictable mood changes, quick-witted, or fickle) similarly derive from the astrological qualities of Jupiter and Mercury, respectively.

The days of the week, which Pierre-Simon Laplace considered "the most ancient monument of astronomical knowledge," might well be the most ubiquitous legacy of astrology.[30] The subdivision of time into a seven-day period is probably an ancient Jewish invention, molded on the six days the Bible says God labored to create the Earth and a seventh day of rest. The original roots of the week appear to go back to even earlier times, with the Sumerians adopting weeks of seven days, one of which was set aside for recreation. The names of the days of the week are, at least in Romance languages, in transparent association with the planets, with a Jewish and Christian twist on Saturday and Sunday. In Italian the week starts on *lunedì* (Monday), the day of *luna*, the Moon; it is followed by *martedì* (Tuesday), the day of *Marte*, Mars; *mercoledì* (Wednesday), the day of *Mercurio* (Mercury); and *giovedì* (Thursday), the day of *Giove* (Jupiter); it ends on *venerdì* (Friday), the day of *Venere* (Venus). *Sabato* (from Shabbat) is the Jewish day of prayer, while *domenica* (from the Latin *dies domini*, "the day of the Lord") is the Christian day of rest. In English, Saturday and Sunday have maintained their designations as the day of Saturn and the day of the Sun, respectively, while gods from Teutonic mythology were swapped in to name Tuesday (from Tiu, the Norse god of war, replacing Mars), Wednesday (Woden, or Odin, the supreme Norse deity), Thursday (Thor, the god of thunder, instead of Jupiter), and Friday (from Frigg, the goddess of love and beauty, standing in for Venus).

So far, Laplace was right. But the deeper astrological connection is hidden in plain sight. Starting from Monday, the sequence Moon-Mars-

Mercury-Jupiter-Venus-Saturn-Sun does not reflect any obvious order, such as their distance from the Earth (in the Ptolemaic system)—until one considers each planet's role as astrological "time lord" for each of the twenty-four hours of a day. Starting from the first hour of Saturday, the planets are assigned each to an hour in reverse-distance order from the Earth: Saturn, Jupiter, Mars, Sun, Venus, Mercury, Moon. The eighth hour of the day goes back to Saturn, and the sequence repeats until the day is completed. After all seven planets are cycled through three times, Saturn, Jupiter, and Mars fill up the last three hours of Saturday, and the remaining four planets are moved over to the next day, with the Sun being first. The scheme continues, so that at each change of day, the first planet in the sequence shifts by three positions. At the end of the week, all hours have been assigned to a planet, and the planet assigned to the first hour of each day (becoming also the astrological ruler for that day) gives the name to the day: Saturn, Sun, Moon, Mars, Mercury, Jupiter, and Venus. Et voilà, the week as we know it has been built.

TO EVERY MAN HIS STAR

Our lives are littered with star signs, not just the zodiac's. Take the five-pointed star, one of the most common symbols in today's world, an instantly recognizable emblem of quality (a five-star hotel), authority (a five-star general), deliciousness (a three-starred Michelin restaurant), and success (the number of football World Cups won by a national team). The potency of this symbol is witnessed by its appearing on at least a quarter of the world's flags, most copiously on the American flag, where the fifty white stars on a blue background represent the fifty US states. The five-pointed, red-filled star used to be the emblem of communism, and it appears today, together with a crescent Moon, on the national flags of countries with an Ottoman past, like Algeria, Turkey, and Tunisia.

Who remembers that, deep down, the five-pointed shape follows the hieroglyph the Egyptians used for *star*? Its origin stretches even further into the distant past, in the form of pentagrams found in the Mesopotamian city of Uruk from at least three millennia BC. A powerful magical symbol of protection, the pentagram has a rich history of meaning and is a staple of occult rites, including preventing Mephistopheles from leaving the room where he had been summoned by Johann Wolfgang von Goethe's Faust. Its power extends into Christian rites: the traditional Christian coffin has a pentagonal section, supposed to protect the body in the perilous passage into the land of the dead.

Faith in the safeguarding power of the five-pointed star led Giuseppe Garibaldi, the Italian revolutionary whose campaign unified Italy, to secretly carry it wherever he went, stitched with golden thread *inside* his beret—unlike Che Guevara, another revolutionary, who wore the star proudly on his cap. Garibaldi avowed that he considered Arcturus, a red giant in Boötes and the third brightest star in the northern sky, to be his special protective star, after he contemplated it at length the night before conquering Palermo with his army. "Every man has his star, and Arcturus is mine," he said.[31]

When Garibaldi had brought his campaign to a successful end, a curious report of a bright "star" surfaced with the inauguration of the Italian parliament, reinstated in Rome on November 27, 1871, after the Eternal City had been unified with the Kingdom of Italy. The crowds assembled in the piazza del Quirinale were stunned to discover a "new star" shining above the palace. "La stella, la stella!" cried men with flat caps and women in peasant clothes, pointing at the portent of their country's bright future. The "star," it appears, was none other than Venus, exceptionally bright because at maximum elongation. But today a five-pointed star shines forth in the official seal of the Italian republic, born in 1946 after the monarchy was abolished by popular vote. We call it, affectionately, *la stellona d'Italia* (Italy's big star).

The star system we use everywhere online to rate purchases, services, and people was first introduced in 1844 by a pioneer of guidebook writing, Karl Baedeker, in his eponymous guides, to single out sights not to be missed.[32] The idea was soon imitated by his competitors. The 1879 English *Handbook for Visitors to Paris* rated attractions "by marking them by stars according to their merit or importance" (actually represented by asterisks on the printed page).[33] The Louvre, Notre Dame, and Versailles achieved the maximum three-star rating, but of the outdoors concerts of the *cafés chantants* on the Champs-Elysées, the no-star review sniffed that "performance tends towards the immoral" and concluded that "respectable people keep aloof."

The Hollywood Walk of Fame immortalizes cinema's greatest "stars" by engraving their names in a five-pointed star in the pavement on Hollywood Boulevard and Vine Street.[34] The phrase *to star* in a play, meaning to play the lead, was first used in 1815, and from 1865 famous entertainers and sportspeople were said to gain *stardom*. Whether we believe in astrology or not, we still describe somebody blessed with consistent good fortune as having been *born under a lucky star*. We bemoan our misfortune when a *disaster* (from the Latin for "bad star") befalls us, perhaps because, like Romeo and Juliet, we feel we are *star-crossed*. If said disaster physically hits us on the head, we are likely to *see stars*. A shiny, eight-pointed copper star identified New York policemen starting in 1845; their first nickname, *star police*, later became *copper* or *cop* from the material the star was made of.[35] From the East Coast, the star as an emblem of authority followed the Sun to become the sheriff's badge across the West. Whether five-, six-, or even seven-pointed, the brass or metal star was highly polished so that it would shine even in the moonlight and identify the bearer from a distance at a time when "shoot first, ask questions later" was the norm.

My son Benjamin's sheriff star is irreplaceable in his games of cops and robbers. He came home from school recently and proudly showed me the result of his error-free spelling test. Along with words of praise, the teacher stamped his workbook page with a radiant symbol of approval and commendation. I looked at the five-pointed star and smiled.

LYING ON MY BACK ON THE POLISHED CONCRETE FLOOR OF TATE Modern, I meditated on the mix of awe and fear that Eliasson's artificial sun had stirred in me. The summer before the opening of the installation, a torrid heatwave had blistered continental Europe, claiming an estimated thirty thousand lives. At the time, I didn't know this was only a warning of things to come, as global temperatures have kept increasing ever since. The benevolent, life-giving force of the Sun is showing us its dangerous, Earth-scorching side—the same ambivalent nature that appears in a myth predating even the Epic of Gilgamesh, stemming from the ancient city of Ugarit, on the coast of today's Syria.

The myth tells the story of the supreme storm god Baal: locked in battle with Mot, the god of death, Baal is tricked into descending into the netherworld, where Mot imprisons him. The Sun goddess Shapshu then falls under the sway of Mot, perhaps because of her daily disappearance under the horizon and therefore presumably into the underworld. A scorching drought follows, with Baal unable to send rain from the land of the dead. Drunk with power, Mot crows,

> *The sun, the lamp of the gods is burning*
> *The heavens are powerless in the hands of Death, the*
> *divine one.*

Eventually Shapshu manages to free Baal from the netherworld and bring him back with her to the surface and to life, restoring him to his magnificent palace where she crowns him king of the gods. Shapshu is both a destroyer, when under the spell of Mot, and a kingmaker, when she helps Baal escape death—two opposing faces of the Sun that are staples of myth the world over and today neatly embody the mortal danger coming from our own star. Human-made climate change has pushed Shapshu into the clutches of Mot once again, from which she is unlikely to return any time soon.

Artwork like Olafur Eliasson's can help in shattering our inflated sense of planetary control and bringing back a sense of reverence for the larger-than-human forces of nature, to which we are subject even in this technology-driven day and age. It is to the future of our relationship with the stars that we now turn.

THE CALIGO TALES

Mist-Catcher's Tale

Mist-Catcher stood, and thus began her story.

When the first snowflakes blanket the hills and the Bear is ready to go to sleep, that's when the Cloud is calling us onto the ice tongues. We light our burning sticks then, and Way-Finder leads us across the silent forest, our faces to the Cloud, feeling its twirling touches. Squirrel sometimes watches us walk by from a snow-covered branch, but we won't use our spears when preparing for Name-Giving. Those with no names walk in front, hands joined so they don't get lost, followed by those whose name they will receive.

The ice tongue awaits at the end of the glow, and we climb it until nothing but ice surrounds us. That's when you must follow Way-Finder closely, for the many mouths of the ice tongue, hidden by the snow, can open any time and swallow you whole. The Mist becomes stronger with every step, puffing out of our lips as we enter more deeply into the land of ice. Everything sparkles with the flames of the burning sticks as we surround in a circle those who are to be named. I step into the circle, a burning stick held high in each hand, as those whose names are to be passed on blow handfuls of snow into the air, their Mist strong in front of them. Tiny flames fill the Dark.

The deer skin is taut on the hollowed tree trunks; their Rumble keeps the Blackness away. The first name-giver comes to the middle: this time, it's Skinner, and she joins hands with the young one. Their noses touch, and I call the Mist forth from their mouths. Skinner's Mist mingles with the young one's, and they become the same.

"Skinner!" I say, "the Mist you received from Elder Skinner is now with the young one. Young one! You receive Skinner's Mist. May you give it new strength and look after it until the Cloud sends us a new young."

"Skinner!" I say, "you shall now be named Elder Skinner. Young one! You shall now be named Skinner the Young. May your Remembering be as strong and good as Elder Skinner can make it!"

One after the other, they come forward, join hands, noses touching, Mists mingling. The young ones become who they are meant to be, so that our people may continue Remembering as the Cloud rolls from one end of the Disc to the other, always different and always the same.

So spoke Mist-Catcher, and my heart grew big at the memory of when I became Cave-Keeper.

10

TO REBEHOLD THE STARS

We mounted up, he first and I the second,
Till I beheld through a round aperture
Some of the beauteous things that Heaven doth bear;
Thence we came forth to rebehold the stars.

—DANTE ALIGHIERI, *The Divine Comedy*

THE BROOM STAR ENCOUNTER

We stretched out on our reclining chairs and passed the popcorn. The streetlamp behind the walnut tree was a nuisance, but we could pretend it wasn't there as long as we didn't look directly north. With all the lights of the house switched off, our eyes took a few minutes to adjust to the darkness, by which time I could hear the rustle of little hands already scraping the bottom of the popcorn bag.

"Can we have the marshmallows, mummy?" my daughter asked.

"Let's at least wait for the show to start!" I protested.

To the west, the last rays of the setting Sun had disappeared behind the spa town of Grado and its golden beaches. The natural lagoon in the foreground was a dark shape against the sky, whose faint glow I knew was caused by lights in the nearby shipyards of Monfalcone. A monstrous cruise ship was being refurbished in the dry docks, and its dozen decks were floodlit day and night. Drifting along the shoreline, my gaze was inevitably hooked by the squat towers of Duino castle, perched on an outcrop overhanging the Mediterranean. I imagined the poet Rainer Maria Rilke gazing out to sea from the elegant stone balcony in 1912, writing in his first Duino Elegy, "Oh and night: there is night, when a wind full of infinite space gnaws at our faces."[1]

I wondered how much darker Rilke's nights would have been over a century ago. Still, from the garden of our new home on the Karst Plateau, I could distinguish the Milky Way—a definite improvement over London!

I turned my attention to the northwest horizon, just above the wild Friuli Dolomite range in the distance. Arcturus, Garibaldi's special star in Boötes, was easy to spot, and I knew that my target was quite a bit lower on the horizon, so much so that it had possibly already been engulfed in the haze blown onshore from the sea. I followed the handle of the Big Dipper downward, toward a region of the sky devoid of bright stars.

"There it is!" I exclaimed. The trick was not to look at it directly but rather to let the peripheral vision of the eye pick up the fuzzy streak of light that was NEOWISE, the first naked-eye comet in over twenty years. It managed to give the impression of great speed without any perceptible motion. I could have sworn that its tail was swaying behind it, though this was a physical impossibility. No wonder the ancient Chinese texts called comets "broom stars"! My wife and children gave a little shout of surprise when they followed my pointing finger. Benjamin's enthusiasm quickly faded, but then revived in a wonder-filled "Oooh!" when he pressed his eye to the small telescope I had rigged up. Neatly framed by the eyepiece, NEOWISE showed its bright, compact core and its bifurcated tail—one stream made of dust reflecting the sunlight, the other of glowing gas. Isaac Newton believed that comets refueled the Sun and that they bestowed on planets gifts of water and "vital spirits."[2] Watching NEOWISE sweep the sky, I was tempted to believe him.

"Dad, is that the International Space Station?!" exclaimed my son, whose attention had wandered away from the eyepiece to a fast-moving, bright dot overhead. There was a time when one could be reasonably certain to have spotted the space station, but this dot was traveling in the wrong direction. I was trying to think of something supportive to say when Emma jumped in: "No, *that* is the space station!" She pointed at another pinprick carving a line through the constellations. There was no point in lying, as I unfortunately knew exactly what they were. But before I could explain, Emma had spotted another moving dot, and then another one. It was an infestation.

"Whatever it is, it's not cool!" exclaimed my daughter, fuming. "It spoils the stars!" She stomped away, heading back to the house. I packed up the telescope and tripod, and we all soon followed her.

A BLACK CANVAS

At the dawn of the space age, the night sky became art's final frontier. The contemporary land art movement had sought to escape the strictures of galleries by creating work in remote locations that could only be experienced on-site and often only from the air. Michael Heizer's *Double Negative*, a "240,000 tons displacement of rhyolite and sandstone," creates two fifteen-meter-deep cuts into Nevada's Mormon Mesa that face each other across a chasm.[3] Robert Smithson's *Spiral Jetty* is a 450-meter-long spiral constructed with black basalt rock and earth, winding counterclockwise from the shore into the Great Salt Lake in Utah. Now half submerged by rising waters and colonized by algae, the fifty-year-old artwork appears as if it could have been left behind by a long-gone civilization. It was only a question of time before artists' attention turned to the sky: What could be more striking than an artwork shining among the stars, embracing the whole of humankind in its orbit, speaking at the same time of the newly achieved powers of our species?

Such were the intentions behind *L'Anneau Lumière*, or *The Ring of Light*, the winning proposal in the competition launched in 1986 to create a modern-day equivalent in space of the Eiffel Tower. When Gustave Eiffel conceived his eponymous monument in 1886, he intended to celebrate with a three-hundred-meter tower of steel the technological prowess of the Industrial Revolution, a beacon of modernity and progress at the heart of the Parisian World Fair in 1889. A century later, *The Ring of Light* was to be a man-made constellation of one hundred reflecting balloons, each six meters in diameter once inflated in orbit. The balloons would be connected by lightweight Kevlar tubes to form a circular shape twenty-four kilometers in diameter, reflecting the Sun's light at night and shining as brightly as first-magnitude stars arranged in a circle larger than the Moon. Like the original Eiffel Tower, which was intended as a temporary exhibition, the new space monument was designed to be ephemeral: it would slowly lose altitude due to friction and eventually burn up in the atmosphere from three months to two years after launch. Unlike the tower, the project was abandoned due to technical difficulties.[4]

The Ring of Light echoed the utopian ideals of the man credited with launching the idea of artworks in "outer space." Five months after astronaut Alan Shepard of Apollo 14 played golf on the Moon in 1971, New York–born artist Albert Notarbartolo felt constrained by painting and drawing in two dimensions, and a move to three-dimensional paper sculptures didn't help. The "tyranny" of his pieces being confined by the walls of the room became so insufferable that he could no longer fall asleep: he began longing for "the space between planets . . . the true freedom of a place that has no sides, no top and no bottom."[5] He conceived of a series of projects, which he called *spaceworks*, that he hoped would give "emotional satisfaction" to the inhabitants of Earth and future weary travelers coming home after long voyages into deep space. *Project Beacon* was an oscillating sculpture reflecting sunlight in geostationary orbit that would welcome space travelers back to Earth as the Statue of Liberty welcomed to the United States those who arrived by sea. Other ideas aimed to provide "aesthetic satisfaction" to a community living on the Moon and to reduce the psychological stress of explorers facing the threats and loneliness of deep space. None of his ideas was ever realized.

After Notarbartolo, many others proposed orbiting artworks that would be visible from Earth: inflatable satellites appearing as bright stars from the ground, artificial constellations illuminated by laser beams, orbiting solar sails, a pair of spheres slowly rotating around each other. Most of the proposals had in common an underlying message of global unity, fraternity, and peace for humankind, visible as they would be from the four corners of the world. They were also, for the most part, controversial, expensive, and technically involved. Already Notarbartolo foresaw the opposition: that *spaceworks* could be seen as merely polluting space and their cost as indefensible given the harsh living conditions of most of humanity on Earth. Others saw them as embodying the purest ideals of "art for art's sake."

This was in the early 1970s, when space was largely devoid of man-made artifacts. But the shine of going to space wore off: already in 1972, the *New York Times* lamented that the footage of barren moonscapes and sauntering astronauts sent back from the Moon had become "ordinary and even tedious."[6] The public agreed and flooded the phonelines of the CBS TV network with complaints when an episode of a popular medical drama series was cut short to switch to coverage of the Apollo 17 launch. After that, the last three Apollo missions were canceled.

By 2018, when space entrepreneur Peter Beck secretly launched what could be described as a giant reflecting disco ball into orbit, the concept had become stale, even offensive. His *Humanity Star* circled the Earth for two months before burning up in the atmosphere, and despite its alleged message of world unity, it was met with ridicule. Astronomers, journalists, and the Twitterati described the three-foot-wide, sixty-five-sided satellite as a "PR stunt," "space graffiti," "satellite vermin," "abusive, disgusting, glittery space garbage," and "a vandalization of the night sky."[7]

With so many man-made mini-stars already up there, what's the point—aesthetic or otherwise—of sending one more? While Beck's *Humanity Star* was a modest ten kilograms in mass, that same year space and electric cars tycoon Elon Musk launched his personal cherry-colored sportscar into an elliptical orbit reaching as far as Mars in a stunt that, while not conceived of as art, Andy Warhol might have applauded. Proposals for orbital billboards periodically resurface, suggesting in all earnestness the tongue-in-cheek idea of Fredric Brown's 1945 short story "Pi in the Sky," in which a wealthy businessman rearranges the 168 brightest stars in the night sky to spell out a giant advertisement for his company's soap. When he realizes that he has misspelled the name of his company, he dies of a heart attack.[8]

Art at its best reimagines the possibilities of our relationships with ourselves and our environment. The commercial imperialism that considers space one more resource to be mined has undercut the field of possibilities artists once saw there. Perhaps the most radical intervention left for the future will be that of reinstating our original view of the sky: to transform the night back into what it would be without our wanton pollution.

The work of French photographer Thierry Cohen comes close to realizing this ideal. Cohen travels to dark sites at the same latitude as the world's largest megalopolises and photographs the night sky there. He then superimposes that image on the corresponding cityscape, after digitally extinguishing all the lights from the urban environment. The results of his *Villes éteintes* (Darkened Cities) project are haunting images that show us "not a fantasy sky as it might be dreamt, but a real one as it should be seen," in the words of art critic Francis Hodgson: the Milky Way blazing over Rio, the New York skyline in silhouette, a Parisian cobbled street flooded with starshine.[9] It's a world where the great cities of man are darkened and presumably abandoned, and the night sky is sovereign once again.

THE LAST GLOBAL COMMONS

The points of lights that had infuriated my daughter weren't artwork. They were rather the metastasis of a new space race, whose main aim is to defy stock market gravity for its proponents. Conquering space has been a matter of pride ever since Sputnik; the race that put Yuri Gagarin into orbit in 1961, sent Neil Armstrong to the Moon in 1969, and built the International Space Station was born of nationalism and fueled by political showmanship. In the twenty-first century, internet magnates and entrepreneurs have redefined space as the final frontier of profit, with astronomical egos to boot. Like the Wild West during the gold rush in nineteenth-century America, space has become a lawless mine, the last remaining commons to be claimed, colonized, and exploited with little to no oversight from governments and no consideration for the impact on communities all around the world.

Access to space used to be the preserve of state governments or international entities like the European Space Agency, which corralled the financial, technological, and scientific resources required to build and operate large rockets. Human spaceflight was particularly challenging, as the cocoon of breathable air that protects a capsule's fragile occupants must withstand intense vibrations at launch, supersonic speeds during ascent, mini-meteorite impacts, deadly radiation and freezing temperatures in space, and furnace-like heat during reentry. It wasn't easy to do, and it didn't come cheaply. The Apollo program that put twelve men on the Moon between 1969 and 1974 cost $700 billion in today's money— that's $1.2 million for each second each astronaut spent walking around the Moon.[10]

Since the early 2000s, some of the richest men on Earth have set their minds and fortunes to reducing the cost of getting to space both for equipment and humans—the necessary prelude, they claim, to colonizing the stars. They have criticized the operations of traditional space agencies like NASA, whose "safety first" approach, they argue, slows down development while inflating costs. They accumulated fabulous riches in Silicon Valley—Elon Musk was cofounder of X.com, an online bank that later became PayPal, before setting up SpaceX and taking over Tesla; Jeff Bezos can bankroll Blue Origin thanks to Amazon; Paul Allen used his Microsoft windfall to fund the first commercial flight into space—and brought the hacker spirit of fast prototyping to bear on space faring. Their trial-and-error approach, common in code development, has enabled them to design cheap, reusable rockets with boosters capable of landing back upright—the space equivalent of low-cost airlines, which twenty years earlier had revolutionized travel by operating airplanes like flying buses. It has worked. The cost of launching a satellite into orbit came down from $60,000 per kilogram with the space shuttle to $3,700 per kilogram with SpaceX's Falcon 9 rocket, named after a *Star Wars* spaceship. Public money contributed, too: the US government has increasingly bought in, allocating more than $7 billion in funding to private space companies between 2000 and 2018.

The new space race is on, with a self-proclaimed lofty humanitarian goal: to connect the whole planet, spread wealth and opportunity, democratize access to knowledge, and much more. To achieve the utopia of global fast internet access, SpaceX has launched thousands of internet satellites, bathing the planet in a permanent rain of radio signals, reaching the four corners of the world. Global internet access could have been achieved with a much smaller number of higher-altitude satellites, but starting in 2018 SpaceX chose instead to deploy a fleet of up to thirty thousand low-altitude satellites, capable of providing much faster internet to two key target communities: bankers (who require it for high-frequency trading) and gamers (who can't bear losing because of a sluggish internet).[11] The financials belie the "democratization of information" rhetoric. The vast majority of people in the least developed countries, who have the greatest need for satellite-based internet access since they have no ground-based alternative, cannot afford the monthly fee SpaceX is charging, not even on a subsidized basis; most people in Western countries live in urban areas where they enjoy fast internet access already. This leaves a few tens of millions in rural areas of North America and hundreds of millions more in China, Brazil, and Thailand as the main potential customers of the service, which also targets cruise ships, private jets, and commercial airlines.[12] Yet we will all be paying for it. We already are.

The astronomical community was blindsided by the unexpectedly rapid pollution of the night sky. Launch after reusable launch, from 2019 onward the night sky has become congested with hundreds, then thousands of internet satellites. Sky watchers initially wondered at the bright trails the satellites etched in their cameras as they were raised toward their final orbits. Videos mushroomed online, showing fast-moving strings of pearls threading the night: swarms of sixty internet satellites being deployed. Their parabolic antennas and solar panels reflect the Sun's light, especially right after sunset and before dawn, turning them into bright mini-stars crisscrossing the night. As the satellites rotate, they act like a mirror and reflect back the Sun's light, even outshining Venus. According to some estimates, by 2030 visible fake stars could outnumber real ones.[13]

This infestation of satellites didn't just spoil the magic of comet spotting for my daughter and countless others. Long exposures of the sky are damaged by their passage, which leaves trails like scars across the picture. This ruins the pictures of deep-sky objects of amateur astronomers, as well as up to half of the data acquired by the billion-dollar telescopes used by professionals—built at great cost in some of the remotest locations on the planet, offering the darkest skies but not immune to the scourge of passing satellites.[14] The Hubble Space Telescope is even more affected, as the satellites are so much closer to its spaceborne eye.[15] Radio astronomers found that the internet signals from overhead satellites can fry their exquisitely sensitive receivers, overwhelmed by a beam ten billion times stronger than the cosmic whispers they are designed to capture. Helplessly scrambling to rein in the excesses of this commercial space rush, astronomers could only bemoan the "special irony that a technology indebted to centuries of study of orbits and electromagnetic radiation from space now holds the power to prevent us permanently from further exploration of the Universe."[16]

One might think that none of this matters in a world where a majority of us live under a permanent blanket of light pollution. Who cares how many artificial dots move above the orange vault of our electric nights, just as invisible as the real stars we can no longer see? But there is a difference. No mountain, desert, or sea is sufficiently remote to evade the tight mesh of fast-moving artificial stars that the space barons are twining around the Earth for their financial gain. Their visibility and impact will be largest for the few remaining indigenous communities, for example, in the depths of the Amazon rainforest, that still rely on a close connection with the night sky.

In 1836, Ralph Waldo Emerson sang the beauty of the night sky thus: "One might think the atmosphere was made transparent with this design, to give man, in the heavenly bodies, the perpetual presence of the sublime."[17] The new, commercial space race will rob us forever of the ancient sublime. Our contemplation of the infinite is diverted to the ephemeral transit of heaps of circuitry that help further shorten our social-media-sized attention spans, the starry messengers shoved aside by instant messaging.

But it is not only our view of the cosmos that's being threatened by the uncontrolled proliferation of satellite mega-constellations. Our access to space is also in danger. Low-Earth orbit is becoming a congested place, a superhighway with no rules and no police, where satellites zip around at twenty-seven thousand kilometers per hour. At that speed, a fragment the size of a grape can blow a hole in the side of the International Space Station; a collision with a defunct satellite would destroy the station on impact. As the number of satellites increases, so does the probability of a collision between them, because of either malfunction or error. And there is no protecting against the impact of one of the over twenty-seven thousand pieces of space junk currently being tracked—some the result of a satellite collision in 2009, and some produced by the testing of anti-satellite weapons. The deployment of such bellicose means is no longer science fiction: Chinese scientists part of the military Strategic Support Force have called for the development of orbital weapons should SpaceX's internet service become a threat to their country's national security.[18] Past a certain density of satellites, the fragments produced by a collision or targeted destruction would hit and destroy further satellites, and so on, in a chain reaction that could obliterate most satellites and fill up the orbit with an impassable belt of space debris—a scenario called the "Kessler effect," named for astronomer Donald Kessler, who first described it in 1978.[19] This would render the passage through low-Earth orbit much more dangerous for any future mission, potentially cutting humanity off from access to space.

The Outer Space Treaty signed in 1967, while defining outer space as "the province of all mankind," does not provide adequate protection for what has become an urgent environmental issue. Space, one of the last remaining virgin territories, is being aggressively targeted for commercial exploitation. The US Congress legalized in 2015 the commercial mining of celestial bodies, with a piece of legislation that exploits a loophole in the treaty and "gives US space firms the right to own, keep, use, and sell the spoils of the cosmos as they deem fit," according to a legal analysis.[20] Several private companies are vying to demonstrate the technical feasibility of asteroid mining. The 2021 satire *Don't Look Up*—starring Jennifer Lawrence and Leonardo DiCaprio as two astronomers whose warning about a catastrophic asteroid impact is overruled by the commercial imperative of mining it for rare metals—is terrifyingly realistic in its depiction of greed and political corruption as the crucial elements of humankind's possible undoing.

BEAM ME UP, SCOTTY!

To reclaim Emerson's "presence of the sublime," scarred by satellites, you might have to become one of them, by buying a ticket on one of the suborbital tourist flights touted by—you guessed it—the same space barons responsible for the blighting of the night. The sky used to be for everybody, but space isn't for all: even in the era of relatively cheap rockets, a few minutes of weightlessness aboard a Virgin Galactic flight in 2023 would set you back $450,000. A seat on the maiden flight of Jeff Bezos's Blue Origin rocket in July 2021 to admire the curvature of the Earth from the edge of space was auctioned off for $28 million (the unnamed customer didn't show up for launch, implausibly citing a "scheduling conflict"), and a couple of weeks vacationing aboard the International Space Station set three private astronauts back $55 million each in April 2022—food was included, but toilets were shared.[21]

During the first wave of space tourism, one man's trip to space transformed fiction into reality. In October 2021, actor William Shatner became the oldest person to go to space, aged ninety, aboard Bezos's *New Shepard* rocket. Shatner rose to fame for his role in the legendary *Star Trek* series and subsequent motion pictures, in which he played Captain James T. Kirk, of the starship *Enterprise*. The original show portrayed the *Enterprise* crew as explorers, traveling at faster-than-light speed through the galaxy "to seek out new lifeforms, and new civilizations," with Kirk often engaging in bare-chested fistfights with quaintly made-up aliens and womanizing blue-skinned alienesses. But the show, shot in the 1960s, also had a progressive character: it featured one of the first interracial kisses on TV, and *Enterprise* crewmembers of Russian, Japanese, and Irish heritage all worked together, under the watchful eye of a Vulcanian first officer, Mr. Spock. A line often repeated by Shatner's character was "Beam me up, Scotty!" to ask his chief engineer to dematerialize him out of danger and back onto the ship. But when, at long last, he got beamed up in the flesh, Captain Kirk wasn't ready for what awaited him out there.

During his ten minutes in space, Shatner reported experiencing a profound feeling, but, contrary to his expectations, it wasn't one of connection with the infinite blackness of space, which he described as a "vicious coldness." Quite the opposite: when turning his gaze back toward the Earth and appreciating the contrast between our beautiful blue planet and the inhospitable cosmic void, he "discovered that the beauty isn't out there, it's down here, with all of us. Leaving that behind made my connection to our tiny planet even more profound. . . . [A] sense of the planet's fragility takes hold in an ineffable, instinctive manner."[22]

Shatner had experienced a version of the *overview effect*, a term coined in 1987 by author Frank White for the sense of awe, connectedness, and even spiritual awakening reported by astronauts and cosmonauts since the beginning of the space age.[23] Apollo 15 astronaut Al Worden, orbiting the Moon alone in the service module while his other two crew members were on the surface, enjoyed a wave of insight:

> At some points in my orbit around the moon, I was sheltered from both the earth and the sun, so I was in complete darkness. And all of a sudden, the star patterns out there became something that I was not ready for. . . . So many stars I couldn't see one. Just a sheet of light. I don't know whether you'd call it spiritual or not, but when I saw the starfield out there in a way that nobody else has ever seen . . . I had some pretty profound thoughts. . . . We are not unique in the universe.[24]

Upon returning to Earth, the feeling only intensified, and Worden felt compelled to process in poetry what he had felt. He captured it thus:

> *Of all the stars, moons, and planets,*
> *Of all I can see or imagine,*
> *This is the most beautiful;*
> *All the colors of the universe*
> *Focused on one small globe;*
> *And it is our home, our refuge*
> *Now I know why I'm here:*
> *Not for a closer look at the moon,*
> *But to look back*
> *At our home*
> *The earth.*[25]

Another Apollo astronaut, Ed Mitchell—"the sixth human to walk on the Moon," as he signed his emails—had what he described as "a subjective visceral experience accompanied by ecstasy" as, during the return flight, he watched the Earth, Moon, and Sun silently pass his rotating capsule's window every two minutes.[26] In an interview in 1974, he explained, "You develop an instant global consciousness, a people orientation, an intense dissatisfaction with the state of the world, and a compulsion to do something about it. From out there on the moon, international politics look so petty. You want to grab a politician by the scruff of the neck and drag him a quarter of a million miles out and say, 'Look at that, you son of a bitch.'"[27]

If Mitchell is right, perhaps the overview effect, even if witnessed from just above the atmosphere, might help other space tourists realize the fragility and fundamental unity of our destiny on Earth. Given the astronomical cost of the commercial experience, the lucky few able to perceive the Earth from space are likely part of the top 1 *permille*, who, if infused with a new environmental awareness, would have the means and the influence to do something about the many threats to our survival, from climate change to water scarcity, from food insecurity to pandemics.

A LOSS OF HAPPINESS

When space tourists gape at the nightside of Earth from their orbital portholes, continents are festooned with strings of fairy lights, entwining the thinning blackness of forests and mountains. Growing at an estimated rate of 6 percent per year, artificial lighting is not just an indicator of our increasing ecological impact: it is in and of itself an ecological menace. Turning the night into permanent twilight is influencing the balance of interconnected ecosystems, and we are just beginning to grasp the pervasive effects.

Birds, bats, fish, insects, and turtles are all affected by artificial light.[28] "City birds" rise earlier and stay up later than their forest cousins and acquire a faster-paced inner clock. Bats tend to avoid lit areas at night, which disrupts their commuting routes to foraging grounds and in some cases reduces their feeding opportunities, as the insects they prey upon are attracted to the very same lights bats dislike. Bright light sources like the beams from the September 11 memorial in New York alter the behavior of migratory birds at night, leading them to circle around the beams in confusion, wasting energy and time, becoming exhausted and ultimately less likely to reproduce.[29] Sea turtle hatchlings, who leave their beach nests at night, are guided to the ocean by its near ultraviolet glare, to which their eyes are very sensitive. Artificial lighting of beaches or even just the skyglow of nearby cities can interfere with their orientation, leading them astray, in one study toward the floodlights of a nearby house, to certain death.[30]

Not only animals are affected. Flowering plants receive fewer visits by nocturnal pollinators such as moths when they are lit, which reduces the number of fruits they develop and, for reasons that are poorly understood, also the number of daytime visitors, such as bees, thus further decreasing pollination.[31] Trees in urban areas retain their foliage for longer, and their leaves can unfold two weeks earlier in the vicinity of LED lights, putting them at increased risk of frost. The veil of light that repels the darkness of space and erases the stars is silently, invisibly changing the subtle balances of entire ecosystems. Its net effect is to reduce biodiversity, increase stress on plants and animals, and otherwise interfere with biological, ecological, and behavioral rhythms.

Light pollution also hides our impact on the appearance of the night sky, and we care less about the loss of what we hardly ever experience. In a collection of memories of World War II collected by the BBC, Londoners speak with nostalgia and lingering amazement of the experience of the night sky during the Blitz: "On a clear night during the blackout there was total darkness. It's difficult to visualise a complete blackness, but there were really no lights at all. . . . And on a clear, say a frosty night, the stars were out in thousands."[32] The tragedy of German carpet bombing brought forth a real-world version of Thierry Cohen's daunting artwork.

I recaptured some of this when I moved to the Karst, the mountainous plateau above Trieste, with the Mediterranean Sea to the west and the forested hills of Slovenia to the northeast. The city of Trieste, some ten kilometers away, is forty times smaller than London. The Milky Way graced every cloudless moonless night. One January evening, I went out on a small errand, and it hit me. Orion was blazing in the black sky; around his feet, his dog leaped faithfully, Sirius dazzling like an exclamation mark in the sky. I looked at the giant's belt, holding his dagger; I marveled at his powerful shoulders, his legs planted wide, the club raised high above him, ready to strike, his left arm outstretched holding his shield. My memory snapped back to that fateful night, so many years ago, when my future wife and I had been looking at Orion when the meteor slashed through it. Even then, I had not experienced the power of Orion so strongly. In that moment, I felt a fraction of the awe that must have possessed our forebears: a palpable sense of the deep time I would never live to witness and a connection with the untold other improbable configurations of atoms who had become aware and raised their eyes to the sky.

Charles Darwin, who in his youth had been a keen lover of the arts, at the end of his life came to regret the price of his unrelenting focus on scientific work:

> My mind seems to have become a kind of machine for grinding general laws out of large collections of facts, but why this should have caused the atrophy of that part of the brain alone, on which the higher tastes depend, I cannot conceive . . . and if I had to live my life again, I would have made a rule to read some poetry and listen to some music at least once every week; for perhaps the parts of my brain now atrophied would thus have been kept active through use. The loss of these tastes is a loss of happiness, and may possibly be injurious to the intellect, and more probably to the moral character, by enfeebling the emotional part of our nature.[33]

Without the stars, I felt like Darwin without music and poetry. The contemplation of the stars, I realized, nurtured for me a humbling sense of smallness and finitude and nourished the emotional part of my nature. If we all could experience every now and then this sense of being a minuscule part of a grand universe, the world would be, it seemed to me then, a much better place.

THE PRICE WE'LL PAY

In a short ten thousand years, guided by the cold light of the stars, the naked ape has built cities that banish the night; space stations that circle the Earth; works of art that move the soul. We have sent twelve of us to the Moon, where some played golf, and created weapons of unimaginable destructive power. But how ephemeral humankind's achievements appear when considered against the deep time that metes out the life of planets and stars!

Writer and geology expert John McPhee offers the following image: if the timespan of life on Earth takes up the space between your outstretched arms (a measure of length that defines "a fathom"), a single pass of a medium-grained nail file would obliterate the entirety of human history.[34] All the beauty and brutality humans bring into the world, the pyramids and the prophets, the Empire State Building and the coal mines, slavery and universal health care, football and Mozart, a mother singing to her baby and an assassin striking in the night, would all dissolve in a minute cloud of impalpable nail dust. Human time, even humankind's time, is nothing but a kitchen clock when measured by the breathing of stars. For all of our scampering like bacteria in a petri dish, our brevity is almost unfathomable.

In such an eye blink, star-inspired science and technology have bestowed upon us great benefits: advances in medicine, food production, and education have increased our quality of life greatly in the last century. Yet we hoard our benefits. Millions of people in the world still suffer from hunger or are undernourished, while millions of others die prematurely due to obesity, heart attacks, and other illnesses related to overconsumption of highly processed food. Income inequalities are higher than they have ever been: the richest 10 percent own three-quarters of all wealth, with the world's ten richest men—including the space barons—seeing their collective wealth double since March 2020, when the Covid pandemic hit. "What the inventive genius of mankind has bestowed upon us in the last hundred years could have made human life care free and happy if the development of the organizing power of man had been able to keep step with his technical advances. . . . As it is, the hardly bought achievements of the machine age in the hands of our generation are as dangerous as a razor in the hands of a three-year-old child."[35] These words, written by Albert Einstein ahead of the 1932 disarmament conference, are even more poignant today than they were over ninety years ago. Einstein surely pondered them later in life, after the devastation brought about in Hiroshima and Nagasaki by the nuclear weapons he played a crucial role in bringing into the world by urging President Franklin Roosevelt to develop them in 1939—something he regretted his whole life.

In the meantime, humans have swarmed the Earth. There are close to eight billion of us, and thanks to science and technology we have lengthened our average lifespan, wiped out diseases, reduced infant mortality, and, for a minority of us, created a world where almost our every material whim can be satisfied at will—in a two-hour, same-day delivery window. In the words of a character in Richard Powers's harrowing novel *The Overstory*, concerning our relationship with trees, "We are cashing in a billion years of planetary savings bonds and blowing it on assorted bling."[36]

The price we will eventually pay is enormous. Forty percent of our world's land is now degraded, according to a 2022 UN report: deforestation continues unabated, destroying irreplaceable ancient ecosystems, while intensive farming creates salinization, soil exhaustion, and erosion.[37] The havoc that the naked ape is wreaking on the planet is striking from space: our beautiful blue planet is scarred in ways that would have been unimaginable a generation ago. By clear-cutting thousand-year-old forests to make space for palm oil plantations that will fail in less than a decade, we are undercutting the basis of all life on Earth. On land, we have tilted the balance of large animals to suit our needs: farmed animals outweigh wild mammals and birds by a staggering ten to one. The oceans, which once seemed an inexhaustible resource, are overexploited: 90 percent of the fish stocks are fully exploited or depleted already. Flying insect numbers in the United Kingdom have crashed by 60 percent since 2004.[38]

We have been here before. Passenger pigeons once numbered in the billions in the eastern United States, gathering in immense flocks that darkened the sky for days when on the move—a colony in Wisconsin in 1871 was 125 miles long and 8 miles wide. They have been described as "a biological storm" or a "feathered hurricane." In the late 1800s, in the space of a few decades they were exterminated, as humans hunted them in their millions for their flesh and often (perhaps a more revelatory motive) just as a pastime. Nobody could imagine that such numbers could ever dwindle to nothing—until it was too late to save them.[39]

The fate of the passenger pigeon is now faced by *one million* animal and plant species, driven to the brink by habitat destruction, poaching, pollution, and climate change. Our carbon-based economy is rapidly increasing the CO_2 in the atmosphere and therefore heating the planet up: as of 2023, the last eight years have been the hottest on record, with global temperatures over 1°C above preindustrial levels. Glaciers are disappearing, the permafrost is melting, the icecap retreating, and sea levels rising. Our planet has entered an out-of-equilibrium phase, whose feedback loop will endanger the lives and livelihoods of billions of people—as has already begun. The passenger pigeon tragedy shows that, once kicked out of equilibrium, the abundance of life can spiral quickly. In the words of Loren Eiseley, the poet paleontologist, we are "a vast black whirlpool spinning faster and faster, consuming flesh, stones, soil, minerals, sucking down the lightning, wrenching power from the atom, until the ancient sounds of nature are drowned in the cacophony of something which is no longer nature."[40]

BECOMING GOOD ANCESTORS

In the face of the human-induced existential threat to life on Earth, the space barons are working to offer us, they say, the means of fleeing to the stars. The mission of Jeff Bezos's Blue Origin is "to build a road to space so our children can build the future," apparently by moving millions of people and heavy industry to space so as to preserve the planet.[41] Whether our children will want to go is another question. Elon Musk aims even higher: to build a modern-era Noah's Ark, not out of wood on a mountaintop but out of steel on top of a rocket, to ensure the survival of the human race against the metaphorical and actual flood that is coming. "You back up your hard drive. . . . Maybe we should back up life, too?," said Musk in 2015. He believes that we should colonize Mars as a lifeboat for humankind and as a stepping-stone to the stars.[42]

The idea is not new. Carl Sagan championed it as an "insurance policy" against the not-unreasonable risk that we will end up wiping ourselves out—a danger that has perhaps never been more sharply defined than today. He wrote in 1994, "If our long-term survival is at stake, we have a basic responsibility to our species to venture to other worlds," a feeling shared by astrophysicist Stephen Hawking.[43] While NASA plans to return humans (and take the first woman and person of color) to the Moon by 2025, SpaceX's schedule for a crewed mission to Mars keeps slipping back: Musk, who had indicated a ten-year time frame for reaching the red planet in 2011, now speaks of 2029. But Mars is a much more difficult goal than the Moon. The voyage takes six to nine months each way, compared to just three days for reaching our satellite, and comes with the additional challenges of prolonged exposure to cosmic rays, the need to either carry supplies for up to two years or else extract power and consumables from Mars, the difficulty of landing a large spacecraft, and the psychological distress of a long journey in cramped conditions—and that's just the tip of the iceberg.

Establishing a colony, especially one that can survive independently from Earth—Musk's long-term ambition—appears an even more fragile prospect. Consider that in 2021, the International Space Station required resupply every six to eight weeks for a crew of seven 250 kilometers up—not a colony of perhaps hundreds 100 million miles away. Some proponents of transhumanism—among them, many space barons—claim that it may not be necessary to send our biological bodies there and that in the next step of evolution we will shed our organic selves in favor of silicon-based simulacra, our minds uploaded to the cloud. Given the current state of artificial intelligence, this is an even more remote prospect, and even if it were to be realized, our synthetic avatars would be fundamentally other. I doubt that the Neanderthals would have taken comfort upon being told that they were to be replaced by a species upgrade!

The question of space colonialism beyond the solar system is even more academic. Other stars in the galaxy likely host Earth-like planets, but the crossing to these hypothetical other habitable worlds would take hundreds or thousands of years in space, not because of the limitations of present-day technology but rather because of the unimaginably vast distances between the stars and the fundamental barrier of the speed of light.[44] It would be easier for an ant to circumnavigate the Earth on a leaf than for humans to reach even the nearest star. And even if we did and discovered a life-bearing planet, its inhabitants might not be happy to share their home with resource-hungry humans. We have seen this sort of aggressive expansion into new territory before, with the colonization of the Americas, Africa, and Oceania by the Europeans in the sixteenth and seventeenth centuries. It didn't end well for the natives.

Mars, too, would be a refuge for one species only. There would be no space for whales, falcons, or butterflies; no meadows full of bluebells, no thousand-year-old redwoods, and no coral reefs. No bees or earthworms or sound of crickets on a warm summer evening. For that matter, there would be no warm summer evenings. As Mars is a desert planet with a thin atmosphere, its surface temperature drops rapidly after sunset, from a chilly –14°C afternoon high to a freezing –90°C at the tropics.[45]

Not that you or I would be invited to join this ultimate gated community, anyway. Writing in the 1960s, at the onset of the space age, philosopher of technology Lewis Mumford described the great pyramids as "the precise static equivalents of our own space rockets. Both [are] devices for securing, at an extravagant cost, a passage to Heaven for the favored few."[46] Indeed only the "favored few," the twenty-first-century internet moguls turned pharaohs, could hope to gain passage on their hypothetical lifeboats—the same men who are today building the rockets and riding them to space for fun. According to one of them, perhaps 50 percent of Silicon Valley billionaires are "doomsday preppers"—people who buy "apocalypse insurance" in the form of well-stocked bunkers guarded by private militias or even vast, self-sufficient private estates in remote parts of New Zealand.[47] When the apocalypse comes (in the form of climate change, a deadly virus, civil disorder, or nuclear war, perhaps precipitated by the planet-devouring technologies they helped build), they want to secure a way out for themselves and their loved ones. If the whole planet is burning, then space is their last escape hatch.

Leaving Earth behind would be the ultimate outcome of what Mumford has termed "the megamachine": the relentless focus of Western civilization on organizing and corralling the entirety of human existence into an ever more efficient, ever more powerful (and destructive) mechanized order of the world, built on the blueprint of the "clockwork universe." To restate the paradox: humans today imagine living among the stars to save themselves from destruction wrought by the technology that looking at the stars brought into the world.

Rather than follow pipe dreams of life on other worlds, our moral imperative is to become stewards of our own planet. Stars guided us to science that gave us technology indistinguishable from magic; they are etched in our psychological makeup and helped us conquer the world. But they are not an escape hatch. Let us focus our attention back on the real issues: how to share our planet's resources more equitably among all humans, how to ensure that nonhuman life can continue to thrive on Earth, how to refashion our civilization on an environmentally sustainable basis, how to bequeath to our children and to their children a planet as diverse and hospitable as the one we inherited. We must learn to become "good ancestors," in the memorable expression of Jonas Salk, inventor of the polio vaccine, who then dismissed the notion of licensing it by asking, "Could you patent the sun?"[48]

"The fault, dear Brutus, is not in our stars," says Cassius in Shakespeare's *Julius Caesar*, spurring him to move against Caesar despite their weakness as underlings. "Men at some time are masters of their fates," he exhorts.[49] In this most important of tasks for the people now alive, that of mastering our fate even as the rudder slips from our hands, the stars can help. The overview effect is available to all of us who seek out a dark sky and look up: up, toward the blackness between the countless burning suns; up, toward the silent shapes of constellations that sang stories to generations of humans; up, toward the infinite expanse that only rare configurations of atoms have been able to appreciate. To truly reconnect with the stars is to give ourselves the perspective necessary to choose our next step wisely: not myopically focused on the next electoral cycle but encompassing the breadth of geological epochs—lest the Anthropocene become the shortest and last of the recorded geological eras. The night sky is the sole aspect of nature shared among all of us on this planet. Other grand landscapes and living things like whales and sequoia trees do engender a similar sense of awe, but the stars are unique in being common to us all. It is to this sense of shared destiny that we must appeal if we are to face as one the mortal dangers confronting us.

"Down how many roads among the stars must man propel himself in search of the final secret? The journey is difficult, immense, at times impossible, yet that will not deter some of us from attempting it," wrote Loren Eiseley.[50] Our road need not be a solitary one. To rebehold the stars is to take the perspective common to all living beings on Earth. One day, after having traveled down the unpaved roads of science from the bosom of the atom to the end of the universe, the time will come when we will turn homeward onto the path not taken—that of love. And that day, Prometheus shall be unbound.

THE CALIGO TALES

The Skeleton's Dance

Mist-Catcher had just finished when a crack so powerful as to shake our hearts ricocheted inside the cave. The Lightning had listened to us! We rose as one, our voices rumbling away with glee.

Shepherd's words somehow dominated the din.

"Cloud-Watcher! The Lightning is here! Let the Trickster's head feel its power once again! Let the Skeleton's Dance begin."

We reached for our spears and rushed out into the rain, fat drops disappearing in little puffs of dust as they hit the ground. Beehive was shaking her spear up and down, calling Lightning to join our dance, her eyes wide, the muscles of her neck bulging. Bison-Seeker was beating our rhythm, sparks flying with each bang of her stones. Others rolled themselves in the dust, calling Lightning to fill them with its power. Spear-Thrower's browridge was already bleeding as he hit his forehead over and over again against the splintered oak tree that had been visited by Lightning long ago.

The rain comes hard and thick now, accompanied by violent Rumbles that shake the earth under our feet, caked in mud and speckled with blood. Lightning has left its nest among the Cloud to visit our people!

Singing emerges from inside the cave, and we freeze: it's Shepherd heading the procession, followed by Cloud-Watcher, Once-Upon-A-Glow, and the others. He holds the Pole high in front of him, and we all fall in line along the sides. My eyes are glued to the Trickster's head, skewered to the top of the pointy pole, swaying as if alive at every step: by the light of the fire, I glimpse some skin clinging onto the Trickster's jaw, like brown leaves when the glows get short.

The sight fills me with an eagerness for war: as I grip my axe, ready to strike, I eye the Darkness all around us, should his companions come back to rescue him—even though no one alive has seen a Trickster in the flesh.

"Trickster-Basher!" Shepherd is now screaming into the rain. "As we follow your lead, may your courage and strength help us face our enemies! May Lightning keep the Tricksters away from our people!"

That's when it happens. Lightning has heard Shepherd and visits him with its full force. A deafening snap splits the ear, and the smell of energy fills the air. My ears ring, my eyes are blinded, and I fall. Screams of agony everywhere. Have the Tricksters returned, I wonder?

I roll onto my side, groping for my spear, the taste of blood on my tongue. My hand clutches something round: it's the Trickster's blackened skull. Shepherd lies next to it, his limbs twisted, his hands smoking. A smell of charred meat reaches my nostrils.

I drag the Trickster's skull next to me and lift it level with my eyes: I feel its burnished, bulging top, its smooth forehead. How could such a weakling ever frighten us, the people of the Cloud?

As I fall back onto the ground, my gaze pierces through a hole at the back of the skull. Through the Trickster's empty eye sockets, I see what nobody ever Remembered: the Cloud rips, revealing the Roof of the Disc. Blackened by fire and encrusted with crystals, it sparkles just like the vault of our cave.

EPILOGUE

SO SPOKE THE SILENT STARS

IN THE TIME IT HAS TAKEN YOU TO READ THIS BOOK, VOYAGER I has slipped away another three hundred thousand kilometers into the darkness of space, edging toward the star Gliese 445, seventeen light-years away. The probe, like its twin, Voyager 2, will roam the galaxy for something close to eternity—it would take ten million times the age of the universe for one of the Voyagers to have an appreciable probability of colliding with a star.

Each of the Voyagers carries mounted on its side a last-minute addition to the mission, slight but packed with the oversize potency of a symbol. Carl Sagan knew that the probes were destined to leave our solar system, and he assembled a cross-disciplinary team to create an interstellar message in a bottle that could be understood by any sentient being that might one day pick up one of the Voyagers. The message takes the shape of a gold-plated disc, known as the golden record.[1]

On its encasing, a set of instructions demonstrate how the record is to be played, together with the number of rotations per minute (16.5) in units of the fundamental frequency of the hydrogen atom (also depicted). The location of our solar system is given with respect to fourteen nearby pulsars. The cover of each Voyager's golden disc is sprayed with a veil of ultrapure Uranium-238, whose slow decay acts as a timer since launch, for the benefit of its hypothetical finders. The radioactive clock winds down by half every 4.5 billion years. By that time, the twin discs—separated at launch and by then half a galaxy away from each other, never to be reunited—will likely be the only testimony that humankind ever existed, long after the Earth will have been obliterated by the swelling, aging Sun.

In a few years' time, the Voyager 1's power source will run out, and the spacecraft will fall silent forever, the soundless decay of its disc's radioactive coating the last barrier to timeless eternity. Unless, that is, an alien hand (if it is a hand) one day removes the disc from its encasing, finding behind it the stylus needed to bring its content back to life. A test picture etched on the cover will allow alien scientists to check that they have the device working correctly. If the Voyager's golden record fulfills its mission, sounds and pictures of 1970s Earth will raise again in the inconceivable future in some distant corner of the galaxy: the song of whales, Mozart's "Queen of the Night" aria, an F-111 flyby, the chirping of long-extinct crickets and frogs; diagrams showing the structure of DNA, an EEG recording, bushmen hunting, and a Titan rocket takeoff; pictures of a traffic jam in Thailand, the Sydney Opera House, a young woman eating grapes in a supermarket aisle, mud houses and skyscrapers, human sex organs, and crocodiles. One can only imagine what kind of impression these and many other sounds and sights, together with greetings in over fifty languages, might make on the sensory organs of another form of life. I like to think that alien scientists will be moved by the sound of a motherly kiss, followed by the wailing of a baby being consoled in hushed tones by a human woman's voice.

Among the sounds, pictures, and greetings on the golden record is a message from then US president Jimmy Carter. Written in 1977, Carter's words ring with even greater urgency today:

Of the 200 billion stars in the Milky Way galaxy, some—perhaps many—may have inhabited planets and space-faring civilizations. If one such civilization intercepts Voyager and can understand these recorded contents, here is our message: This is a present from a small distant world, a token of our sounds, our science, our images, our music, our thoughts, and our feelings. We are attempting to survive our time so we may live into yours. We hope someday, having solved the problems we face, to join a community of galactic civilizations. This record represents our hope and our determination and our goodwill in a vast and awesome universe.[2]

Should anybody come looking for Earth in a distant future, guided by the fourteen pulsars on the golden record's cover, will they be disappointed to find a dead planet, a cosmic tombstone marking the hubris of the naked ape? Or will they marvel from the orbit of Jupiter at our beautiful blue dot, sparkling against the dark? Whether or not we believe that we will one day join a "community of galactic civilizations," it is our urgent task today to halt the march of the megamachine; to repurpose its planetary power so it may serve the needs of all life on Earth; to fortify ourselves not merely to survive our time but to create a new age.

There is something profound in the gesture of launching a disc the size of a dinner plate into the void between the stars, hoping against hope and reason that sometime, somewhere, somebody—or something—will pick it up and remember us.

We existed, says the disc. We are starborn, and to the stars we entrust our memory.

ACKNOWLEDGMENTS

WHEN I PITCHED A BOOK ON A "WORLD WITHOUT STARS" TO T. J. Kelleher on a sunny June afternoon in London, I did not imagine that from that seed of an idea would sprout a life-changing journey of discovery. I am grateful to T. J. for believing in this project from the very beginning.

To all the dedicated people who nurtured that seed until it grew root, thank you: to my agents, Peter Tallack and Louisa Pritchard at Curious Minds Agency, to Lara Heimert and Sarah Caro at Basic Books, and to their teams, whose passion for creating gorgeous books shines through. Special thanks to Meghan Houser, whose insightful comments and incisive editing were crucial in giving the text its current shape, and to Jennifer Kelland, whose copyediting further improved flow and clarity.

Thank you to the friends, colleagues, and family who supported, encouraged, and inspired me in more ways than I can express here: Laura Cameron and Andrew Eaton-Lewis, Loretta Gianettoni and Fausto Pagnamenta, Gianfranco Bertone, Ivan Cabrillo, Eliel Camargo-Molina, Aifric Campbell, David Cunial, Ed Dark, Stephen Follows, Gigi Funcis and Giulia Carollo, Alessandro Laio, Louis Lyons, Guido Sanguinetti, Tereza Stehlikova, Elisabetta Tola, and Richard Watson. Thanks to Nastja

Gartner and Gregor Višnar of the Golden Beaver Ranch for providing a writer's heaven in their Slovenian Walden.

Thanks to the experts who have helped: David Benqué, Jimena Canales, Arnaud Czaja, Edward Gryspeerdt, Marc McCaughrean, Felicity Mellor, Roger Kneebone, Ed Krupp, Andy Lawrence, Tyler Nordgren, Lala Rolls, Steve Warren, Michael Weatherburn, and Rebecca Wragg Sykes. Of course, any mistakes are entirely my own.

I am grateful to the library staff who sourced rare or difficult-to-find material, which was crucial for my research: my thanks to the librarians at Imperial College London (especially Ann Brew and Rosemary Russell), the Huntington Library, the Yerkes Observatory, and the International School for Advanced Study in Trieste (especially Stefania Cantagalli, Gerardina Cargnelutti, Barbara Corzani, and Marina Picek).

To my friend and colleague Fabio Iocco, a heartfelt thank-you: your presence, support, and insightful suggestions made a difference—starting with the cover.

To my dad, whose eyes no longer behold the stars, thank you for everything: the music may have stopped, but the dance never ends.

To my wife, Elisa, and my children, Benjamin and Emma: this book wouldn't exist without your encouragement, love, and patience. Know that even though my head was often in the stars, you have been and always will be the brightest lights in my universe.

NOTES

PROLOGUE

1. Homer, *The Odyssey*, 11.567.

2. Einstein recalled "a paradox upon which I had already hit at the age of sixteen: If I pursue a beam of light with the velocity c (velocity of light in a vacuum), I should observe such a beam of light as an electromagnetic field at rest though spatially oscillating. There seems to be no such thing, however, neither on the basis of experience nor according to Maxwell's equations. . . . One sees in this paradox the germ of the special relativity theory is already contained" (quoted in Norton, "Chasing the Light," 123).

3. Leopardi, *La storia dell'astronomia*, 731. My own translation from Italian does not do justice to the elegance of the poet: "Dacché la Terra ebbe degli uomini, il cielo ebbe degli ammiratori."

4. Mumford, *Technics and Civilization*, 47.

5. Eliade, *Patterns*, 39.

6. Quoted in Krupp, *Beyond the Blue Horizon*, 25.

7. Alighieri, *The Divine Comedy*, Paradise Canto 33, 145.

8. Bridgman, "Who Were the Cimmerians?," 39–40.

9. Homer, *The Odyssey*, 11.11–13.

CHAPTER 1: A PALE BLUE DOT

1. Sagan, *Pale Blue Dot*, 6.

2. Randall and Reece, "Dark Matter as a Trigger for Periodic Comet Impacts."

3. Eliade, *Patterns*, 39.

4. Poincaré, *The Value of Science*, 84. He then continues with a hubris that we should since have learned to rue: "[Astronomy] shows us how small is man's body, how great his mind, since his intelligence can embrace the whole of this dazzling immensity, where his body is only an obscure point, and enjoy its silent harmony. Thus we attain the consciousness of our power, and this is something which can not cost too dear, since this consciousness makes us mightier."

5. Poincaré, *The Value of Science*, 84–85.

6. Quoted in Crawford, *Atlas of AI*, 227.

7. Poincaré, *The Value of Science*, 85.

8. Mumford, *Technics and Civilization*, 14.

CHAPTER 2: THE LOST SKY

My main sources for ancient beliefs around solar eclipses are Krupp, *Beyond the Blue Horizon*; Kelley and Milone, *Exploring Ancient Skies*; Close, *Eclipses*. On contemporary knowledge, see Nordgren, *Sun Moon Earth*. A good source on popular lore on comets is Schechner, *Comets*.

1. Emerson, *Nature*, 9.

2. Joel 2:31 (King James Version).

3. Solar eclipses don't occur every new Moon because the orbit of the Moon is inclined by five degrees with respect to the plane of the orbit of the Earth; so in most cases, when the Moon is between us and the Sun—at new Moon—it is not aligned in front of it.

4. Herodotus, *Histories*, I.74.2–3.

5. Quoted in Krupp, *Beyond the Blue Horizon*, 162.

6. Humphreys and Waddington, "Dating the Crucifixion"; Schaefer, "Lunar Visibility and the Crucifixion"; Schaefer, "Lunar Eclipses That Changed the World."

7. Amos 8:9 (King James Version).

8. Quoted in Chambers, *The Story of Eclipses*, chap. 12.

9. Shayegan, "Aspects of History and Epic in Ancient Iran."

10. There is no indication as to whether the eclipse was total, although it may well have been observed in the capital city of Ashur, in present-day northern Iraq (Stephenson, "How Reliable Are Archaic Records?"). Humphreys and Waddington ("Solar Eclipse of 1207 BC") suggested that a passage in the Old Testament could be interpreted as describing a total solar eclipse in 1207 BC, and there are claims of a solar eclipse recorded as early as 1223 BC (de Jong and van Soldt, "The Earliest Known Solar Eclipse").

11. Krupp, *Beyond the Blue Horizon*, 51–53; Frazer, *The Worship of Nature*, 556, 559–560, 596.

12. It may be that the large size of our Moon is not such a coincidence after all: a ponderous satellite might be needed to ensure the long-term stability of the Earth's rotation, a prerequisite for life—and therefore, our existence (see Laskar, Joutel, and Robutel, "Stabilization of the Earth's Obliquity"; Lissauer, Barnes, and Chambers, "Obliquity Variations").

13. Numbers 24:17 (King James Version).

14. Matthew 2:2 and 2:9 (King James Version).

15. Another supernova explosion in the galaxy on February 23, 1987, was studied in great detail with modern instruments. The 1987 supernova was, however, of a different type than Kepler's.

16. The definitive account of what Kepler thought (and how his position was misrepresented in the nineteenth century) is given by Burke-Gaffney, "Kepler and the Star of Bethlehem"; quotes on 420–421. See also Kidger, *The Star of Bethlehem*.

17. Quoted in Schechner, *Comets*, 51.

18. Dio Cassius, *Roman History*, 8:66.

19. Aquinas, *Summa Theologiae*, Question 73, article 1.

20. Luke 21:25 (King James Version).

21. Botley and White, "Halley's Comet in 1066," 4–6.

22. Legend has it that William's wife, Queen Matilda, crafted it with her ladies-in-waiting, although the actual author is unknown.

23. Westfall, *Never at Rest*, 104.

24. Quoted in Chambers, *The Story of Eclipses*, chap. 12.

25. Milton, *Paradise Lost*, VII, 580.

26. Dunkin, *The Midnight Sky*, 116.

27. DeLillo, *Underworld*, 623.

28. This anecdote was apparently embellished in a 2008 article on light pollution in the *New York Times*, according to which "numerous calls came into emergency centers and even the Griffith Observatory," anxiously demanding explanations for a "giant silvery cloud" in the sky, which, the astronomers allegedly assured, was merely the Milky Way (Sharkey, "Helping the Stars"). The original source of the story, Dr. Edward Krupp, the director of Griffith Observatory, who was personally involved in the events, told me that "people simply responded to the dark sky and the profusion of stars, but not to the Milky Way, which was very close to the horizon and not particularly conspicuous or even visible." Callers, probably around a dozen in total, "were inquisitive and puzzled, not upset or worried. They seemed to think the earthquake might have caused the 'odd' sky, but they didn't realize it was just an electrical shutdown that revealed stars few had ever seen" (Edward Krupp, emails to author, November 24–25, 2020).

29. Quoted in MacCarthy, *Gropius*, 239.

30. Hintz, Hintz, and Lawler, "Prior Knowledge Base of Constellations."

31. Thoreau, *Walden*, chap. 9.

32. Emerson, *Nature*, 9.

CHAPTER 3: LIFE UNDER A CLOUD

Details on Callanish can be found in Ponting, *Callanish*; Sawyer Hogg, "Out of Old Books." Example of clouds on other planets are described in Helling, "Clouds in Exoplanetary Atmospheres"; Moses, "Cloudy with a Chance of Dustballs"; Kipping and Spiegel, "Detection of Visible Light from the Darkest World"; Libby-Roberts et al., "The Featureless Transmission Spectra." On clouds and climate on Earth, see Still et al., "Influence of Clouds"; Hartmann, Ockert-Bell, and Michelsen, "The Effect of Cloud Type." The burgeoning science of exoplanet climatology is reviewed in Shields, "The Climates of Other Worlds." On the figure of Luke Howard, see Hamblyn, *The Invention of Clouds*.

1. Ponting, *Callanish*, 10.

2. Captain Sommerville, quoted in Sawyer Hogg, "Out of Old Books," 86.

3. The phenomenon of major lunar standstills is caused by the Moon's orbit being inclined by five degrees with respect to the plane of the Earth's orbit around the Sun. For about three years around the major standstill peak, the Moon rises and sets at more southerly (around summer solstice) and more northerly (around winter solstice) locations than the Sun ever reaches. The cycle of 18.6 years between major standstills is caused by the precession of the Moon's nodes (the intersection of the lunar orbital plane with the plane of the ecliptic) and is distinct from the Metonic cycle of 19 years (encompassing 235 full Moons), which is described in Chapter 5. The next opportunity to witness a major standstill will be in 2025.

4. Olson, Doescher, and Olson, "When the Sky Ran Red."

5. Gryspeerdt, "Where Is the Cloudiest Place on Earth?"

6. An instrument made by Campbell himself around 1876, on display at the National Maritime Museum in London, features a highly polished brass bowl, with dark engravings showing the cardinal points and degrees of the circle, like on a navigational compass. A Latin inscription in sans serif letters declares, "I count only the serene hours," giving the misleading impression that the instrument is a time-measuring device. But the heart of the sunshine recorder is a glass sphere sitting in the bowl, looking exactly like the crystal ball of clairvoyants; it captures the light from the Sun as it moves across the sky and concentrates the rays so they produce a scorching on a paper strip at the bottom of the bowl. The resulting charred lines show the path of the Sun and record the cloudiness of the sky.

7. "Niels Ryberg Finsen—Facts."

8. London, *The People of the Abyss*.

9. Quoted in Robson-Mainwaring, "The Great Smog of 1952."

10. Quoted in Robson-Mainwaring, "The Great Smog of 1952."

11. Robinson, "15 Most Polluted Cities in the World."

12. Abbot, "The Habitability of Venus," 170.

13. Quoted in Launius, "Venus-Earth-Mars," 257.

14. Bradbury, "The Long Rain."

15. Burroughs, *Pirates of Venus.*

16. Barlow, *The Immortals' Great Quest*, 81.

17. Barlow, *The Immortals' Great Quest*, 114.

18. The novel was later republished under his real name in Barlow, *The Immortals' Great Quest.*

19. Sagan, "The Planet Venus," 849.

20. In a recent twist in the search for life on Venus, a team of astronomers (Greaves et al., "Phosphine Gas in the Cloud Decks of Venus") reported in 2020 the detection of phosphine in the planet's clouds—a toxic gas that on Earth is produced by bacteria in marshlands and bogs, found in penguin dung and the bowels of badgers. To find phosphine is to find (smelly) life. This exciting prospect was later put to rest, as a reanalysis of the data indicated that the phosphine detection was spurious (Villanueva et al., "No Evidence of Phosphine in the Atmosphere of Venus").

21. Kreidberg et al., "Clouds in the Atmosphere."

22. Poincaré, *The Value of Science*, 84.

23. Quoted in Beck, "The Caves of Forgotten Times."

24. Quoted in Hooper, "Three Years in a Cave."

25. Kaiho et al., "Global Climate Change Driven by Soot."

26. On the asteroid impact that led to the dinosaurs' extinction, see Renne et al., "Time Scales of Critical Events"; on the cometary impact theory around the transition between the Paleolithic and Neolithic periods, see Powell, "Premature Rejection"; Sweatman, "The Younger Dryas Impact Hypothesis."

27. Gould, "The Evolution of Life on the Earth," 100.

CHAPTER 4: THE WEIGHT OF STARLIGHT

On Neanderthals' life and evidence from paleontology, see Wragg Sykes, *Kindred*. On Aboriginal sky lore, see Hamacher, *The First Astronomers*; Norris, "Dawes Review 5." On Inuit knowledge of the sky, see MacDonald, *The Arctic Sky*. On legends surrounding the Pleiades, see Krupp, *Beyond the Blue Horizon*, 241ff; Kelley and Milone, *Exploring Ancient Skies*, 141ff. On animal stellar orientation, see Foster et al., "How Animals Follow the Stars."

1. Wragg Sykes, *Kindred*, 38.

2. Price, "Africans Carry Surprising Amount of Neanderthal DNA."

3. Wragg Sykes, *Kindred*, 377.

4. Gibbons, "Neanderthals Carb Loaded."

5. d'Errico et al., "The Origin and Evolution of Sewing Technologies."

6. Knight, *Blood Relations*, 344.

7. At mid to high northerly latitudes, there is an exception to this sequence: in the evenings following the full Moon closest to the autumn equinox, the so-called Harvest Moon, our satellite rises farther north on the horizon each night, which reduces or even cancels the time lag to the Sun setting. As a consequence, the dusk-to-dawn fully moonlit period can stretch to several nights. The same happens in the Southern Hemisphere around the spring equinox in March.

8. Colagè and d'Errico, "Culture."

9. Hare and Woods, *Survival of the Friendliest*.

10. Knight, *Blood Relations*.

11. Marshack, *The Roots of Civilization*.

12. For a review of Upper Paleolithic lunar notation claims, see Hayden and Villeneuve, "Astronomy in the Upper Palaeolithic?"

13. Another artifact, an engraved tooth from an extinct Australian marsupial dating to twenty thousand years ago, displays twenty-eight notches that have been claimed to represent a lunar calendar (Vanderwal and Fullagar, "Engraved Diprotodon Tooth"). A recent reanalysis shows that the markings were made by a small animal (Langley, "Re-analysis of the 'Engraved' Diprotodon Tooth").

14. The association of the planet Venus with the goddess of love might also have been inspired by astronomical lore: Venus is visible as the "evening star" for 263 days, approximately the length of a human pregnancy, before disappearing behind the Sun and reemerging, fifty days later, as the "morning star" for another 265 days.

15. Its name has nothing to do with the color; the *Maine Farmers' Almanac*, entry for August 21–22, 1937, explains, "This extra moon had a way of coming in each of the seasons so that it could not be given a name appropriate to the time of year like the other moons. It was usually called the Blue Moon."

16. Some studies appear to show that women living at close quarters tend to synchronize their menstrual cycles—dubbed "the McClintock effect," for the researcher who first posited it in 1971 (McClintock, "Menstrual Synchrony and Suppression")—although the evidence is disputed (Gosline, "Do Women Who Live Together"). Ethnographic accounts of modern hunter-gatherers generally show no trace of synchronization of women's cycles with the Moon phase, with the exception of the well-documented case of the Nuu-chah-nulth Native American people of Vancouver Island (Knight, "Menstruation and the Origins of Culture," 211). There are also indications that menses might align, at least intermittently, with the lunar luminous or gravitational cycle, but here, too, the devil is in the statistical detail (Helfrich-Forster et al., "Women Temporarily Synchronize"). In any case, if such correspondences between the lunar cycle and women's fertility are observable in our modern society, Chris Knight argues, it

is possible that they were much stronger—even if only on a symbolic level—in a prehistoric setting, when women lived together in a tight-knit group and there was no artificial lighting, except that provided by fire.

17. Knight, *Blood Relations*, 97.

18. Glaz, "Enheduanna," 33.

19. While Marshack's work has been fiercely criticized as plagued by "loose generality, skimpy delineation of methodology, and assertions offered as demonstrated truth" (King, "Reviewed Works," 1897), it did raise questions about the origin of numerical notation that are still being debated today (Robinson, "Not Counting on Marshack").

20. O'Connell, Allen, and Hawkes, "Pleistocene Sahul and the Origins of Seafaring."

21. Quoted in Fuller, Norris, and Trudgett, "The Astronomy of the Kamilaroi and Euahlayi Peoples," 10.

22. Hamacher, "On the Astronomical Knowledge," 82.

23. Norris and Harney, "Songlines and Navigation," 143.

24. MacDonald, *The Arctic Sky*, 169.

25. Quoted in MacDonald, *The Arctic Sky*, 167.

26. Norris and Harney, "Songlines and Navigation," 143.

27. Hamacher ("On the Astronomical Knowledge," chap. 5) demonstrates that many Aboriginal traditions and stories connected with the sky cannot be older than ten thousand years, for before that point, the association with the sky was lost due to the precession of the equinoxes.

28. Quoted in MacDonald, *The Arctic Sky*, 167.

29. Norris, "Dawes Review 5," 22–23.

30. Fuller et al., "Star Maps and Travelling to Ceremonies," relates a *bora* ceremony in 1894 where attendees traveled up to 160 kilometers by foot to take part.

31. Hayden and Villeneuve, "Astronomy in the Upper Palaeolithic?"

32. From *Hymn to Taurus*, quoted in Allen, *Star Names*, 392.

33. Andrews, *The Seven Sisters of the Pleaides*, 179ff; Allen, *Star Names*, 392.

34. Alfred Tennyson, "Locksley Hall," quoted in Allen, *Star Names*, 396.

35. "Origin of the Name Subaru."

36. Rappenglück, "The Pleiades in the 'Salle des Taureaux.'"

37. Quoted in Allen, *Star Names*, 407.

38. Aratus, *Phenomena*, 253.

39. Norris and Norris, "Why Are There Seven Sisters?"; see also Hamacher, *The First Astronomers*, 147ff.

40. Johnson, "Interpretations of the Pleiades," 293.

41. The six stars that are easily observable with the naked eye take their names from five of the seven mythological sisters (Alcyone, Merope, Electra, Maia, and Taygeta), while the sixth is called Atlas, actually the sisters' father in

the Greek legend. The seventh star, which is too close to Atlas to be seen with the naked eye, is called Pleione, the mother of the seven sisters. Under exceptional circumstances, and for people with great visual acuity, it is possible to discern more stars: Michael Mästlin (one of Kepler's teachers) counted and described fourteen in 1579 without a telescope. Hipparchus speaks of seven; the Barasana of Colombia, eight; the Peruvian Quechua people, ten, thirteen, or sixteen. An Aztec codex shows nine; an Australian Aboriginal bark, thirteen.

42. Norris and Norris, "Why Are There Seven Sisters?"; Norris and Norris, *Emu Dreaming*.

43. Eiseley, *The Immense Journey*, 50.

44. Since the solar year is not exactly 365 days and a quarter, even the Julian reform didn't entirely eliminate the drift between seasons and civil year. Another tweak was needed, and Pope Gregory XIII reformed the calendar again in 1582, eliminating leap years in all centuries not divisible by four hundred. To make up for the accumulated drift, October 5, 1582, became October 15. The Gregorian reform was adopted by Britain and its colonies only in 1752, when the loss of eleven days and the ensuing confusion over wages and payments caused riots in London.

45. Krupp, *Beyond the Blue Horizon*, 67–68.

46. Norris, "Dawes Review 5," 27. The Inuit named the following thirteen Moons after the dominant natural events at that time: Sun is possible; Sun gets higher; premature birth of seal pups; seal pups; tenting month; caribou calves; eggs; caribou hair sheds; caribou hair thickens; velvet peels from caribou antlers; winter starts; hearing (news from neighboring camps); great darkness. This last one is omitted when necessary to sync with the solar year, as it is a month of indefinite duration (MacDonald, *The Arctic Sky*, 194ff).

47. Mithen, *The Prehistory of the Mind*, 149.

48. Keith, "Whence Came the White Race?" Proponents of scientific racism and white supremacy, like the anthropologist Arthur Keith, saw the disappearance of Neanderthals as part of the same natural order that justified, in their eyes, the extermination of what they saw as inferior races.

49. Mathews, "Message-Sticks," 292–293.

50. Bird et al., "Early Human Settlement of Sahul."

51. Caveat: only about three hundred Neanderthal fossils have been discovered to date, a preciously small trove of evidence, given that millions of their kind must have walked the Earth.

CHAPTER 5: CELESTIAL CLOCKS

On constellation myths, see Ridpath, *Star Tales*. On the Antikythera mechanism, see de Solla Price, "Gears from the Greeks"; Freeth et al., "A Model of the Cosmos." On the Egyptian decans, see Neugebauer, "The Egyptian 'Decans'"; van der Waerden, "Babylonian Astronomy."

1. Mitchell, *Gilgamesh*, 162.

2. Ossendrijver, "Ancient Babylonian Astronomers."

3. A modern-day genetic explanation has been suggested by Ashrafian, "Ancient Genetics."

4. Bickel and Gautschy, "Eine Ramessidische Sonnenuhr."

5. The Egyptians recognized that there are stars that never set or rise, which we call "circumpolar"; they considered these blessed souls of "yonder people of whom it has been said: you have not died the death" (Krauss, "Egyptian Calendars and Astronomy," 133).

6. Krupp, *Beyond the Blue Horizon*, 220.

7. Quoted in Krauss, "Egyptian Calendars and Astronomy," 131.

8. Quoted in Duke, "Hipparchus' Coordinate System," 428.

9. Quoted in Burton, *The History of Mathematics*, 26.

10. The rationalists of the French Revolution attempted to reform the sexagesimal time-measuring tradition by introducing decimal time in 1794: ten hours a day, one hundred decimal minutes per hour, one hundred decimal seconds per decimal minute. Its mandatory use ended after just seven months.

11. Bedini and Maddison, "Mechanical Universe," 18.

12. Quoted in Bedini and Maddison, "Mechanical Universe," 18.

13. de Solla Price, *Science Since Babylon*, 28.

14. Despite his ingenuity, Hooke was considered little more than a servant by the haughty Royal Society fellows who employed him. At a time when doing science was a highbrow pastime for the affluent and otherwise idle gentleman, an "experimental philosopher" who was on the payroll didn't fit easily into the social landscape (Shapin, "Who Was Robert Hooke?").

15. Bennett, "Robert Hooke as Mechanic," 36. Hooke also perfected the air pump and used it to experiment on birds, fish, and himself to investigate what would happen to the subject inside an air-tight vessel when the air was pumped out (the animals died, and he became dizzy before stopping the experiment). His thirst for experimentation was not confined to the world of physics: his first report that he kept a dog alive while cutting away all the ribs and diaphragm of his poor victim was met with incredulity, something he resented to the point of repeating the gruesome vivisection with the Royal Society's "noble company" as witnesses (Shapin, "Who Was Robert Hooke?," 284).

16. Quoted in Shapin, "Who Was Robert Hooke?," 274.

17. Lawson Dick, *Aubrey's Brief Lives*, 164.

18. Quoted in Bedini and Maddison, "Mechanical Universe," 15.

19. Quoted in Bedini and Maddison, "Mechanical Universe," 25.

20. Quoted in Bedini and Maddison, "Mechanical Universe," 26.

21. de Solla Price, "Leonardo Da Vinci and the Clock."

22. de Solla Price, *Science Since Babylon*, 29.

23. de Solla Price, *Science Since Babylon*, 12.

24. Quoted in Frank, *About Time*, 85.

25. Quoted in Bedini and Maddison, "Mechanical Universe," 20.

26. All the fascinating details of the finding are reported in Throckmorton, *Shipwrecks and Archaeology*, chap. 4.

27. de Solla Price, "The Prehistory of the Clock," 157.

28. Cicero, *Tuscalan Disputation*, I, 63, quoted in de Solla Price, "Gears from the Greeks," 57.

29. Equation of Time apart, the maximum precision of a sundial is about a minute, owing to the lack of sharpness of the gnomon's shade due to the finite size of the Sun's disc.

30. Quoted in Bedini, "Along Came a Spider, Part 2," 7.

31. Quoted in Turner, "Spiders in the Crosshairs," 10.

32. Quoted in Turner, "Spiders in the Crosshairs," 12. Seeking to mass-produce spider silk, in the early 2000s a biotech start-up successfully implanted the silk-producing gene into mountain goats, with the aim of extracting the valuable thread from the goats' milk. It went bankrupt in 2009 (Levy, "The Race to Put Silk").

33. Charlot et al., "The Third Realization of the International Celestial Reference Frame."

34. Time reckoning is primitive on Caligo—beyond the alternation of day and night and the yearly cycles of nature, its inhabitants have no other natural means (and perhaps no need) to measure time. The People of Foucault take matters to the opposite extreme, becoming obsessed by the pendulum and thereby accidentally discovering the rotation of the Earth. The plane of oscillation of a pendulum remains constant as the Earth rotates under it, which French physicist Léon Foucault demonstrated in 1851 with a pendulum sixty-seven meters long hanging from the dome of the Panthéon in Paris, where it can still be seen today.

CHAPTER 6: TRIPLE BRONZE AND OAK

On the figure of Tupaia, see Salmond, "Tupaia, the Navigator-Priest"; Druett, *Tupaia*. On the Longitude Prize, apart from the excellent account of Sobel, *Longitude*, see also Perkins, "Edmond Halley, Isaac Newton." The first-person accounts of Cook, *Captain Cook's Journal*, and Banks, *Journal*, are fascinating and revealing about the nature of their interactions with the people they encountered. In relation to Cook's navigation skills and his transit-of-Venus expedition, see Deacon and Deacon, "Captain Cook as a Navigator"; Woolley, "Captain Cook"; Beaglehole, "On the Character." On Maskelyne, see Howse, *Nevil Maskelyne*. On traditional Polynesian navigation, see Low, *Hawaiki Rising*, and Lewis, *We, the Navigators*. On the making and interpretation of Tupaia's Map, see Finney,

"Nautical Cartography"; Eckstein and Schwarz, "The Making of Tupaia's Map"; Di Piazza and Pearthree, "A New Reading." On the establishment of the Prime Meridian, see Howse, *Greenwich Time.*

1. Horace, *The Odes*, Book I:III.

2. It is possible that the very names of Europe and Asia derive from the Phoenician words for "the region of the setting Sun" (*Ereb*) and "sunrise" (*aṣū*), respectively.

3. Callimachus, *Iambus* I, 52, quoted in Kirk, Raven, and Schofield, *The Presocratic Philosophers*, 84. Thales was so engrossed in the stars that, according to Plato, he once fell into a well while walking with his eyes to the sky.

4. Homer, *The Odyssey*, Book V, translated in Boitani, "Poetry of the Stars," 289. Homer, incorrectly, singles out the Great Bear as the only constellation that never dips under the horizon ("she alone is denied a plunge"). In reality, other constellations, like the Little Bear, or Cassiopeia, are also circumpolar as seen from a Mediterranean location, now and at the time when the myth is set. Perhaps Homer simply takes the Great Bear to represent all of the constellations that can be used as wayfinding aids all year round.

5. Bryant, "Hymn to the North Star."

6. Shakespeare, *Julius Caesar*, act 3, scene 1.

7. Allen, *Star Names*, 454.

8. Even today Polaris is not precisely fixed, for it is still one degree away (twice the diameter of the full Moon) from the actual north celestial pole. Hence it executes a small circle around it.

9. Polaris, contrary to what is sometimes believed, is not an especially prominent star, but finding it is simple: look for the two stars at the end of the Big Dipper's cup (or, if you choose to see a Big Wagon, look for the rear side of the chariot), follow their direction for about six times their separation, and you'll get to Polaris.

10. "Aotearoa" was the name that the Māori gave to the North Island, and today it designates New Zealand in the Māori language.

11. Rodman and Stokes, "The Sacred Calabash," claimed that Hawaiian seafarers used a gourd filled with water as a mirror to measure the altitude of Polaris so as to determine the latitude when sailing back from Tahiti. This, however, is disputed (Richards-Jones, "The Myth of the Sacred Calabash"). Another interpretation is that the sacred calabash had a magical rather than navigational function: blocking out all but the favorable winds (Makemson, *The Morning Star Rises*, 147).

12. Andía y Varela, "An Account of Traditional," 2:282. Joseph Banks (*Journal*, 159ff) describes the voyaging canoes of the Society Islands, called *pahi*, in detail: one of them was fifty-one feet long and three feet at its widest. Two canoes were fastened together for stability and equipped with one or two masts, fitted

with triangular sails. A pennant made of feathers ran the length of the mast, twenty-five feet, spectacular when blown out by the wind.

13. Lewis et al., "Voyaging Stars," 133.

14. A blind Tongan master navigator was said to have saved a lost flotilla by asking his son to tell him the position of some stars, and after plunging his hand in the water, he correctly indicated the position of Fiji, just beyond the horizon (Lewis et al., "Voyaging Stars").

15. Lewis et al., "Voyaging Stars," 141.

16. Andía y Varela, "An Account of Traditional," 2:284.

17. Quoted in Turnbull, "(En)-countering Knowledge Traditions," 68.

18. For example, observations of lunar eclipses he made while anchored in the Caribbean yielded enormous errors of longitude of twenty-two degrees (in 1494) and thirty-eight degrees (in 1504), corresponding to thousands of miles (Randles, "Portuguese and Spanish Attempts," 236).

19. Costa Canas, "The Astronomical Navigation in Portugal"; Laguarda Trìas, "Las longitudes geográficas," 172–173.

20. The Longitude Prize was set up after William Whiston (Newton's successor as Lucasian professor) and mathematician Humphrey Ditton proposed in all earnestness to anchor a fleet of signal boats at six-hundred-mile intervals across the oceans, firing guns at midnight so that mariners could set their clocks by them (Perkins, "Edmond Halley, Isaac Newton," 128; Sobel, *Longitude*, 48).

21. Gingerich, "Cranks and Opportunists," 135.

22. Quoted in Sobel, *Longitude*, 52.

23. For the story of Galileo's attempt to solve the longitude problem, see de Grijs, "European Longitude Prizes." While it never worked at sea, on terra firma Galileo's method eventually delivered superb results, but it required that skilled observers timed simultaneously the eclipses of Jupiter's moons at different locations. When astronomer Gian Domenico Cassini presented Louis XIV with a new, more accurate map of France he had produced in 1693 using Galileo's method, the French king reportedly bemoaned that he was losing more territory to his astronomers than to his enemies (Van Helden, "Longitude and the Satellites of Jupiter").

24. The king himself had bankrolled the expedition with £4,000 of his own money, tickled in his pride by the Royal Society's shrewdly worded plea: "That the British Nation has been justly celebrated in the learned World, for their Knowledge of Astronomy, to which they are Inferior to no nation upon Earth ancient or Modern; and it would cast Dishonour upon them should they neglect to have correct observations made of this Important Phaenomenon" (Carter, "The Royal Society," 251).

25. Hamilton, *Captain James Cook*, n.p.

26. Cook, *Captain Cook's Journal*, 317 (August 23, 1770).

27. Banks, *Journal*, 73 (April 13, 1769). Banks considered the *Endeavour* voyage the ultimate grand tour and spent the equivalent of £2 million in current money in salaries, equipment, and supplies for his round-the-world adventure on Cook's ship. As he was always one to travel in style, his entourage aboard the *Endeavour* consisted of a botanist, a draughtsman, two artists, and four servants (plus two greyhounds); delicacies for his table included fine ale, port, wine, apple pies, and Cheshire cheese.

28. Cook, *Captain Cook's Journal*, 76 (June 3, 1769).

29. "Secret Instructions to Captain Cook," 1.

30. Cook, *Captain Cook's Journal*, 87 (July 13, 1769).

31. Banks, *Journal*, 109 (July 12, 1769). In 1774, Cook brought back with him from his second voyage a young man from Tupaia's same island, called Mai, whom Banks took under his protection, as he had intended to do with Tupaia. Banks introduced him to London society, and Mai became a celebrity: the exotic visitor was often received at court by King George and Queen Charlotte, invited to the opera and Royal Society dinners, and included on fox hunting expeditions; he learned to play chess and backgammon and was the subject of a play performed at the Theatre Royal. After the initial excitement, Banks became bored of his charge, so much so that a scholar wrote that Banks seemed "to keep him as an Object of Curiosity, to observe the Workings of an untutored, unenlighted Mind" (quoted in Salmond, *The Trial of the Cannibal Dog*, 298; see also Shapin, "Keep Him as a Curiosity").

32. Banks, *Journal*, 110 (July 13, 1769).

33. Cook, *Captain Cook's Journal*, 118 (July 31, 1769).

34. Quoted in Williams, "Tupaia, Polynesian Warrior," 40.

35. Marra, *Journal*, 217.

36. Banks, *Journal*, 162 (Chapter VII).

37. Banks, *Journal*, 124 (August 9, 1769).

38. Cook, *Captain Cook's Journal*, 121 (August 15, 1769).

39. All that is known about the process by which Tupaia's Map was created comes from the secondhand description of Johann Forster, the naturalist who replaced Banks on Cook's second voyage, according to whom "Tupaia had perceived the meaning and use of charts, he gave directions for making one according to his account, and always pointed to the part of the heavens, where each isle was situated, mentioning at the same time that it was either larger or smaller than [Tahiti], and like-wise whether it was high or low, whether it was peopled or not, adding now and then some curious account relative to some of them" (quoted in Finney, "Nautical Cartography," 447).

40. Turnbull, "(En)-countering Knowledge Traditions," 69.

41. But Tupaia's Map is not the only surviving testimony of his skills. Tupaia also took up painting, crafting some astonishing watercolors—among them, Australian natives paddling in their bark canoes, musicians in Tahiti playing the

flute with their nostrils, the Chief Mourner in his colorful traditional dress, and a delightful caricature of Banks, in his frock coat, cocked hat, and buckled shoes, warily trading a handkerchief for a huge red crayfish off a local dressed only in bark cloth (Smith, "Tupaia's Sketchbook").

42. Quoted in O'Brian, *Joseph Banks*, 109.

43. Cook, *Captain Cook's Journal*, 229 (March 1770).

44. Quoted in Williams, "Tupaia," 40.

45. Quoted in Salmond, *The Trial of the Cannibal Dog*, 188.

46. Hornsby, "LIII. The Quantity of the Sun's Parallax."

47. Quoted in Howse and Hutchinson, *Clocks and Watches*, 194.

48. James Cook undertook one last voyage around the world, during which he died in Hawaii in February 1779, killed in a skirmish with locals he had done much to provoke. His body was dismembered, his flesh roasted in the traditional ritual for vanquished chiefs. His legacy is today being questioned by Polynesian people. The K-1 stopped working within two months of Cook's death.

49. Brumfiel, "U.S. Navy Brings Back Navigation."

50. Quoted in Low, "Polynesian Navigation."

51. Lodestone is a rare, naturally magnetized mineral that attracts iron and can be used as a magnetic compass, hence its name (meaning "course stone," from Old English *lode*, for "way"). Lodestone was first reported in China in the fourth century BC, although archeological findings suggest its magnetic properties may have been recognized a thousand years earlier (Carlson, "Lodestone Compass"). Lodestone minerals are often mistaken for meteorites, given their black appearance.

CHAPTER 7: FROM BEAUTY, ORDER

On the figures of Brahe and Kepler, see the double biography of Ferguson, *The Nobleman*. On Copernicus's biography, see O'Connor and Roberts, "Nicolaus Copernicus." On the relationship between Galileo and Kepler, see Peterson, *Galileo's Muse*, chap. 8. On the figure of Newton, see Westfall, *Never at Rest*; More, *Isaac Newton*. On the role of women astronomers at Harvard Observatory, see Sobel, *The Glass Universe*. On the aesthetics of space telescope pictures, see Kessler, *Picturing the Cosmos*; Castello, "The Art Behind."

1. Prescod-Weinstein et al., "The James Webb," argues that Webb was co-responsible for policy discriminating against homosexuals, a period that became known as "the lavender scare"; a report by the NASA chief historian found no evidence involving Webb directly with the lavender scare (Odom, "NASA Historical Investigation").

2. The combination of a concave (thinner at the center) and a convex (fatter at the center) lens to make "both distant and near objects larger than they would

otherwise appear" is already mentioned in a 1589 book by the Neapolitan poly-math Giambattista della Porta, who, however, never put the idea into practice. When Lippershey submitted his invention to the States-General of the Netherlands seeking to patent it, two other glassmakers, James Metius and Zacharias Jansen, advanced futile claims of having independently come up with the same idea (King, *The History of the Telescope*, 30–32).

3. Quoted in Rosen, "Galileo and the Telescope," 180.

4. Psalm 104:5 (King James Version).

5. Qur'an 21:33 (Yusuf Ali translation).

6. Quoted in Hoskin, *The Cambridge Illustrated History*, 92.

7. Quoted in Hoskin, *The Cambridge Illustrated History*, 97.

8. Copernicus, *De revolutionibus*, 528.

9. Of the fabulous Uraniborg—its fountain, its gardens, its observatory, its paper mill, its workshops, its visitors' quarters—nothing remains. After Brahe went into exile in 1597 because of a dispute with the new king he had done much to provoke, the inhabitants of Hven celebrated their restored freedom from their loathed lord by quarrying his magnificent castle, the symbol of all their forced toiling. Within a century, oblivious sheep grazed once again on the remains of what had been the world's first and best astronomical center.

10. Quoted in Ferguson, *The Nobleman*, 183.

11. Johannes Kepler, quoted in Ferguson, *The Nobleman*, 189.

12. "Tycho Brahe Wasn't Poisoned After All."

13. Ferguson, *The Nobleman*, 279.

14. Ferguson, *The Nobleman*, 320.

15. Galilei, *The Sidereal Messenger*, 44–45.

16. Quoted in Sobel, *Galileo's Daughter*, 35.

17. Galilei, *The Sidereal Messenger*, 69–70.

18. Quoted in Partridge and Whitaker, "Galileo's Work," 411.

19. The Saturn discovery Galileo similarly scrambled as "smaismrmilme-poetaleumibunenugttaurias," which stood for "Altissimum planetam tergemi-num observavi" (remembering that *u* and *v* are the same letter in Latin), or "I have observed the highest of the planets, triple-bodied." The coded announcement reached Kepler at the court of Rudolph II, and both the astronomer and the emperor "went crazy trying to understand the cipher" (Marcus and Findlen, "Deciphering Galileo," 965). Three months later, Kepler decrypted it as "Salve umbistineum geminatum Martia proles," or "Hail, satellites, children of Mars" (ibid., 966), believing it to be the announcement that Mars had two moons—something that was actually discovered by the astronomer Asaph Hall in 1877 (Bodifée, "La découverte").

20. Joshua 10:13 (King James Version).

21. Quoted in Rosen, "Galileo and Kepler," 264.

22. Abbot, "Discovery of Galileo's Long-Lost Letter."

23. Wisan, "Galileo and God's Creation," 479.

24. D'Amico, *Giordano Bruno*, 385.

25. Quoted in Wisan, "Galileo and God's Creation," 482.

26. The famous story, according to which Galileo muttered, "It moves nevertheless," is regarded as a legend. However, a portrait of Galileo painted just thirteen years after the facts shows precisely these legendary words (Eve, "Galileo and Scientific History").

27. Gattei, *On the Life of Galileo*, 47.

28. Newton, "Two Letters from Humphrey Newton." Isaac Newton's assistant was called Humphrey Newton (no relation), a young man from the same grammar school in Grantham (Westfall, *Never at Rest*, 343).

29. Macomber, "Glimpses of the Human Side," 304.

30. Toward the end of his life, Newton told the story of the apple to at least four different people. This is the version related by Richard Conduitt, Newton's half niece's husband, in a memorandum on Newton's life written in 1727–1728 (Westfall, *Never at Rest*, 154).

31. Westfall, *Never at Rest*, 403.

32. Westfall, *Never at Rest*, 405.

33. Westfall, *Never at Rest*, 406.

34. Westfall, *Never at Rest*, 460.

35. Byron, *Lord Byron*, Canto 10.

36. Newton, "Two Letters from Humphrey Newton," 1.

37. Quoted in Westfall, *Never at Rest*, 470.

38. Quoted in Westfall, *Never at Rest*, 473.

39. Seneca, *Naturales quæstiones*, 7, XXVII, 2.

40. Quoted in Broughton, "The First Predicted Return," 125.

41. Seneca, *Naturales quæstiones*, 7, XXVI, 1.

42. Letter from Tommaso Campanella to Galileo Galilei, August 5, 1632; quoted in Lipking, *What Galileo Saw*, 11. The Italian original can be found at Archivio Tommaso Campanella, letter N. 92, https://www.iliesi.cnr.it/ATC /testi.php?tp=1&iop=Lettere&pg=92 (accessed April 23 2023).

43. Galilei, *The Sidereal Messenger*, 42–43.

44. Descartes, *The World*, chap. 6, para. 26.

45. Hoskin, *The Cambridge Illustrated History*, 211.

46. Quoted in Hoskin, "Newton, Providence," 82–84.

47. Wright, *An Original Theory*, 76.

48. Herschel, "Address Delivered," 453.

49. Bessel, "A Letter from Prof. Bessel," 71.

50. Hafez, "Abd Al-Rahman Al-Ṣūfi," 351.

51. Quoted in Editors of Encyclopedia Britannica, "Simon Marius."

52. Quoted in Sobel, *The Glass Universe*, 13.

53. Leavitt, "1777 Variables," 107.

54. Among the proponents of the latter theory was the newly appointed director of Harvard Observatory, Harlow Shapley—who ironically had perfected the calibration of Leavitt's law. So it was only fitting that when the letter from Hubble arrived, triumphantly announcing his definitive discovery ("Dear Shapley, you will be interested to hear that I have found a Cepheid variable in the Andromeda Nebula," it began), Cecilia Payne, a Cambridge graduate who had taken over Leavitt's old desk, found herself witnessing Shapley's reaction to it: "Here is the letter that has destroyed my universe," he declared (Sobel, *The Glass Universe*, 204).

55. Quoted in Hoskin, "Newton, Providence," 96.

56. Newton, *Principia*, quoted in Snobelen, "The Myth of the Clockwork Universe," 160.

57. Quoted in Snobelen, "The Myth of the Clockwork Universe," 165.

58. Galileo's letter to G. Gallanzoni of July 16, 1611, quoted in Olschki, "Galileo's Philosophy," 354.

59. Quoted in Lubbock, *The Herschel Chronicle*, 310–311.

60. It turns out that Laplace's conclusion wasn't actually correct: Poincaré introduced at the end of the nineteenth century the notion of "dynamical chaos," the idea that tiny differences in initial conditions can lead to drastically different final outcomes for certain physical systems—the solar system being a case in point: chance perturbations get amplified over time.

61. Kornei, "How Many of the Moon's Craters."

CHAPTER 8: THE DEMON UNLEASHED

On Laplace's demon (and many other kinds), see Canales, *Bedeviled*. On the figure of Babbage, see Snyder, *The Philosophical Breakfast Club*; Swade, *The Difference Engine*. On how the Victorian era might have been changed had Babbage succeeded, see the counterfactual novel by Gibson and Sterling, *The Difference Engine*. On the figure of Laplace, see Gillispie, *Pierre-Simon Laplace*; on his contribution to statistics, see Stiegler, *The History of Statistics*. On the links between a mechanistic conception of the cosmos, geology, and Darwin's theory of evolution, see Eiseley, *The Firmament of Time*.

1. Quoted in Newcomb, *Navaho Folk Tales*, 83.

2. For in-depth research about the origin of the phrase, see "Lies, Damned Lies and Statistics."

3. Laplace, *A Philosophical Essay*, 2.

4. Quoted in Burton, *The History of Mathematics*, 433.

5. Lovering, "The Méchanique Céleste," 186.

6. Quoted in Simmons, *Calculus Gems*, 161.

7. Quoted in Hoskin, *The Cambridge Illustrated History*, 187.

8. Quoted in Foderà Serio, Manara, and Sicoli, "Giuseppe Piazzi," 21.

9. Quoted in Foderà Serio, Manara, and Sicoli, "Giuseppe Piazzi," 20–21.

10. Teets and Whitehead, "The Discovery of Ceres," 84.

11. Quoted in Dunnington, *Carl Friedrich Gauss*, 44.

12. Stiegler, *The History of Statistics*, 158.

13. Jahoda, "Quetelet and the Emergence," 1.

14. Dickens writing in the magazine *Household Works*, quoted in Porter, "From Quetelet to Maxwell," 354.

15. Quoted in Stiegler, *The History of Statistics*, 171.

16. Quoted in Jahoda, "Quetelet and the Emergence," 3.

17. Quoted in Jahoda, "Quetelet and the Emergence," 3.

18. Laplace, *A Philosophical Essay*, 4.

19. Fourier, *Oeuvres de Fourier*, Discours préliminaire, XIV.

20. Locke, *The Conduct*, 7–8.

21. Quoted in Burton, *The History of Mathematics*, 311.

22. Lardner, "Babbage's Calculating Engine," 274.

23. Quoted in Swade, *The Difference Engine*, 13.

24. Babbage, *Passages from the Life*, 42.

25. Babbage, *Passages from the Life*, 34.

26. "Fellows Deceased: Charles Babbage," 107.

27. Dreyer, *History*, 6.

28. Quoted in Buxton and Hyman, *Memoir of the Life*, 46. This is how Babbage recalled the incident in November 1839, although in earlier versions of the story he wasn't sure whether Herschel or he had proposed a mechanical solution. For details, see Collier, "The Little Engines," chap. 2.

29. Quoted in Snyder, *The Philosophical Breakfast Club*, 92.

30. Quoted in Wilkes, "Herschel, Peacock," note 29.

31. Quoted in Schaffer, "Babbage's Intelligence," 207.

32. Swade, *The Difference Engine*.

33. In 1834, Whewell wrote, "There was no general term by which these gentlemen [members of the British Association for the Advancement of Science] could describe themselves with reference to their pursuits. *Philosophers* was felt to be too wide and too lofty a term . . . ; *savans* was rather assuming, beside being French instead of English; some ingenious gentleman proposed that, by analogy with *artist*, they might form *scientist* . . . but this was not generally palatable" (Whewell, "On the Connexion," 59). The "ingenious gentleman" was Whewell himself; he had put forward "scientist" after the poet and philosopher Samuel Coleridge expressed his dismay at "prostituting the name of Philosopher . . . to every Fellow who has made a lucky experiment" (Schaffer, "Scientific Discoveries," 409–410). "Scientman," proposed in 1661, never caught on. For an account of Babbage's soirées, see Ticknor, *Life, Letters*, 144;

Schweber, "The Origin," 286; Somerville, *Personal Recollections*, 140; Goldman, *Victorians and Numbers*, 45.

34. Seven letters from Darwin to Babbage have survived for the period March 14, 1837, to May 26, 1840, in which Darwin either sends his regrets ("I am very much obliged for your kind invitation for tomorrow evening, and whether for beauty or for shells, I should have had great pleasure in accepting if I had not happened to be engaged." "Letter no. 349," Darwin Correspondence Project, accessed on February 8, 2023, https://www.darwinproject.ac.uk/letter/?docId =letters/DCP-LETT-349.xml) or asks for permission to bring guests along ("My sister is at present staying with us, will you be so kind as to allow me to bring her to your party on Saturday, that she may see *the World*." "Letter no. 479," Darwin Correspondence Project, accessed on February 8, 2023, https://www.darwinproject .ac.uk/letter/?docId=letters/DCP-LETT-479.xml).

35. Babbage, *The Ninth Bridgewater Treatise*, 44–46.

36. Darwin, *Charles Darwin's Notebooks*, Red Notebook, 127–130. The notebook entry is undated but is likely to have been written around mid-March 1837 (Darwin, *Charles Darwin's Notebooks*, introduction, 18). According to his correspondence (Darwin, "To Caroline Darwin, 27 February 1837"), Darwin attended his first party at Babbage's on March 4, 1837.

37. "Charles Babbage," 28.

38. Snyder, *The Philosophical Breakfast Club*, 354.

39. Quoted in Woodhouse, "Eugenics," 129.

40. "Documenting the Numbers of Victims."

41. Quoted in Canales, "Exit the Frog," 178.

42. Quoted in Mollon and Perkins, "Errors of Judgement," 102.

43. Canales, "Exit the Frog," 181–186.

44. Quoted in Mumford, *Technics and Civilization*, 146.

45. Poincaré, *The Value of Science*, 88.

46. Quoted in McPhee, *Annals*, 73.

47. The importance of catastrophic events, such as asteroid or cometary impacts, in shaping geology and biology on Earth is today being reevaluated in light of recent discoveries on the frequency of major collisions with extraterrestrial objects (Sweatman, "The Younger Dryas Impact Hypothesis").

48. Herschel to Lyell, February 20, 1836, quoted in note 5 in Darwin, "To Caroline Darwin."

49. Darwin, "To Caroline Darwin."

50. Darwin, "To W. D. Fox."

51. Joly, *The Birth-Time of the World*, 3.

52. Quoted in Becker, "Halley on the Age of the Ocean," 461.

53. Stacey, "Kelvin's Age of the Earth"; Lord Kelvin quoted in Darwin, "Radio-Activity and the Age of the Sun," 496.

54. Darwin, "Radio-Activity and the Age of the Sun."

55. Galilei, *The Assayer*, 238.

56. Quoted in Rosenthal-Schneider, Braun, and Miller, *Reality and Scientific Truth*, 74.

57. Dirac, "Quantised Singularities," 60.

58. Wigner, "The Unreasonable Effectiveness."

59. Anderson, "The End of Theory."

60. Poincaré, *The Value of Science*, 84.

CHAPTER 9: A MIRROR TO OURSELVES

On the history of astrology, see Campion, *A History of Western Astrology*. On mythology tied to the Sun, see Frazer, *The Worship of Nature*; in connection to the Moon, see Cashford, *The Moon*. On the cult of Mithra in Rome, see Adrych et al., "Reconstructions."

1. Frazer, *The Worship of Nature*, 465ff.

2. Frazer, *The Worship of Nature*, 491. For the etymology, see Curtius and Windisch, *Grundzüge*, 399–400.

3. St. Augustine, "Sermon 190," 24.

4. Barnes, "The First Christmas Tree."

5. Berger, *Palace of the Sun*, 2–3.

6. Jung, "The Archetypes," paragraph 87.

7. Quoted in Hirsch, "Coming Out into the Light," 13.

8. Quoted in Cashford, *The Moon*, 65.

9. Quoted in Cashford, *The Moon*, 65.

10. Gospel of Matthew 12:40 (King James Version).

11. Frazer, *The Worship of Nature*, 561.

12. Anaxagoras, fragment 18.

13. Quoted in Cashford, *The Moon*, 179.

14. Revelation 12:1 (King James Version).

15. Jung, "The Archetypes," paragraph 6.

16. Cashford, *The Moon*, 67.

17. Quoted in Cashford, *The Moon*, 158.

18. Emerson, *Nature*, 9–10.

19. Jung, *Jung on Astrology*, 23–24.

20. Ripat, "Expelling Misconceptions," 117n9.

21. Campion, *Astrology and Cosmology*, 95.

22. Tacitus, *Historiæ*, I, 22.

23. Walker, *Spiritual and Demonic Magic*, 206–208.

24. Quoted in Kollerstrom, "Galileo's Astrology," 427. The Latin original can be found in Galilei, "Astrologica Nonnulla," 119–120. See also Favaro, "Galileo, Astrologer."

25. Campion, *Astrology and Cosmology*, 169.

26. As related by Conduitt, cited in Whiteside, "Isaac Newton," 58.

27. Related by Conduitt, in Whiteside, Hoskin, and Prag, *The Mathematical Papers*, 1:15–19.

28. Tacitus, *Historiæ*, I, 22.

29. Ciardi and Williams, *How Does a Poem Mean?*, 9.

30. Quoted in Copeland, "Sources of the Seven-Day Week," 175.

31. Quoted in Lista, *La stella d'Italia*, 42–43.

32. Mendelson, "Baedeker's Universe."

33. *Handbook for Visitors to Paris*, 37.

34. The origin of the Walk of Fame is lost in myth. The *Los Angeles Magazine* (Rozbrook, "The Real Mr. Hollywood," 19–20) claims that it was the brainchild of Hollywood businessman and theater manager Harry Sugarman, inspired in 1953 by the drinks menu of one of his bars, which featured celebrities' headshots framed by golden stars. The number of stars dedicated to date, twenty-seven hundred, is approximately the same as the number of real stars one can hope to see in a very dark sky, and so many more than those actually visible from light-polluted Los Angeles.

35. "Early New York City Police 'Badges.'"

CHAPTER 10: TO REBEHOLD THE STARS

1. Rilke, *Duino Elegies*, Elegy 1.

2. Schechner, *Comets*, 151–152.

3. "Double Negative. 1969."

4. Gillieron, "La Tour Eiffel de l'espace."

5. Notarbartolo, "Some Proposals for Art Objects," 140.

6. O'Connor, "Apollo 17 Coverage."

7. Tuckner, "One Man's Mission."

8. Brown, "Pi in the Sky."

9. Cohen, *Villes éteintes*.

10. Extravehicular activity on the Moon by the crews of Apollo 11 to 17 totaled eighty hours and thirty-two minutes ("Extravehicular Activity").

11. Lawrence, *Losing the Sky*.

12. Rawls et al., "Satellite Constellation."

13. Lawrence et al., "The Case for Space Environmentalism," 5; McDowell, "The Low Earth Orbit."

14. Responding to the astronomers' concerns, SpaceX reoriented its satellites' solar panels to minimize reflection and added a deployable solar shade to new ones to reduce glare. These measures were only partially successful (Mallama, "The Brightness of VisorSat-Design"; Horiuchi et al., "Multicolor and Multi-spot Observations").

15. The JWST, orbiting as it does 1.5 million kilometers away, is mercifully being spared.

16. Venkatesan et al., "The Impact of Satellite Constellations," 1043.

17. Emerson, *Nature*, 9.

18. Turner, "Chinese Scientists."

19. Kessler and Cour-Palais, "Collision Frequency." Today's models for the collision frequency for the case of forty thousand orbiting satellites show that the number of disabling collisions would likely exceed the speed at which satellites are being replenished (Lawrence et al., "The Case for Space Environmentalism").

20. Mallick and Rajagopalan, "If Space Is 'the Province of Mankind,'" 7.

21. Roulette, "How Much Does a Ticket to Space on New Shepard Cost?"

22. Shatner, "William Shatner."

23. Yaden et al., "The Overview Effect."

24. Quoted in Weibel, "The Overview Effect," 11.

25. Worden, *Hello Earth*, 27–30.

26. Quoted in Homans, "The Lives They Lived."

27. Quoted in Weibel, "The Overview Effect," 20.

28. Sanders et al., "A Meta-analysis of Biological Impacts."

29. Van Doren et al., "High-Intensity Urban Light Installation."

30. Berry, Booth, and Limpus, "Artificial Lighting and Disrupted Sea-Finding Behaviour."

31. Knop et al., "Artificial Light at Night."

32. Quoted in Lister, "Seeing the Northern Lights."

33. Darwin, *The Autobiography*, 139.

34. McPhee, *Annals*, 89.

35. Einstein, "The 1932 Disarmament Conference."

36. Powers, *The Overstory*, 482.

37. Harvey, "UN Says."

38. Carrington, "Flying Insect Numbers."

39. Bodio, "A Feathered Tempest."

40. Eiseley, *The Firmament of Time*, 203.

41. "About Blue Origin."

42. Heath, "How Elon Musk."

43. Sagan, *Pale Blue Dot*, 312. Advocates of "longtermism" consider that the highest purpose of human existence is to colonize the galaxy and provide "net-positive lives" to the largest possible number of sentient beings, whether biological, cybernetic, or simulated in a computer. In pursuit of the goal of fulfilling humanity's "potential," longtermists contend that shorter-term human sufferance (lasting for a few millennia) and even the destruction of the whole ecosystem are a price worth paying. Space expansionism and the creation of an all-powerful, benevolent superintelligence are the central planks of their plan (Torres, "Against Longtermism").

44. Theoretical physicists have put forward speculative ideas for circumventing the speed-of-light barrier, such as the so-called Alcubierre drive, in which spacetime is warped around the hypothetical spaceship, thus achieving faster-than-light displacement (Alcubierre, "Letter to the Editor"; Van Den Broek, "A 'Warp Drive'"). This *Star Trek*–inspired engine, requiring an amount of "negative energy" equivalent to the mass of the Sun to create a warp bubble with a one-hundred-meter radius, is just as fantastical as the imaginary dilithium reactor of the starship *Enterprise* (Sternbach and Okuda, *Star Trek: The Next Generation*).

45. Not easily discouraged, the techno-optimists retort with the terraforming fantasy—to make Mars hospitable by reengineering the entire biosphere. This is just that—a fantasy.

46. Mumford, *The Myth of the Machine*, 11–12.

47. Osnos, "Doomsday Prep."

48. Salk, "Are We Being Good Ancestors?"; "Could You Patent the Sun?"

49. Shakespeare, *Julius Caesar*, act 1, scene 2.

50. Eiseley, *The Immense Journey*, 10.

EPILOGUE

1. On the creation of the Voyager's record, see Sagan, *Murmurs of the Earth*. Some of the golden record's contents are available from goldenrecord.org.

2. Carter, "Voyager Spacecraft Statement."

BIBLIOGRAPHY

Abbot, Alison. "Discovery of Galileo's Long-Lost Letter Shows He Edited His Heretical Ideas to Fool the Inquisition." *Nature* 561 (2018): 441–442. https://doi.org/10.1038/d41586-018-06769-4.

Abbot, C. G. "The Habitability of Venus, Mars, and Other Worlds." Annual Report of the Board of Regents of the Smithsonian Institution for 1920. Washington, DC: Government Printing Office, 1922.

"About Blue Origin." Blue Origin. Accessed February 11, 2023. https://www.blueorigin.com/about-blue.

Adrych, Philippa, Robert Bracey, Dominic Dalglish, Stefanie Lenk, and Rachel Wood. "Reconstructions: Mithras in Rome." In *Images of Mithra*, edited by Jas Elsner. Oxford: Oxford University Press, 2017.

Alcubierre, Miguel. "Letter to the Editor: The Warp Drive: Hyper-Fast Travel Within General Relativity." *Classical and Quantum Gravity* 11 (1994): L73–L77. https://doi.org/10.1088/0264-9381/11/5/001.

Alighieri, Dante. *The Divine Comedy*. Translated by Henry Longfellow (Boston: Ticknor and Fields, 1867). Originally published in 1321.

Allen, Richard H. *Star Names: Their Lore and Meaning*. New York: Dover, 1963. Originally published in 1899.

Anaxagoras. Fragment 18. In David Warmflash. "An Ancient Greek Philosopher Was Exiled for Claiming the Moon Was a Rock, Not a God." *Smithsonian Magazine*. June 20, 2019. https://www.smithsonianmag.com/science-nature/ancient-greek-philosopher-was-exiled-claiming-moon-was-rock-not-god-180972447.

Anderson, Chris. "The End of Theory: The Data Deluge Makes the Scientific Method Obsolete." *Wired*. June 23, 2008. https://www.wired.com/2008/06/pb-theory.

Andía y Varela, Ignacio. "An Account of Traditional Tahitian Navigation (Journal 1774)." In *The Quest and Occupation of Tahiti by Emissaries of Spain During the Years 1772–6*, edited by B. G. Corney. 3 vols. London: Hakluyt Society, 1913–1919. Accessed February 4, 2023. https://archive.hokulea.com/ike/hookele/ancient_tahitian_navigation.html.

Andrews, M. *The Seven Sisters of the Pleiades: Stories from Around the World*. North Geelong, Australia: Spinifex Press, 2004.

Antonello, Elio. "The Palaeolithic Sky." In *The Light, the Stones and the Sacred: Proceedings of the XVth Italian Society of Archaeoastronomy Congress*, edited by Andrea Orlando. Cham: Springer, 2017. https://doi.org/10.1007/978-3-319-54487-8.

Aratus. *Phenomena*. In *Callimachus: Hymns and Epigrams, Lycophron and Aratus*. Translated by A. W. Mair and G. R. Loeb. Classical Library 129. London: William Heinemann, 1921.

Arianrhod, Robyn. *Seduced by Logic: Émilie du Châtelet, Mary Somerville and the Newtonian Revolution*. Oxford: Oxford University Press, 2012.

Ashrafian, H. "Ancient Genetics—Was Gilgamesh a Mosaic?" *Genetics in Medicine* 10, no. 11 (2008): 843.

Ashworth, William J. "The Calculating Eye: Baily, Herschel, Babbage and the Business of Astronomy." *British Journal for the History of Science* 27 (1994): 409–441. https://doi.org/10.1017/s0007087400032428.

Asimov, Isaac. *Nightfall and Other Stories*. Garden City, NY: Doubleday & Company, 1969. Originally published in 1941.

Babbage, C. *Passages from the Life of a Philosopher*. London: Longman, Green, Longman, Roberts, & Green, 1864. https://books.google.it/books?id=Fa1JAAAAMAAJ.

Babbage, Charles. *The Ninth Bridgewater Treatise*. 2nd ed. London: John Murray, 1838. http://darwin-online.org.uk/converted/Ancillary/1838_Bridgewater_A25/1838_Bridgewater_A25.html.

Bailey, Regina. "Extremophiles—Extreme Organisms." *ThoughtCo*. Last modified April 7, 2020. https://www.thoughtco.com/extremophiles-extreme-organisms-373905.

Banks, Joseph. *Journal of the Right Hon. Sir Joseph Banks During Captain Cook's First Voyage in H.M.S. Endeavour in 1768–71*, edited by J. D. Hooker. Cambridge: Cambridge University Press, 2011.

Barlow, James William. *The Immortals' Great Quest. Translated From an Unpublished Manuscript in the Library of a Continental University*. London: Forgotten Books, 2017. Originally published in 1909; https://archive.org/details/immortalsgreatqu00barl.

Barnes, Alison. "The First Christmas Tree." *History Today* 56, no. 12 (December 2006). https://www.historytoday.com/archive/history-matters/first-christmas-tree.

Bauer, S. W. *The History of the Ancient World*. New York: W. W. Norton & Company, 2007.

Beaglehole, J. C. *The Life of Captain James Cook*. Palo Alto, CA: Stanford University Press, 1992.

Beaglehole, J. C. "On the Character of Captain James Cook." *Geographical Journal* 122, no. 4 (1956): 417–429. https://doi.org/10.2307/1790186.

Beck, Julie. "The Caves of Forgotten Time." *The Atlantic*. November 9, 2015. https://www.theatlantic.com/health/archive/2015/11/the-caves-of-forgotten-time/414894.

Becker, George F. "Halley on the Age of the Ocean." *Science* 31, no. 795 (1910): 459–461. https://doi.org/10.1126/science.31.795.459.b.

Bedini, Silvio A. *The Pulse of Time: Galileo Galilei, the Determination of Longitude, and the Pendulum Clock*. Florence: Olshki, 1991.

Bedini, Silvio A., and Francis R. Maddison. "Mechanical Universe: The Astrarium of Giovanni De' Dondi." *Transactions of the American Philosophical Society* 56, no. 5 (1966): 1–69. https://doi.org/10.2307/1006002.

Bedini, Silvio. "Along Came a Spider—Spinning Silk for Cross-Hairs: The Search for Cross-Hairs for Scientific Instrumentation, Part 1." *American Surveyor* (March/April 2005).

Bedini, Silvio. "Along Came a Spider—Spinning Silk for Cross-Hairs: The Search for Cross-Hairs for Scientific Instrumentation, Part 2." *American Surveyor* (May 2005).

Bennett, J. A. "Robert Hooke as Mechanic and Natural Philosopher." *Notes and Records of the Royal Society of London* 35, no. 1 (1980): 33–48. https://doi.org/doi:10.1098/rsnr.1980.0003.

Berger, R. W. *Palace of the Sun: The Louvre of Louis XIV*. University Park: Pennsylvania State University Press, 2010. https://books.google.it/books?id=1IkUHp8efo4C.

Berry, Megan, David T. Booth, and Colin J. Limpus. "Artificial Lighting and Disrupted Sea-Finding Behaviour in Hatchling Loggerhead Turtles (*Caretta caretta*) on the Woongarra Coast, South-East Queensland, Australia." *Australian Journal of Zoology* 61, no. 2 (2013): 137–145.

Beson, Michael. *Cosmigraphics: Picturing Space Through Time*. New York: Abrams, n.d.

Bessel, Friedrich William. "A Letter from Prof. Bessel to Sir J. Herschel, Bart., Dated Königsberg, Oct 23, 1838." In *The London and Edinburgh Philosophical Magazine and Journal of Science* 14 (January–June 1839): 68–72. https://books.google.it/books?id=UdChCg32-a0C.

Bickel, Susanne, and Rita Gautschy. "Eine Ramessidische Sonnenuhr Im Tal Der Könige." *Zeitschrift für Ägyptische Sprache und Altertumskunde* 141, no. 1 (2014): 3–14. https://doi.org/doi:10.1515/zaes-2014-0001.

Bigdeli, Mohammad, Rajat Srivastava, and Michele Scaraggi. "Dynamics of Space Debris Removal: A Review." arXiv. April 12, 2023. https://doi.org/10.48550/arXiv.2304.05709.

"Bio: Edgar Mitchell's Strange Voyage." *People Weekly*, April 8, 1974, 20–23.

Bird, Michael I., Scott A. Condie, Sue O'Connor, Damien O'Grady, Christian Reepmeyer, Sean Ulm, Mojca Zega, Frédérik Saltré, and Corey J. A. Bradshaw. "Early Human Settlement of Sahul Was Not an Accident." *Scientific Reports* 9, no. 1 (2019): 8220. https://doi.org/10.1038/s41598-019-42946-9.

Bodifée, G. "La découverte des satellites de Mars." *L'Astronomie* 91 (1977): 235. https://ui.adsabs.harvard.edu/abs/1977LAstr..91..235B.

Bodio, Stephen J. "A Feathered Tempest: The Improbable Life and Sudden Death of the Passenger Pigeon." Cornell Lab of Ornithology. April 15, 2010. https://www.allaboutbirds.org/news/a-feathered-tempest-the-improbable-life-and-sudden-death-of-the-passenger-pigeon/#.

Boitani, P. "Poetry of the Stars." In *The Inspiration of Astronomical Phenomena VI*, edited by E. M. Corsini, 289–309. Astronomical Society of the Pacific Conference Series 441. San Francisco, CA: Astronomical Society of the Pacific, 2011.

Bonechi, Sara. *How They Make Me Suffer . . . A Short Biography of Galileo Galilei*. Translated by Anna Teicher. Florence: Istituto e Museo di Storia della Scienza, 2008.

Botley, C. M., and R. E. White. "Halley's Comet in 1066." Leaflet of the Astronomical Society of the Pacific 10 (1967): 9–16.

Bradbury, Ray. "The Long Rain." In *The Illustrated Man*. New York: Doubleday, 1951.

Bridgman, Tim. "Who Were the Cimmerians?" *Hermathena*. No. 164 (1998): 31–64. http://www.jstor.org/stable/23041189.

Brooke-Hitching, Edward. *The Sky Atlas*. London: Simon & Schuster, 2019.

Broughton, Peter. "The First Predicted Return of Comet Halley." *Journal for the History of Astronomy* 16, no. 2 (1985): 123–133. https://doi.org/10.1177/002182868501600203.

Brown, Fredric. "Pi in the Sky." In *The Best of Fredric Brown*. Edited by Robert Bloch, 86–112. New York: Nelson Doubleday, 1977. Originally published in 1945.

Brumfiel, Geoff. "U.S. Navy Brings Back Navigation by the Stars for Officers." *NPR*. Last modified February 22, 2016. https://www.npr.org/2016/02/22/467210492/u-s-navy-brings-back-navigation-by-the-stars-for-officers.

Brush, Stephen G. "Poincaré and Cosmic Evolution." *Physics Today* 33, no. 3 (1980): 42–49. https://doi.org/10.1063/1.2913996.

Bryant, William Cullen. "Hymn to the North Star." https://quod.lib.umich.edu/a/amverse/BAD0508.0001.001/1:95?rgn=div1;view=fulltext.

Burke-Gaffney, W. "Kepler and the Star of Bethlehem." *Journal of the Royal Astronomical Society of Canada* 31 (1937): 417. https://ui.adsabs.harvard.edu/abs/1937JRASC..31..417B.

Burroughs, Edgar Rice. *Pirates of Venus*. London: New English Library, 1972. Originally published in 1932.

Burton, David M. *The History of Mathematics: An Introduction*. New York: McGraw-Hill, 2007.

Buxton, H. W., and A. Hyman. *Memoir of the Life and Labours of the Late Charles Babbage Esq. F.R.S.* Cambridge, MA: MIT Press, 1988. https://books.google.it/books?id=_EDYswEACAAJ.

Byron, George Gordon. *Lord Byron: The Complete Poetical Works*. Vol. 5: *Don Juan*. Edited by Jerome J. McGann. Oxford: Oxford University Press, 1986.

Campion, Nicholas. *A History of Western Astrology*. Vol. 1: *The Ancient and Classical Worlds*. New York: New York University Press, 2012.

Campion, Nicholas. *Astrology and Cosmology in the World's Religions*. New York: Continuum, 2008.

Campion, N., and N. Kollerstrom, eds. "Galileo's Astrology." Special issue, *Culture and Cosmos* 7, no. 1 (spring/summer 2003).

Canales, Jimena. "Exit the Frog, Enter the Human: Physiology and Experimental Psychology in Nineteenth-Century Astronomy." *British Journal for the History of Science* 34, no. 2 (2001): 173–197. https://doi.org/10.1017/S0007087401004356.

Canales, Jimena. *Bedeviled: A Shadow History of Demons in Science*. Princeton, NJ: Princeton University Press, 2020.

Carlson, John B. "Lodestone Compass: Chinese or Olmec Primacy?" *Science* 189, no. 4205 (1975): 753–760. https://doi.org/doi:10.1126/science.189.4205.753.

Carrington, Damian. "Flying Insect Numbers Have Plunged by 60% Since 2004, GB Survey Finds." *The Guardian*. May 4, 2022. https://www.theguardian.com/environment/2022/may/05/flying-insect-numbers-have-plunged-by-60-since-2004-gb-survey-finds.

Carter, Harold B. "The Royal Society and the Voyage of HMS 'Endeavour' 1768–71." *Notes and Records of the Royal Society of London* 49, no. 2 (1995): 245–260. http://www.jstor.org/stable/532013.

Carter, Jimmy. "Voyager Spacecraft Statement by the President." American Presidency Project. July 29, 1977. https://www.presidency.ucsb.edu/node/243563.

Cashford, Jules. *The Moon: Symbols of Transformation*. Carterton, UK: Greystones Press, 2003.

Castello, Jay. "The Art Behind NASA's Scientific Space Photos." *The Verge*. October 10, 2022. https://www.theverge.com/2022/10/10/23393194/nasa-image-processing-jwst-astrophotography.

Chambers, G. F. *The Story of Eclipses Simply Told for General Readers*. London: George Newnes Ltd., 1899. https://www.gutenberg.org/files/24222/24222-h/24222-h.htm.

"Charles Babbage." *Nature* 5, no. 106 (1871): 28–29. https://doi.org/10.1038/005028a0.

Charlot, P., C. S. Jacobs, D. Gordon, S. Lambert, A. de Witt, J. Böhm, A. L. Fey, et al. "The Third Realization of the International Celestial Reference Frame by Very Long Baseline Interferometry." *Astronomy & Astrophysics* 644 (2020): A159. https://doi.org/10.1051/0004-6361/202038368.

Chatley, H. "Ancient Egyptian Star Tables and the Dekans." *The Observatory* 65 (December 1, 1943): 121. https://ui.adsabs.harvard.edu/abs/1943Obs....65..121C.

Chevalier, Jean, and Alain Gheerbrant, eds. *The Penguin Dictionary of Symbols*. London: Penguin Books, 1996.

Christianson, Gale E. *Edwin Hubble: Mariner of the Nebulae*. Chicago: University of Chicago Press, 1996.

Christianson, Gale E. *The Wild Abyss: The Story of the Men Who Made Modern Astronomy*. New York: The Free Press, 1978.

Ciardi, John, and Miller Williams. *How Does a Poem Mean?* Boston: Houghton Mifflin, 1959.

Close, Frank. *Eclipses: What Everyone Needs to Know*. New York: Oxford University Press, 2019.

Cohen, Thierry. *Villes éteintes*. Paris: Marval, n.d.

Colagè, Ivan, and Francesco d'Errico. "Culture: The Driving Force of Human Cognition." *Topics in Cognitive Science* 12 (July 22, 2018). https://doi.org/10.1111/tops.12372.

Collier, Bruce. "The Little Engines That Could've: The Calculating Machines of Charles Babbage." PhD diss., Harvard University, 1970. http://robroy.dyndns.info/collier/index.html (accessed February 8, 2023).

Cook, James. *Captain Cook's Journal During His First Voyage Round the World, Made in H.M. Bark Endeavour, 1768–71*, edited by W. J. Lloyd Wharton. Cambridge: Cambridge University Press, 2014.

Cooper, J. C. *An Illustrated Encyclopaedia of Traditional Symbols*. London: Thames & Hudson, 2017.

Copeland, Leland S. "Sources of the Seven-Day Week." *Popular Astronomy* 47 (April 1, 1939): 175. https://ui.adsabs.harvard.edu/abs/1939PA.....47..175C.

Copernicus, Nicolaus. *De revolutionibus orbium coelestium*. Translated by Charles Glenn Wallis. Britannica Great Books 16. Chicago: Encyclopaedia Britannica, 1955. Originally published in 1543.

Costa Canas, António. "The Astronomical Navigation in Portugal in the Age of Discoveries." *Cahiers François Viète* III-3 (2017): 15–36. https://doi.org/10.4000/cahierscfv.752.

"Could You Patent the Sun?" Video posted to YouTube by Global Citizen. April 12, 1955. https://www.youtube.com/watch?v=erHXKP386Nk.

Crawford, Kate. *Atlas of AI*. New Haven, CT: Yale University Press, 2021.

Croarken, Mary. "Providing Longitude for All." *Journal for Maritime Research* 4, no. 1 (2002): 106–126.

Curtius, G., and E. Windisch. *Grundzüge der griechischen Etymologie*. Leipzig: B. G. Teubner, 1879. https://books.google.it/books?id=40ITAAAAYAAJ.

D'Amico, Matteo. *Giordano Bruno: Avventure e misteri del grande mago nell'Europa del Cinquecento*. Casale Monferrato: Edizioni Piemme, 2003.

"Dark and Quiet Skies for Science and Society." United Nations Office for Outer Space Affairs. 2020. https://www.iau.org/static/publications/dqskies-book -29-12-20.pdf.

d'Errico, Francesco, Luc Doyon, Shuangquan Zhang, Malvina Baumann, Martina Lázničková-Galetová, Xing Gao, Fuyou Chen, and Yue Zhang. "The Origin and Evolution of Sewing Technologies in Eurasia and North America." *Journal of Human Evolution* 125 (2018): 71–86. https://doi.org/10.1016/j .jhevol.2018.10.004.

Darwin, Charles. "To Caroline Darwin, 27 February 1837." Darwin Correspondence Project. Accessed February 8, 2023. https://www.darwinproject.ac.uk /letter/?docId=letters/DCP-LETT-346.xml.

Darwin, Charles. "To W. D. Fox [9–12 August] 1835." Darwin Correspondence Project. Accessed February 8, 2023. https://www.darwinproject.ac.uk/letter /?docId=letters/DCP-LETT-282.xml.

Darwin, Charles. *Charles Darwin's Notebooks, 1836–1844: Geology, Transmutation of Species, Metaphysical Enquiries*, edited by Paul H. Barrett, Peter J. Gautrey, Sandra Herbert, David Kohn, and Sydney Smith. Cambridge: Cambridge University Press, 1998.

Darwin, Charles. *The Autobiography of Charles Darwin, 1809–1882*, edited by Nora Barlow. New York: W. W. Norton & Co, 1993.

Darwin, George. H. "Radio-Activity and the Age of the Sun." *Nature* 68, no. 1769 (1903): 496–496.

de Grijs, Richard. "European Longitude Prizes. I: Longitude Determination in the Spanish Empire." *Journal of Astronomical History and Heritage* 23 (2020): 465–494. https://ui.adsabs.harvard.edu/abs/2020JAHH...23..465D.

de Jong, T., and W. H. van Soldt. "The Earliest Known Solar Eclipse Record Redated." *Nature* 338, no. 6212 (1989): 238–240. https://doi.org/10 .1038/338238a0.

de Solla Price, Derek J. "Gears from the Greeks. The Antikythera Mechanism: A Calendar Computer from Ca. 80 B. C." *Transactions of the American Philosophical Society* 64, no. 7 (1974): 1–70. https://doi.org/10.2307/1006146.

de Solla Price, Derek J. "The Prehistory of the Clock." *Discovery* 17 (1957): 153–157.

de Solla Price, Derek J. *Science Since Babylon*. New Haven, CT: Yale University Press, 1961.

de Solla Price, Derek J. "Leonardo Da Vinci and the Clock of Giovanni De Dondi." *Antiquarian Horology*. No. 2. June (1958): 127–128.

Deacon, G. E. R., and Margaret Deacon. "Captain Cook as a Navigator." *Notes and Records of the Royal Society of London* 24, no. 1 (1969): 33–42. http://www.jstor .org/stable/530739.

DeLillo, Don. *Underworld*. New York: Scribner, 1997.

Department of Ancient Near Eastern Art. "The Phoenicians (1500–300 B.C.)." In *Heilbrunn Timeline of Art History*. October 2004. http://www.metmuseum .org/toah/hd/phoe/hd_phoe.htm.

Descartes, René. *The World, or Treatise on Light*. Translated by Michael S. Mahoney. 1629–1633. https://www.princeton.edu/~hos/mike/texts/descartes /world/worldfr.htm.

Di Piazza, Anne, and Erik Pearthree. "A New Reading of Tupaia's Chart." *Journal of the Polynesian Society* 116, no. 3 (2007): 321–340. http://www.jstor.org /stable/20707400.

Di Piazza, Anne, and Erik Pearthree. "Il Cartografo Tupaia, James Cook e il Confronto Tra Due Saperi Geografici." *Quaderni Storici* (2008): 575–592. https://hal.archives-ouvertes.fr/hal-00412208.

Dio, Cassius. *Roman History*. Translated by Earnest Cary. Cambridge MA: Harvard University Press, 1925.

"Documenting the Numbers of Victims of the Holocaust and Nazi Persecution." United States Holocaust Memorial Museum Holocaust Encyclopaedia. Last modified December 8, 2020. https://encyclopedia.ushmm.org /content/en/article/documenting-numbers-of-victims-of-the-holocaust -and-nazi-persecution.

"Double Negative. 1969." Museum of Contemporary Art. Accessed February 10, 2023. https://www.moca.org/collection/work/double-negative-2.

Dirac, Paul Adrien Maurice. "Quantised Singularities in the Electromagnetic Field." *Proceedings of the Royal Society of London. Series A, Containing Papers of a Mathematical and Physical Character* 133, no. 821 (1931): 60–72. https://doi .org/doi:10.1098/rspa.1931.0130.

Dreyer, J. L. E. *History of the Royal Astronomical Society*. London: Royal Astronomical Society, 1923.

Druett, Joan. *Tupaia: Captain Cook's Polynesian Navigator*. Westport, CT: Praeger, 2011.

Duke, Dennis W. "Hipparchus' Coordinate System." *Archive for History of Exact Sciences* 56 (2002): 427–433. https://ui.adsabs.harvard.edu/abs/2002AHES ...56..427D.

Dunkin, Edwin. *The Midnight Sky: Familiar Notes on the Stars and Planets with Star-Maps and Other Illustrations*. London: Religious Tract Society, 1869.

Dunnington, Guy Waldo. *Carl Friedrich Gauss: Titan of Science*. Whitefish, MT: Literary Licensing, 2012.

Eagleton, Katie. "An Islamic Astrolabe." Whipple Museum of the History of Science. University of Cambridge. Last updated 1999. http://www.sites.hps.cam .ac.uk/starry/isaslabe.html.

"Early New York City Police 'Badges' & Emblems of Office—1800–1845." NYP DHistory.com. October 11, 2020. https://nypdhistory.com/staves.

Eckstein, Lars, and Anja Schwarz. "The Making of Tupaia's Map: A Story of the Extent and Mastery of Polynesian Navigation, Competing Systems of Wayfinding on James Cook's *Endeavour*, and the Invention of an Ingenious Cartographic System." *Journal of Pacific History* 54, no. 1 (2019): 1–95. https://doi.org/10.1080/00223344.2018.1512369.

Editors of Encyclopedia Britannica. "Simon Marius." *Britannica*. January 6, 2023. https://www.britannica.com/biography/Simon-Marius.

Einstein, Albert. "The 1932 Disarmament Conference." *The Nation*. August 23, 2001. https://www.thenation.com/article/archive/1932-disarmament-conference-0.

Eiseley, Loren. *The Firmament of Time*. In *Loren Eiseley: Collected Essays on Evolution, Nature and the Cosmos*, Vol. 1. New York: Library of America, 2016. Originally published in 1946.

Eiseley, Loren. *The Immense Journey*. In *Loren Eiseley: Collected Essays on Evolution, Nature and the Cosmos*, Vol. 1. New York: Library of America, 2016. Originally published in 1946.

Eliade, Mircea. *Patterns in Comparative Religion*. New York: Sheen and Ward, 1958.

Emerson, Ralph Waldo. *Nature*. Boston: James Munroe and Company, 1836.

Emerson, Ralph Waldo. "Fragments on Nature and Life, Nature." In *The Complete Works of Ralph Waldo Emerson: Poems*. Boston: Houghton, Mifflin, 1903–1904.

Enheduanna. *Princess, Priestess, Poet: The Sumerian Temple Hymns of Enheduanna*, edited by Betty De Shong Meador. Austin: University of Texas Press, 2009.

Eve, A. S. "Galileo and Scientific History: The Leaning Tower and Other Stories." *Nature* 137, no. 3453 (936): 8–10. https://doi.org/10.1038/137008a0.

"Extravehicular Activity." NASA. https://history.nasa.gov/SP-4029/Apollo_18-30 _Extravehicular_Activity.htm (accessed April 25, 2023).

Favaro, A. "Galileo, Astrologer." In "Galileo's Astrology." Special issue, *Culture and Cosmos* 7, no 1 (spring/summer 2003): 9–19. Originally published in 1881.

"Fellows Deceased: Charles Babbage, F. R. S." *Monthly Notices of the Royal Astronomical Society* 32 (1872): 101. https://doi.org/10.1093/mnras/32.4.101.

Ferguson, Kitty. *The Nobleman and His Housedog: Tycho Brahe and Johannes Kepler: The Strange Partnership That Revolutionised Science*. London: Review, 2002.

Finney, Ben R. "Nautical Cartography and Traditional Navigation in Oceania." In *The History of Cartography: Cartography in the Traditional African, American, Arctic, Australian, and Pacific Societies*, edited by D. Woodward and

G. Malcolm Lewis, Vol. 2, Bk. 3, 443–494. Chicago: University of Chicago Press, 1998.

Foderà Serio, G., A. Manara, and Piero Sicoli. "Giuseppe Piazzi and the Discovery of Ceres." In *Asteroids III*, edited by W. F. Bottke Jr., A. Cellino, P. Paolicchi, and R. P. Binzel, 17–24. Tucson: University of Arizona Press, 2002. https://ui.adsabs.harvard.edu/abs/2002aste.book...17F.

Foster, James J., Jochen Smolka, Dan-Eric Nilsson, and Marie Dacke. "How Animals Follow the Stars." *Proceedings of the Royal Society B: Biological Sciences* 285, no. 1871 (2018): 20172322. https://doi.org/10.1098/rspb.2017.2322.

Fourier, Jean-Baptiste Joseph. *Oeuvres de Fourier: Théorie analytique de la chaleur*. Paris: Gauthier-Villars et fils, 1888. https://books.google.com.bn/books?id=JZNWAAAAMAAJ.

Frank, Adam. *About Time: Cosmology and Culture at the Twilight of the Big Bang*. New York: Free Press, 2012.

Frazer, James George. *The Worship of Nature*. London: Macmillan, 1926.

Freeth, T., Y. Bitsakis, X. Moussas, J. H. Seiradakis, A. Tselikas, H. Mangou, M. Zafeiropoulou, et al. "Decoding the Ancient Greek Astronomical Calculator Known as the Antikythera Mechanism." *Nature* 444, no. 7119 (2006): 587–591. https://doi.org/10.1038/nature05357.

Freeth, Tony, David Higgon, Aris Dacanalis, Lindsay MacDonald, Myrto Georgakopoulou, and Adam Wojcik. "A Model of the Cosmos in the Ancient Greek Antikythera Mechanism." *Scientific Reports* 11, no. 1 (2021): 5821. https://doi.org/10.1038/s41598-021-84310-w.

Fuller, R. S., R. P. Norris, and M. Trudgett. "The Astronomy of the Kamilaroi and Euahlayi Peoples and Their Neighbours." *Australian Aboriginal Studies*. No. 2 (2014): 3–27.

Fuller, Robert S., Michelle Trudgett, Ray P. Norris, and Michael G. Anderson. "Star Maps and Travelling to Ceremonies: The Euahlayi People and Their Use of the Night Sky." *Journal of Astronomical History and Heritage* 17 (2014): 149–160. https://ui.adsabs.harvard.edu/abs/2014JAHH...17..149F.

Galilei, Galileo. *The Assayer*. In *Discoveries and Opinions of Galileo*, translated by Stillman Drake, 229–280. Garden City, NY: Doubleday Anchor, 1957. Originally published in 1623.

Galilei, Galileo. "Astrologica Nonnulla." In *Le opere di Galileo Galilei: Appendice, Volume III*, edited by Andrea Battistini, Michele Camerota, Germana Ernst, Romano Gatto, Mario Otto Helbing, and Patrizia Ruffo, 108–193. Firenze: Giunti Editore, 2017.

Galilei, Galileo. *The Sidereal Messenger*, translated by E. S. Carlos. London: Rivingtons, 1880. Originally published in 1610.

Galluzzi, Paolo, ed. *Galileo: Images of the Universe from Antiquity to the Telescope*. Florence: Giunti, 2009.

Galway-Witham, Julia, and Chris Stringer. "How Did *Homo sapiens* Evolve?" *Science* 360, no. 6395 (2018): 1296. https://doi.org/10.1126/science.aat6659.

Gattei, Stefano. *On the Life of Galileo: Viviani's Historical Account and Other Early Biographies*. Princeton, NJ: Princeton University Press, 2019.

Gibbons, Ann. "Neanderthals Carb Loaded, Helping Grow Their Big Brains." *Science* (2021). https://doi.org/10.1126/science.abj4012.

Gibson, Bruce, and William Sterling. *The Difference Engine*. New York: Bantam Books, 1992.

Gillieron, Philippe. "La Tour Eiffel de l'espace." *La Jaune et la Rouge* 425 (1989): 13–20.

Gillispie, Charles Coulston. *Pierre-Simon Laplace, 1749–1827: A Life in Exact Science*. Princeton, NJ: Princeton University Press, 2021.

Gingerich, Owen. "Cranks and Opportunists: 'Nutty' Solutions to the Longitude Problem." In *The Quest for Longitude: The Proceedings of the Longitude Symposium, Harvard University, Cambridge, Massachusetts, November 4–6, 1993*, edited by William J. H. Andrewes. Cambridge, MA: Collection of Historic and Scientific Instruments, Harvard University, 1996.

Glaz, Sarah. "Enheduanna: Princess, Priestess, Poet, and Mathematician." *Mathematical Intelligencer* 42, no. 2 (2020): 31–46. https://doi.org/10.1007/s00283-019-09914-7.

Goldman, Lawrence. *Victorians and Numbers: Statistics and Society in Nineteenth Century Britain*. Oxford: Oxford University Press, 2022.

Gosline, Anna. "Do Women Who Live Together Menstruate Together?" *Scientific American*. Last modified December 7, 2007. https://www.scientificamerican.com/article/do-women-who-live-together-menstruate-together.

Gould, Stephen J. "The Evolution of Life on the Earth." *Scientific American* 271, no. 4 (October 1994): 84–91. doi: 10.1038/scientificamerican1094-84 .PMID: 7939569.

Grafton, Anthony. "Some Uses of Eclipses in Early Modern Chronology." *Journal of the History of Ideas* 64, no. 2 (2003): 213–229. https://doi.org/10.2307/3654126.

Greaves, J. S., A. M. S. Richards, W. Bains, P. B. Rimmer, H. Sagawa, D. L. Clements, S. Seager, et al. "Phosphine Gas in the Cloud Decks of Venus." *Nature Astronomy* 5 (2021): 655–664. https://doi.org/10.1038/s41550-020-1174-4.

Green, Judith A. *Henry I, King of England and Duke of Normandy*. Cambridge: Cambridge University Press, 2006.

Gryspeerdt, Edward. "Where Is the Cloudiest Place on Earth?" *Clouds and Climate*. Last modified January 24, 2021. https://www.cloudsandclimate.com/blog/where_is_the_cloudiest.

Gryspeerdt, Edward. "Where Is the Cloudiest Place on Earth? (Part 2—Satellites)." *Clouds and Climate*. Last modified March 30, 2021. https://www.cloudsandclimate.com/blog/where_is_cloudiest_part2.

Hafez, Ihsan. "Abd Al-Rahman Al-Ṣūfi and His Book of the Fixed Stars: A Journey of Re-discovery." PhD diss., James Cook University, 2010. https://research online.jcu.edu.au/28854.

Hamacher, Duane Willis. "On the Astronomical Knowledge and Traditions of Aboriginal Australians." PhD diss., Macquarie University, 2012. http://hdl .handle.net/1959.14/268547.

Hamacher, Duane. *The First Astronomers*. Crows Nest, Australia: Allen & Unwin, 2022.

Hamblyn, Richard. *The Invention of Clouds: How an Amateur Meteorologist Forged the Language of the Skies*. London: Picador, 2001.

Hamilton, J. C. *Captain James Cook and the Search for Antarctica*. Barnsley, UK: Pen & Sword Books, 2020. https://books.google.it/books?id=oRnhDwAAQBAJ.

Handbook for Visitors to Paris. London: John Murray; Paris: Galignani, 1879. https:// archive.org/details/handbookforvisit00lond/page/n5/mode/2up.

Hardy, Karen, Stephen Buckley, Matthew J. Collins, Almudena Estalrrich, Don Brothwell, Les Copeland, Antonio García-Tabernero, et al. "Neanderthal Medics? Evidence for Food, Cooking, and Medicinal Plants Entrapped in Dental Calculus." *Naturwissenschaften* 99, no. 8 (2012): 617–626. https://doi .org/10.1007/s00114-012-0942-0.

Hare, Brian, and Vanessa Woods. *The Survival of the Friendliest: Understanding Our Origins and Rediscovering Our Common Humanity*. New York: Random House, 2020.

Hartmann, Dennis L., Maureen E. Ockert-Bell, and Marc L. Michelsen. "The Effect of Cloud Type on Earth's Energy Balance: Global Analysis." *Journal of Climate* 5, no. 11 (1992): 1281–1304.

Harvey, Fiona. "UN Says Up to 40% of World's Land Now Degraded." *The Guardian*. April 27, 2022. https://www.theguardian.com/environment/2022 /apr/27/united-nations-40-per-cent-planet-land-degraded.

Hawkins, Richard. "Barlow, James William." In *Dictionary of Irish Biography*. Last modified June 2021. https://doi.org/10.3318/dib.000376.v1.

Hayden, B., and S. Villeneuve. "Astronomy in the Upper Palaeolithic?" *Cambridge Archaeological Journal* 21, no. 3 (2011): 331–355.

Heath, Chris. "How Elon Musk Plans on Reinventing the World (and Mars)." *GQ*. December 12, 2015. https://www.gq.com/story/elon-musk-mars-spacex -tesla-interview.

Helfrich-Forster, C., S. Monecke, I. Spiousas, T. Hovestadt, O. Mitesser, and T. A. Wehr. "Women Temporarily Synchronize Their Menstrual Cycles with the Luminance and Gravimetric Cycles of the Moon." *Science Advances* 7, no. 5 (2021): https://doi.org/10.1126/sciadv.abe1358.

Helling, Christiane. "Clouds in Exoplanetary Atmospheres." In *Exofrontiers: Big Questions in Exoplanetary Science*, edited by N. Madhusudhanm, 20-1–20-7. Bristol, UK: IOP Publishing, 2021.

Herodotus. *Histories*. Translated by R. Godley. Cambridge, MA: Harvard University Press, 1920.

Herschel, John. "Address Delivered at the General Meeting of the Royal Astronomical Society, February 12, 1842, on Presenting the Honorary Medal to M. Bessel." *Memoirs of the Royal Astronomical Society* 12 (1842): 442–454.

Hintz, Eric G., Maureen L. Hintz, and Jeannette M. Lawler. "Prior Knowledge Base of Constellations and Bright Stars Among Non–Science Majoring Undergraduates and 14–15 Year Old Students." *Journal of Astronomy & Earth Sciences Education* 2 (2015). https://doi.org/10.19030/jaese.v2i2.9515.

Hirsch, Edward. "Coming Out into the Light: W. B. Yeats's 'The Celtic Twilight' (1893, 1902)." *Journal of the Folklore Institute* 18, no. 1 (1981): 1–22. https://doi.org/10.2307/3814184.

Homans, Charles. "The Lives They Lived: Edgar Mitchell." *New York Times Magazine*. December 21, 2016. https://www.nytimes.com/interactive/2016/12/21/magazine/the-lives-they-lived-edgar-mitchell.html.

Homer. *The Odyssey*. Translated by A. T. Murray. Cambridge, MA: Harvard University Press; London: William Heinemann, Ltd., 1919.

Hooper, John. "Three Years in a Cave— and Trying for Six." *The Guardian*. October 13, 2006. https://www.theguardian.com/world/2006/oct/13/italy.mainsection.

Horace. *The Odes*, translated by A. S. Kline. Poetry in Translation. https://www.poetryintranslation.com/PITBR/Latin/HoraceOdesBkIV.php (accessed April 25, 2023).

Horiuchi, Takashi et al. "Multicolor and Multi-spot Observations of Starlink's Visorsat." Publications of the Astronomical Society of Japan. April 8, 2023. https://doi.org/10.1093/pasj/psad021.

Hornsby, Thomas. "LIII. The Quantity of the Sun's Parallax as Deduced from the Observations of the Transit of Venus, on June 3, 1769." *Philosophical Transactions of the Royal Society of London* 61 (1771): 574–579. https://doi.org/10.1098/rstl.1771.0054.

Hoskin, M. A. "Newton, Providence and the Universe of Stars." *Journal for the History of Astronomy* 8 (1977): 77. https://doi.org/10.1177/002182867700800203.

Hoskin, Michael, ed. *The Cambridge Illustrated History of Astronomy*. Cambridge: Cambridge University Press, 1996.

Howse, Derek. *Greenwich Time and the Longitude*. London: Philip Wilson Publishers, 2003.

Howse, Derek. *Nevil Maskelyne: The Seaman's Astronomer*. Cambridge: Cambridge University Press, 1989.

Howse, Derek, and Beresford Hutchinson. *Clocks and Watches of Captain James Cook, 1769–1969*. London: Antiquarian Horological Society, 1970.

Humphreys, Colin, and Graeme Waddington. "Solar Eclipse of 1207 BC Helps to Date Pharaohs." *Astronomy & Geophysics* 58, no. 5 (2017): 5.39–5.42. https://doi.org/10.1093/astrogeo/atx178.

Humphreys, Colin J., and W. G. Waddington. "Dating the Crucifixion." *Nature* 306, no. 5945 (December 22, 1983): 743–746.

Hunt, Lucas R., Megan C. Johnson, Phillip J. Cigan, David Gordon, and John Spitzak. "Imaging Sources in the Third Realization of the International Celestial Reference Frame." *Astronomical Journal* 162, no. 3 (2021): 121. https://doi.org/10.3847/1538-3881/ac135d.

Irwin, Geoffrey. *The Prehistoric Exploration and Colonisation of the Pacific*. Cambridge: Cambridge University Press, 1992.

Jahoda, G. "Quetelet and the Emergence of the Behavioral Sciences." *SpringerPlus* 4 (2015): 473. https://doi.org/10.1186/s40064-015-1261-7.

Jeguès-Wolkiewiez, C. "Aux racines de l'astronomie ou l'ordre caché d'une oeuvre paléolithique." *Antiquités Nationales* 37 (2005): 43–52.

Jensen, F., and S. Mullen. *C. G. Jung, Emma Jung and Toni Wolff: A Collection of Remembrances*. San Francisco, CA: Analytical Psychology Club of San Francisco, 1982. https://books.google.it/books?id=ETcQAQAAIAAJ.

Johnson, Dianne. "Interpretations of the Pleiades in Australian Aboriginal Astronomies." *Proceedings of the International Astronomical Union* 7, no. S278 (2011): 291–297. https://doi.org/10.1017/s1743921311012725.

Johnson, George. *Fire in the Mind: Science, Faith and the Search for Order*. New York: Vintage: 1996.

Joly, John. *The Birth-Time of the World and Other Scientific Essays*. London: T. Fisher Unwin Ltd., 1915.

Jung, C. G., *Jung on Astrology*, edited by S. Rossi, and K. L. Grice. Milton Park, UK: Taylor & Francis, 2017.

Jung, Carl Gustav. "The Archetypes and the Collective Unconscious" (1936). In *Collected Works of C. G. Jung*, Vol. 9, Part 1. Translated by Gerhard Hadler and R. F. C. Hull. Princeton, NJ: Princeton University Press, 1959.

Kaiho, Kunio, Naga Oshima, Kouji Adachi, Yukimasa Adachi, Takuya Mizukami, Megumu Fujibayashi, and Ryosuke Saito. "Global Climate Change Driven by Soot at the K-Pg Boundary as the Cause of the Mass Extinction." *Scientific Reports* 6, no. 1 (2016): 28427.

Keith, Arthur. "Whence Came the White Race?" *New York Times*. October 12, 1930.

Kelley, David H., and Eugene F. Milone. *Exploring Ancient Skies: A Survey of Ancient and Cultural Astronomy*. New York: Springer, 2011.

Kepler, Johannes. *Kepler's Conversation with Galileo's Sidereal Messenger*. Edited and translated by Edward Rosen. The Sources of Science, no. 5. New York: Johnson Reprint Corporation, 1965. https://gwern.net/doc/science/1965-kepler-keplersconversationwithgalileossiderealmessenger.pdf.

Kessler, Donald J., and Burton G. Cour-Palais. "Collision Frequency of Artificial Satellites: The Creation of a Debris Belt." *Journal of Geophysical Research* 83 (1978): 2637–2646. https://doi.org/10.1029/JA083iA06p02637.

Kessler, Elizabeth. *Picturing the Cosmos: Hubble Space Telescope Images and the Astronomical Sublime*. Minneapolis: University of Minnesota Press, 2012.

Kidger, Mark. *The Star of Bethlehem*. Princeton, NJ: Princeton University Press, 1999.

King, Arden R. "Review: [Untitled]: Reviewed Works: *The Roots of Civilization: The Cognitive Beginnings of Man's First Art, Symbol and Notation* by Alexander Marshack; *Notation dans les gravures du Paléolithique supérieur: Nouvelles méthodes d'analyse* by Alexander Marshack." *American Anthropologist* 75, no. 6 (December 1973): 1897–1900. http://www.jstor.org/stable/673696.

King, Henry C. *The History of the Telescope*. Mineola, NY: Dover Publications, 2003. Originally published in 1955.

Kipping, David M., and David S. Spiegel. "Detection of Visible Light from the Darkest World." *Monthly Notices of the Royal Astronomical Society: Letters* 417, no. 1 (2011): L88–L92. https://doi.org/10.1111/j.1745-3933.2011.01127.x.

Kirk, G. S., J. Raven, and Malcolm Schofield. *The Presocratic Philosophers: A Critical History with a Selection of Texts*. Cambridge: Cambridge University Press, 1983.

Klugler, Jeffrey. "Why the SpaceX Falcon Heavy Rocket Is Such a Big Deal for Elon Musk." *Time*. February 6, 2018. https://time.com/5133813/elon-musk-spacex-falcon-heavy-launch/.

Knight, Chris. *Blood Relations*. New Haven, CT: Yale University Press, 1995.

Knight, Chris. "Menstruation and the Origins of Culture." PhD diss., University College London, 1987.

Knop, Eva, Leana Zoller, Remo Ryser, Christopher Gerpe, Maurin Hörler, and Colin Fontaine. "Artificial Light at Night as a New Threat to Pollination." *Nature* 548, no. 7666 (2017): 206–209. https://doi.org/10.1038/nature23288.

Kollerstrom, Nick. "Galileo's Astrology." In *Largo campo di filosofare: Eurosyposium Galileo 2001*, edited by J. Montesinos and C. Solís, 421–431. La Orotava: Fundación Canaria Orotava de Historia de la Ciencia, 2001.

Kornei, Katherine. "How Many of the Moon's Craters Are Named for Women?" *The Independent*. May 3, 2021. https://www.independent.co.uk/news/science/moon-crater-names-women-space-b1840157.html.

Krauss, Rolf. "Egyptian Calendars and Astronomy." In *The Cambridge History of Science*. Vol. 1: *Ancient Science*, edited by Alexander Jones and Liba Taub, 131–143. Cambridge: Cambridge University Press, 2018.

Kreidberg, Laura, Jacob L. Bean, Jean-Michel Désert, Björn Benneke, Drake Deming, Kevin B. Stevenson, Sara Seager, et al. "Clouds in the Atmosphere of the Super-Earth Exoplanet Gj 1214b." *Nature* 505, no. 7481 (January 1, 2014): 69–72. https://doi.org/10.1038/nature12888.

Krupp, Edward. *Beyond the Blue Horizon*. Oxford: Oxford University Press, 1991.

Laguarda Trìas, Rolando. "Las longitudes geográficas de la membranza de Magallanes y del primer viaje de circum navegación." In *A viagem de Fernão de Magalhães e a questão das Molucas: Actas do II Colóquio Luso-Espanhol de história ultramarina*, edited by Avelino Teixeira da Mota, 135–178. Lisbon: Junta de Investigações Científicas do Ultramar, 1975.

Langley, M. "Re-analysis of the 'Engraved' Diprotodon Tooth from Spring Creek, Victoria, Australia." *Archaeology in Oceania* 55 (2020): 1–9.

Laplace, Pierre Simon. *A Philosophical Essay on Probabilities*. Translated by F. W. Truscott and F. L. Emory. London: Chapman & Hall, 1902. Originally published in 1825.

Laplace, Pierre Simon. *Théorie analytique des probabilités*. Paris: Courcier, 1820. https://gdz.sub.uni-goettingen.de/download/pdf/PPN585523401/PPN585523401.pdf.

Lardner, Dyonisus. "Babbage's Calculating Engine." *Edinburgh Review* (July 1834): 263–327.

Laskar, J., F. Joutel, and P. Robutel. "Stabilization of the Earth's Obliquity by the Moon." *Nature* 361, no. 6413 (1993): 615–617. https://doi.org/10.1038/361615a0.

Launius, Roger D. "Venus-Earth-Mars: Comparative Climatology and the Search for Life in the Solar System." *Life* 2, no. 3 (2012): 255–273. https://www.mdpi.com/2075-1729/2/3/255.

Launius, Roger D. "Visions of Venus at the Dawn of the Space Age." *Roger Launius's Blog*. November 7, 2014. https://launiusr.wordpress.com/2014/11/07/visions-of-venus-at-the-dawn-of-the-space-age.

Lauterjung, Isabel. "Powders of Sympathy." *The Royal Society Blog*. Last modified January 25, 2022. https://royalsociety.org/blog/2022/01/powders-of-sympathy.

Lawrence, Andy, Meredith L. Rawls, Moriba Jah, Aaron Boley, Federico Di Vruno, Simon Garrington, Michael Kramer, et al. "The Case for Space Environmentalism." *Nature Astronomy* (2022). https://doi.org/10.1038/s41550-022-01655-6.

Lawrence, Andy. *Losing the Sky*. Edinburgh: Photon Productions, 2021.

Lawson Dick, Oliver. *Aubrey's Brief Lives*. London: Secker and Warburg, 1950.

Leavitt, Henrietta S. "1777 Variables in the Magellanic Clouds." *Annals of Harvard College Observatory* 60 (1908): 87–108.3. https://ui.adsabs.harvard.edu/abs/1908AnHar..60...87L.

Leavitt, Henrietta S., and Edward C. Pickering. "Periods of 25 Variable Stars in the Small Magellanic Cloud." *Harvard College Observatory Circular* 173 (1912): 1–3. https://ui.adsabs.harvard.edu/abs/1912HarCi.173....1L.

Leopardi, Giacomo. *La storia dell'astronomia*. In *Tutte le opere di Giacomo Leopardi. Le poesie e le prose*, edited by W. Binni, 2:585–750. Florence: Sansoni, 1969.

Levy, Max G. "The Race to Put Silk in Nearly Everything." *Wired*. June 28, 2021. https://www.wired.com/story/the-race-to-put-silk-in-nearly-everything.

Lewis, D., P. W. Gathercole, David George Kendall, S. Piggott, Desmond George King-Hele, I. E. S. Edwards, and F. R. Hodson. "Voyaging Stars: Aspects of Polynesian and Micronesian Astronomy." *Philosophical Transactions of the Royal Society of London. Series A, Mathematical and Physical Sciences* 276, no. 1257 (1974): 133–148. https://doi.org/doi:10.1098/rsta.1974.0015.

Lewis, David. *We, the Navigators: The Ancient Art of Landfinding in the Pacific.* Honolulu: University of Hawaii Press, 1972.

Libby-Roberts, Jessica E., Zachory K. Berta-Thompson, Jean-Michel Désert, Kento Masuda, Caroline V. Morley, Eric D. Lopez, Katherine M. Deck, et al. "The Featureless Transmission Spectra of Two Super-Puff Planets." *Astronomical Journal* 159, no. 2 (2020): 57. https://doi.org/10.3847/1538-3881/ab5d36.

"Lies, Damned Lies and Statistics." Department of Mathematics, University of York. Last modified July 19, 2012. https://www.york.ac.uk/depts/maths/histstat/lies.htm.

Linge, Mary Kay. "How Ronald Reagan's Wife Nancy Let Her Astrologer Control the Presidency." *New York Post.* October 18, 2021. https://nypost.com/article/ronald-reagans-wife-nancy-astrologer-joan-quigley.

Lipking, L. *What Galileo Saw: Imagining the Scientific Revolution.* Ithaca, NY: Cornell University Press, 2014.

Lissauer, Jack J., Jason W. Barnes, and John E. Chambers. "Obliquity Variations of a Moonless Earth." *Icarus* 217, no. 1 (2012): 77–87. https://doi.org/10.1016/j.icarus.2011.10.013.

Lista, G. *La stella d'Italia.* Milan: Mudima, 2010.

Lister, Elizabeth. "Seeing the Northern Lights over East London." *WW2 People's War. BBC.* July 4, 2005. https://www.bbc.co.uk/history/ww2peopleswar/stories/48/a4354148.shtml.

Locke, John. *The Conduct of the Understanding.* London: Scott, Webster, and Geary, 1838. Originally published in 1706.

London, Jack. *The People of the Abyss.* London: Pluto Press, 2001. Originally published in 1903.

"'Lone' Longitude Genius May Have Had Help." *New Scientist.* Last modified May 12, 2009. https://www.newscientist.com/gallery/dn17119-lone-longitude-pioneer-had-help.

Lovering, Joseph. "The Méchanique Céleste by Laplace, and Its Translation with a Commentary by Bow-Ditch." In *Proceedings of the American Academy of Arts and Sciences* (1846): 185–201.

Low, Sam. *Hawaiki Rising: Hōkūle‘a, Nainoa Thompson, and the Hawaiian Renaissance.* Honolulu: University of Hawaii Press, 2018.

Low, Sam. "Polynesian Navigation." *Soundings Magazine.* November 2003. http://www.samlow.com/sail-nav/starnavigation.htm.

Lubbock, Constance Ann. *The Herschel Chronicle: The Life-Story of William Herschel and His Sister Caroline Herschel*. Cambridge: Cambridge University Press, 2013.

MacCarthy, Fiona. *Gropius: The Man Who Built the Bauhaus*. Cambridge, MA: Harvard University Press, 2019.

MacDonald, John. *The Arctic Sky: Inuit Astronomy, Star Lore, and Legend*. Toronto and Iqaluit: Royal Ontario Museum and Nunavut Research Institute, 1998.

Macomber, Henry P. "Glimpses of the Human Side of Sir Isaac Newton." *Scientific Monthly* 80, no. 5 (1955): 304–309. http://www.jstor.org/stable/21590.

Makemson, Maud W. *The Morning Star Rises*. New Haven, CT: Yale University Press 1941.

Mallama, Anthony. "The Brightness of VisorSat-Design Starlink Satellites." arXiv. January 2, 2021. https://doi.org/10.48550/arXiv.2101.

Mallick, Senjuti, and Rajeswari Pillai Rajagopalan. "If Space Is 'the Province of Mankind,' Who Owns Its Resources? An Examination of the Potential of Space Mining and Its Legal Implications." ORF Occasional Paper 182. Observer Research Foundation. January 2019. https://www.orfonline.org/research/if-space-is-the-province-of-mankind-who-owns-its-resources-47561.

Mann, G. S. "The Polynesian, Master Mariner and Astronomer." *Irish Astronomical Journal* 1 (1950): 114. https://ui.adsabs.harvard.edu/abs/1950IrAJ....1..114M.

Marchant, Jo. "Archimedes' Legendary Sphere Brought to Life." *Nature* 526, no. 7571 (2015): 19. https://doi.org/10.1038/nature.2015.18431.

Marcus, Hannah, and Paula Findlen. "Deciphering Galileo: Communication and Secrecy Before and After the Trial." *Renaissance Quarterly* 72, no. 3 (2019): 953–995. https://www-jstor-org.iclibezp1.cc.ic.ac.uk/stable/26845908.

Marra, John. *Journal of the Resolution's Voyage, in 1772, 1773, 1774, and 1775*, edited by D. Henry and F. Newbery. Rex Nan Kivell Collection; NK913. London: Printed for F. Newbery, 1775.

Marshack, Alexander. "Lunar Notation on Upper Paleolithic Remains." *Science* 146, no. 3645 (1964): 743–745. https://doi.org/10.1126/science.146.3645.743.

Marshack, Alexander. *The Roots of Civilization: The Cognitive Beginnings of Man's First Art, Symbol and Notation*. New York: McGraw-Hill, 1971.

Marshack, Alexander. "The Taï Plaque and Calendrical Notation in the Upper Palaeolithic." *Cambridge Archaeological Journal* 1, no. 1 (1991): 25–61. https://doi.org/10.1017/S095977430000024X.

Marshack, Alexander. "Upper Paleolithic Notation and Symbol." *Science* 178, no. 4063 (1972): 817–828. http://www.jstor.org/stable/1734899.

Maslin, Mark. *The Cradle of Humanity*. Oxford: Oxford University Press, 2019.

Mathews, R. H. "Message-Sticks Used by the Aborigines of Australia." *American Anthropologist* 10, no. 9 (1897): 288–298. http://www.jstor.org/stable/658501.

McClintock, Martha K. "Menstrual Synchrony and Suppression." *Nature* 229, no. 5282 (1971): 244–245. https://doi.org/10.1038/229244a0.

McDowell, Jonathan C. "The Low Earth Orbit Satellite Population and Impacts of the SpaceX Starlink Constellation." *Astrophysical Journal Letters* 892:L36, no. 2 (2020). https://doi.org/10.3847/2041-8213/ab8016.

McPhee, John. *Annals of the Former World*. New York: Farrar, Straus and Giroux, 1998.

Mendelson, Edward. "Baedeker's Universe." *Yale Review* 74 (spring 85), 386–483.

Milton, John. *Paradise Lost*. London: Samuel Simmons, 1667.

Mitchell, Stephen, trans. *Gilgamesh: A New English Version*. New York: Atria, 2013.

Mithen, Stephen J. *After the Ice Age: A Global Human History, 20,000–5,000 BC*. Cambridge, MA: Harvard University Press, 2006.

Mithen, Stephen J. *The Prehistory of the Mind: A Search for the Origins of Art, Religion and Science*. London: Thames & Hudson, 1996.

Mollon, J. D., and A. J. Perkins. "Errors of Judgement at Greenwich in 1796." *Nature* 380, no. 6570 (1996): 101–102. https://doi.org/10.1038/380101a0.

More, Louis Trenchard. *Isaac Newton: A Biography*. London: Charles Scribner's Sons, 1934.

Moses, Julianne. "Cloudy with a Chance of Dustballs." *Nature* 505, no. 7481 (2014): 31–32. https://doi.org/10.1038/505031a.

Mumford, Lewis. *The Myth of the Machine: Technics and Human Development*. San Diego, CA: Harcourt, Brace & World, 1967.

Mumford, Lewis. *Technics and Civilization*. Chicago: University of Chicago Press, 2010. Originally published in 1934.

Murray, Andrew, and Derek Howse. "Lieutenant Cook and the Transit of Venus, 1769." *Astronomy & Geophysics* 38, no. 4 (1997): 27–30. https://doi.org/10.1093/astrog/38.4.27.

Neugebauer, O. "The Egyptian 'Decans.'" In *Astronomy and History Selected Essays*, 205–209. New York: Springer, 1983.

Neugebauer, O. "The History of Ancient Astronomy Problems and Methods." *Journal of Near Eastern Studies* 4, no. 1 (1945): 1–38. https://doi.org/10.1086/370729.

Newcomb, F. J. *Navaho Folk Tales*. Albuquerque: University of New Mexico Press, 1990.

Newton, Humphrey. "Two Letters from Humphrey Newton to John Conduitt, Dated 17 January and 14 February 1727/8." Keynes Ms. 135, King's College, Cambridge, UK, The Newton Project. Accessed February 7, 2023. https://www.newtonproject.ox.ac.uk/view/texts/normalized/THEM00033.

Nichols, Peter. *Evolution's Captain: The Story of the Kidnapping That Led to Charles Darwin's Voyage Aboard the Beagle*. New York: Harper Perennial, 2004.

"Niels Ryberg Finsen—Facts." The Nobel Prize. Accessed February 1, 2023. https://www.nobelprize.org/prizes/medicine/1903/finsen/facts.

Nordgren, Tyler. *Sun Moon Earth: The History of Solar Eclipses from Omens of Doom to Einstein and Exoplanets*. New York: Basic Books, 2016.

Norris, Ray P. "Dawes Review 5: Australian Aboriginal Astronomy and Navigation." *Publications of the Astronomical Society of Australia* 33 (2016). https://doi
.org/10.1017/pasa.2016.25.

Norris, Ray P., and Barnaby R. M. Norris. "Why Are There Seven Sisters?" In
Advancing Cultural Astronomy: Studies in Honour of Clive Ruggles, edited by
Efrosyni Boutsikas, Stephen C. McCluskey, and John Steele, 223–235.
Cham: Springer International Publishing, 2021.

Norris, Ray P., and Cilla Norris. *Emu Dreaming: An Introduction to Australian
Aboriginal Astronomy*. Sydney: Emu Dreaming, 2009.

Norris, Ray P., and Bill Yidumduma Harney. "Songlines and Navigation in Wardaman and Other Australian Aboriginal Cultures." *Journal of Astronomical
History and Heritage* 17, no. 2 (2014): 141–148.

Norton, John D. "Chasing the Light: Einstein's Most Famous Thought Exper-
iment." In *Thought Experiments in Philosophy, Science and the Arts*, edited by
James Robert Brown, Mélanie Frappier, and Letitia Meynell, 123–140. New
York: Routledge, 2013.

Notarbartolo, Albert. "Some Proposals for Art Objects in Extraterrestrial Space."
Leonardo 8, no. 2 (1975): 139–141. https://doi.org/10.2307/1572957.

O'Brian, P. *Joseph Banks: A Life*. Chicago: University of Chicago Press, 1997.
https://books.google.it/books?id=p0lB4qYDvGgC.

O'Connell, James F., and Jim Allen. "The Restaurant at the End of the Universe:
Modelling the Colonisation of Sahul." *Australian Archaeology*, no. 74 (2012):
5–17. http://www.jstor.org/stable/23621508.

O'Connell, James F., Jim Allen, and Kristen Hawkes. "Pleistocene Sahul and the
Origins of Seafaring." In *The Global Origins and Development of Seafaring*, ed-
ited by A. Anderson, J. H. Barrett, and K. V. Boyle, 57–68. Cambridge, UK:
McDonald Institute for Archaeological Research, 2010.

O'Connor, John J. "Apollo 17 Coverage Gets Little Viewer Response." *New York
Times*. December 14, 1972.

O'Connor, J. J., and E. F. Robertson. "Nicolaus Copernicus." MacTutor, University of St Andrews. Last modified November 2002. https://mathshistory
.st-andrews.ac.uk/Biographies/Copernicus.

Odom, Brian C. "NASA Historical Investigation into James E. Webb's Relationship to the Lavender Scare." Accessed February 6, 2023. https://www.nasa
.gov/sites/default/files/atoms/files/nasa_historical_investigation_james
_webb_0.pdf.

Olschki, Leonardo. "Galileo's Philosophy of Science." *Philosophical Review* 52, no.
4 (1943): 349–365. https://doi.org/10.2307/2180669.

Olson, Donald W., Russell L. Doescher, and Marilynn S. Olson. "When the Sky
Ran Red: The Story Behind *The Scream*." *Sky & Telescope* (February 2004):
29–35.

Orchiston, Wayne. "Cook, Green, Maskelyne and the 1769 Transit of Venus: The Legacy of the Tahitian Observations." *Journal of Astronomical History and Heritage* 20 (2017): 35–68.

Orchiston, Wayne. "James Cook's 1769 Transit of Venus Expedition to Tahiti." *Proceedings of the International Astronomical Union*, Issue. IAUC196 (June 2004): 52–66. doi:10.1017/S1743921305001262.

"Origin of the Name Subaru." Web Archive. Accessed February 2, 2023. https://web.archive.org/web/20100411083646/http:/www.subaru-global.com/origin_name.html.

Osnos, Evan. "Doomsday Prep for the Super-rich." *New Yorker*. January 22, 2017. https://www.newyorker.com/magazine/2017/01/30/doomsday-prep-for-the-super-rich.

Ossendrijver, M. "Ancient Babylonian Astronomers Calculated Jupiter's Position from the Area Under a Time-Velocity Graph." *Science* 351, no. 6272 (2016): 482–484.

Panek, Richard. *Seeing and Believing: How the Telescope Opened Our Eyes and Minds to the Heavens*. New York: Viking, 1998.

Partridge, E. A., and H. C. Whitaker. "Galileo's Work on Saturn's Rings." *Popular Astronomy* 3 (1896): 408–414. https://ui.adsabs.harvard.edu/abs/1896PA......3..408P.

Peggy V. Beck, Anna Lee Walters, and Nia Francisco. *The Sacred*. Tsaile, AZ: Navajo Community College Press, 1992.

Perkins, Adam J. "Edmond Halley, Isaac Newton and the Longitude Act of 1714." In *The History of Celestial Navigation: Rise of the Royal Observatory and Nautical Almanacs*, edited by P. Kenneth Seidelmann and Catherine Y. Hohenkerk, 69–143. Cham: Springer, 2020. https://doi.org/10.1007/978-3-030-43631-5.

Perryman, Michael. "The History of Astrometry." *European Physical Journal H* 37, no. 5 (2012): 745–792. https://doi.org/10.1140/epjh/e2012-30039-4.

Peterson, Mark A. *Galileo's Muse*. Cambridge, MA: Harvard University Press, 2011.

Poincaré, Henri. *The Value of Science*. Translated by George Halsted. New York: The Science Press, 1907. Originally published in 1905.

Ponting, Gerald. *Callanish and Other Megalithic Sites of the Outer Hebrides*. Glastonbury, UK: Wooden Books, 2007.

Porter, Theodore M. "From Quetelet to Maxwell: Social Statistics and the Origins of Statistical Physics." In *The Natural Sciences and the Social Sciences: Some Critical and Historical Perspectives*, edited by I. Bernard Cohen, 345–362. Boston Studies in the Philosophy of Science 150. Dordrecht: Springer Netherlands.

Porter, Theodore M. "The Mathematics of Society: Variation and Error in Quetelet's Statistics." *British Journal for the History of Science* 18, no. 1 (1985): 51–69. https://doi.org/10.1017/S0007087400021695.

Powell, James Lawrence. "Premature Rejection in Science: The Case of the Younger Dryas Impact Hypothesis." *Science Progress* 105, no. 1 (2022). https://doi.org/10.1177/00368504211064272.

Powers, Richard. *The Overstory*. London: Vintage, 2019.

Prescod-Weinstein, Chanda, Sarah Tuttle, Lucianne Walkowicz, and Brian Nord. "The James Webb Space Telescope Needs to Be Renamed." *Scientific American*. Last modified March 1, 2021. https://www.scientificamerican.com/article/nasa-needs-to-rename-the-james-webb-space-telescope.

Price, Michael. "Africans Carry Surprising Amount of Neanderthal DNA." *Science*. January 30, 2020. https://www.doi.org/10.1126/science.abb0984.

Quetelet, Adolphe. "Des lois concernant le développement de l'homme." *Bulletin de l'Académie Royale des Sciences, des Lettres et des Beaux-Arts de Belgique* 29 (1870): 669–680.

Randall, Lisa, and Matthew Reece. "Dark Matter as a Trigger for Periodic Comet Impacts." *Physical Review Letters* 112, no. 16 (2014): 161301. https://doi.org/10.1103/PhysRevLett.112.161301.

Randles, W. G. L. "Portuguese and Spanish Attempts to Measure Longitude in the 16th Century." *Vistas in Astronomy* 28 (January 1, 1985): 235–241. https://doi.org/10.1016/0083-6656(85)90031-5.

Rappenglück, M. "The Pleiades in the 'Salle des Taureaux,' Grotte de Lascaux. Does a Rock Picture in the Cave of Lascaux Show the Open Star Cluster of the Pleiades at the Magdalénien Era (Ca 15.300 BC)?" In *Actas del IV Congreso de la SEAC "Astronomía en la Cultura" Held in Salamanca (1996)*, edited by C. Jaschek and F. Atrio Barandela, 217–225. Salamanca: Univesidad de Salamanca, 1997. https://ui.adsabs.harvard.edu/abs/1997ascu.conf..217R.

Rawls, Meredith L., Heidi B. Thiemann, Victor Chemin, Lucianne Walkowicz, Mike W. Peel, and Yan G. Grange. "Satellite Constellation Internet Affordability and Need." *Research Notes of the AAS* 4, no. 10 (2020): 189. https://doi.org/10.3847/2515-5172/abc48e.

Reiser, Oliver L. "The Evolution of Cosmologies." *Philosophy of Science* 19, no. 2 (1952): 93–107. http://www.jstor.org/stable/185818.

Renne, Paul R., Alan L. Deino, Frederik J. Hilgen, Klaudia F. Kuiper, Darren F. Mark, William S. Mitchell, Leah E. Morgan, Roland Mundil, and Jan Smit. "Time Scales of Critical Events Around the Cretaceous-Paleogene Boundary." *Science* 339, no. 6120 (2013): 684–687. https://doi.org/doi:10.1126/science.1230492.

Richards-Jones, P. "The Myth of the Sacred Calabash." *Journal of Navigation* 26, no. 4 (1973): 480–481. https://doi.org/10.1017/S0373463300021603.

Ridpath, Ian. *Star Tales: Revised and Expanded Edition*. Cambridge, UK: Lutterworth Press, 2018.

Rietbergen, Peter. "Urban VII Between White Magic and Black Magic, or Holy and Unholy Power." In *Power and Religion in Baroque Rome: Barberini Cultural Policies*, 336–376. Leiden: Brill, 2006.

Rigaud, Stephen Peter. *Some Account of Halley's Astronomiae cometicae synopsis*. Oxford and London: J. H. Parker and Whittaker & Co., 1835. https://doi.org/10.3931/e-rara-1299.

Rilke, Rainer Maria. *Duino Elegies*. Translated by Stephen Mitchell. Berkeley, CA: Shambhala Publications, 1992.

Ripat, Pauline. "Expelling Misconceptions: Astrologers at Rome." *Classical Philology* 106, no. 2 (2011): 115–154. https://doi.org/10.1086/659835.

Robinson, Deena. "15 Most Polluted Cities in the World." Earth.org. Last modified March 26, 2022. https://earth.org/most-polluted-cities-in-the-world.

Robinson, Judy. "Not Counting on Marshack: A Reassessment of the Work of Alexander Marshack on Notation in the Upper Palaeolithic." *Journal of Mediterranean Studies* 2 (1992): 1–17. https://muse.jhu.edu/article/669979.

Robson-Mainwaring, Laura. "The Great Smog of 1952." *The National Archives Blog*. Last modified July 19, 2022. https://blog.nationalarchives.gov.uk/the-great-smog-of-1952.

Rodman, Hugh, and John F. G. Stokes. "The Sacred Calabash." *Journal of the Polynesian Society* 37, no. 1 (145) (1928): 75–87. http://www.jstor.org/stable/20702185.

Rosen, Edward. "Galileo and Kepler: Their First Two Contacts." *Isis* 57, no. 2 (1966): 262–264. https://doi.org/10.1086/350119.

Rosen, Edward. "Galileo and the Telescope." *Scientific Monthly* 72, no. 3 (1951): 180–182. http://www.jstor.org/stable/20225.

Rosenthal-Schneider, I., T. Braun, and A. I. Miller. *Reality and Scientific Truth: Discussions with Einstein, Von Laue, and Planck*. Detroit, MI: Wayne State University Press, 1980. https://books.google.it/books?id=7tHaAAAAMAAJ.

Roulette, Joey. "How Much Does a Ticket to Space on New Shepard Cost? Blue Origin Isn't Saying." *New York Times*. October 13, 2021. https://www.nytimes.com/2021/10/13/science/space/blue-origin-ticket-cost.html.

Rozbrook, Roslyn. "The Real Mr. Hollywood." *Los Angeles Magazine* 43, no. 2 (February 1998): 19–20.

Russell, Stuart. "Artificial Intelligence: The Future Is Superintelligent." *Nature* 548, no. 7669 (2017): 520–521. https://doi.org/10.1038/548520a.

Sagan, Carl. *Murmurs of the Earth—The Voyager Interstellar Record*. New York: Ballantine Books, 1979.

Sagan, Carl. *Pale Blue Dot*. New York: Ballantine Books, 1997. Originally published in 1994.

Sagan, Carl. "The Planet Venus." *Science* 133, no. 3456 (1961): 849–858.

Salk, Jonas. "Are We Being Good Ancestors?" *World Affairs: The Journal of International Issues* 1, no. 2 (1992): 16–18. http://www.jstor.org/stable/45064193.

Salmond, Anne. "Tupaia, the Navigator-Priest." In *Tangata O Le Moana: New Zealand and the People of the Pacific*, edited by S. Mallon, K. U. Mahina-Tuai, and D. I. Salesa. Wellington, NZ: Te Papa Press, 2012.

Salmond, Anne. *The Trial of the Cannibal Dog: Captain Cook in the South Seas*. New Haven, CT: Yale University Press, 2003.

Sanders, Dirk, Enric Frago, Rachel Kehoe, Christophe Patterson, and Kevin J. Gaston. "A Meta-analysis of Biological Impacts of Artificial Light at Night." *Nature Ecology & Evolution* 5 (2020): 74–81. https://doi.org/10.1038/s41559-020-01322-x.

Sawyer Hogg, Helen. "Out of Old Books (the Callanish Stones)." *Journal of the Royal Astronomical Society of Canada* 60 (1966): 80. https://ui.adsabs.harvard.edu/abs/1966JRASC..60...80S.

Schaefer, Bradley E. "Lunar Eclipses That Changed the World." *Sky and Telescope* (December 1992): 639–642.

Schaefer, Bradley E. "Lunar Visibility and the Crucifixion." *Quarterly Journal of the Royal Astronomical Society* 31 (1990): 53–67.

Schaffer, Simon. "Babbage's Intelligence: Calculating Engines and the Factory System." *Critical Inquiry* 21, no. 1 (1994): 203–227. http://www.jstor.org/stable/1343892.

Schaffer, Simon. "Scientific Discoveries and the End of Natural Philosophy." *Social Studies of Science* 16, no. 3 (1986): 387–420. http://www.jstor.org/stable/285025.

Schechner, Sara. *Comets, Popular Culture, and the Birth of Modern Cosmology*. Princeton, NJ: Princeton University Press, 1997.

Schweber, Silvan S. "The Origin of the 'Origin' Revisited." *Journal of the History of Biology* 10, no. 2 (1977): 229–316. http://www.jstor.org/stable/4330676.

Seager, Sara. *Exoplanet Atmospheres: Physical Processes*. Princeton, NJ: Princeton University Press, 2010.

"Secret Instructions to Captain Cook, 30 June 1768." Museum of Australian Democracy. Accessed February 5, 2023. https://www.foundingdocs.gov.au/resources/transcripts/nsw1_doc_1768.pdf.

Seneca. *Naturales quæstiones*. In John Clarke, *Physical Science in the Time of Nero: Being a Translation of the Quaestiones naturales of Seneca*. London: Macmillan and Co., 1910.

Shapin, Steven. "Keep Him as a Curiosity." *London Review of Books* 16, no. 42 (August 13, 2020).

Shapin, Steven. "Who Was Robert Hooke?" In *Robert Hooke: New Studies*, edited by M. Hunter and S. Schaffer, 253–285, Woodbridge, UK: Boydell Press, 1989. http://nrs.harvard.edu/urn-3:HUL.InstRepos:3415435.

Sharkey, Joe. "Helping the Stars Take Back the Night." *New York Times*. August 30, 2008. https://www.nytimes.com/2008/08/31/business/31essay.html.

Shatner, William. "William Shatner: My Trip to Space Filled Me with 'Overwhelming Sadness.'" *Variety*. October 6, 2022. https://variety.com/2022/tv/news/william-shatner-space-boldly-go-excerpt-1235395113.

Shayegan, M. Rahim. "Aspects of History and Epic in Ancient Iran: From Gaumāta to Wahnām." Hellenic Studies Series 52. Washington, DC: Center for Hellenic Studies, 2012.

Sheynin, O. B. "A. Quetelet as a Statistician." *Archive for History of Exact Sciences* 36, no. 4 (1986): 281–325. http://www.jstor.org/stable/41133805.

Shields, A. L. "The Climates of Other Worlds: A Review of the Emerging Field of Exoplanet Climatology." *Astrophysical Journal Supplement Series* 243, no. 2 (2019).

Simmons, George F. *Calculus Gems: Brief Lives and Memorable Mathematics*. Providence: American Mathematical Society, 2007.

Smith, Keith Vincent. "Tupaia's Sketchbook." *Electronic British Library Journal* 10 (2005): 1–6.

Snobelen, Stephen. "The Myth of the Clockwork Universe: Newton, Newtonianism, and the Enlightenment." In *The Persistence of the Sacred in Modern Thought*, edited by Chris L. Firestone and Nathan A. Jacobs, 149–184. Notre Dame, IN: University of Notre Dame Press, 2012.

Snyder, Laura J. *The Philosophical Breakfast Club: Four Remarkable Friends Who Transformed Science and Changed the World*. New York: Broadway Books, 2011.

Sobel, Dava. *Galileo's Daughter: A Historical Memoir of Science, Faith, and Love*. London: Walker Books, 1999.

Sobel, Dava. *The Glass Universe: How the Ladies of the Harvard Observatory Took the Measure of the Stars*. New York: Penguin, 2016.

Sobel, Dava. *Longitude*. London: Harper Perennial, 2011.

Soltis, Joseph, Robert Boyd, and Peter J. Richerson. "Can Group-Functional Behaviors Evolve by Cultural Group Selection? An Empirical Test." *Current Anthropology* 36, no. 3 (1995): 473–494. https://doi.org/10.1086/204381.

Somerville, M. *Personal Recollections, from Early Life to Old Age, of Mary Somerville with Selections from Her Correspondence*. Charleston, SC: BiblioBazaar, 2010. Originally published in 1873; https://books.google.ws/books?id=3AaIcg AACAAJ.

Soressi, Marie, Shannon P. McPherron, Michel Lenoir, Tamara Dogandžić, Paul Goldberg, Zenobia Jacobs, Yolaine Maigrot, et al. "Neandertals Made the First Specialized Bone Tools in Europe." *Proceedings of the National Academy of Sciences* 110, no. 35 (2013): 14186. https://doi.org/10.1073/pnas.1302730110.

St. Augustine. "Sermon 190." In *Sermons on the Liturgical Seasons*. The Fathers of the Church 38, translated by Sister Mary Sarah Muldowney, 3–48. Washington, DC: Catholic University of America Press, 1959. https://www.jstor.org/stable/j.ctt32b3nc.

Stacey, Frank D. "Kelvin's Age of the Earth Paradox Revisited." *Journal of Geophysical Research: Solid Earth* 105, no. B6 (June 10, 2000): 13155–13158. https://doi.org/10.1029/2000JB900028.

Stahl, Saul. "The Evolution of the Normal Distribution." *Mathematics Magazine* 79, no. 2 (2006): 96–113. https://doi.org/10.2307/27642916.

Stephenson, F. Richard. "How Reliable Are Archaic Records of Large Solar Eclipses?" *Journal for the History of Astronomy* 39, no. 2 (May 1, 2008): 229–250. https://doi.org/10.1177/002182860803900205.

Sternbach, Rick, and Michael Okuda. *Star Trek: The Next Generation: Technical Manual*. New York: Pocket Books, 1991.

Stiegler, Stephen M. *The History of Statistics: The Measurement of Uncertainty Before 1900*. Cambridge, MA: Belknap Press, 1990.

Still, C. J., W. J. Riley, S. C. Biraud, D. C. Noone, N. H. Buenning, J. T. Randerson, M. S. Torn, et al. "Influence of Clouds and Diffuse Radiation on Ecosystem-Atmosphere CO_2 and $CO18O$ Exchanges." *Journal of Geophysical Research* 114, no. G1 (2009). https://doi.org/10.1029/2007jg000675.

Swade, Doron. *The Difference Engine: Charles Babbage and the Quest to Build the First Computer*. New York: Viking, 2001.

Sweatman, Martin B. "The Younger Dryas Impact Hypothesis: Review of the Impact Evidence." *Earth-Science Reviews* 218 (July 2021): 103677. https://doi.org/10.1016/j.earscirev.2021.103677.

"Sympathetic Vibrations." *Royal Museums Greenwich Blog*. Last modified March 3, 2011. https://www.rmg.co.uk/stories/blog/sympathetic-vibrations.

Tacitus. *Historiæ*. Translated by Clifford H. Moore. Loeb Classical Library 111. Cambridge, MA: Harvard University Press, 1925.

Taylor, E. G. R. "Navigation in the Days of Captain Cook." *Journal of Navigation* 21, no. 3 (1968): 256–276. https://doi.org/10.1017/S0373463300024735.

Teets, Donald, and Karen Whitehead. "The Discovery of Ceres: How Gauss Became Famous." *Mathematics Magazine* 72, no. 2 (1999): 83–93. https://doi.org/10.2307/2690592.

Teilhard de Chardin, Pierre. "The Evolution of Chastity." In *Toward the Future*. Translated by René Hague. New York: Harcourt Brace Jovanovich, 1975. https://books.google.it/books?id=LqNqqOH3LqYC.

Thoreau, Henry David. *Walden, or: Life in the Woods*. Boston: Ticknor and Fields, 1854.

Thoren, Victor E. *The Lord of Uraniborg: A Biography of Tycho Brahe*. Cambridge: Cambridge University Press, 2007.

Throckmorton, Peter. *Shipwrecks and Archaeology*. Boston: Little, Brown and Co., 1969.

Ticknor, G. *Life, Letters, and Journals of George Ticknor.* London: Osgood, 1876. https://books.google.it/books?id=7jppAAAAcAAJ.

Torres, Émile P. "Against Longtermism." *Aeon*. October 19, 2021. https://aeon.co /essays/why-longtermism-is-the-worlds-most-dangerous-secular-credo.

Tracy, Gene. "Sky Readers." *Aeon*. December 23, 2015. https://aeon.co/essays /what-have-we-lost-now-we-can-no-longer-read-the-sky.

Trinkhaus, Erik, and Pat Shipman. *The Neandertals: Changing the Image of Mankind*. New York: Alfred A. Knopf, 1993.

Tuckner, Ian. "One Man's Mission to Conquer Space." *The Guardian*. February 11, 2018. https://www.theguardian.com/science/2018/feb/11/one-mans -mission-to-conquer-space-peter-beck-humanity-star.

Turnbull, David. "(En)-countering Knowledge Traditions. The Story of Cook and Tupaia." *Humanities Research* 7, no. 1 (2000): 55–76.

Turner, Ben. "Chinese Scientists Call for Plan to Destroy Elon Musk's Starlink Satellites." *Live Science*. May 28, 2022. https://www.livescience.com /china-plans-ways-destroy-starlink.

Turner, H. D. "Robert Hooke and Boyle's Air Pump." *Nature* 184, no. 4684 (1959): 395–397. https://doi.org/10.1038/184395a0.

Turner, Steven. "Spiders in the Crosshairs: Cobwebs, Instrument Makers, and the Search for the Perfect Line." *Journal of the Antique Telescope Society* 1 (1992): 10. https://ui.adsabs.harvard.edu/abs/1992JATSo...1...10T.

"Tycho Brahe Wasn't Poisoned After All." *The History Blog*. Last modified November 19, 2012. http://www.thehistoryblog.com/archives/21535.

Van Den Broeck, Chris. "A 'Warp Drive' with More Reasonable Total Energy Requirements." *Classical and Quantum Gravity* 16, no. 12 (1999): 3973. https:// dx.doi.org/10.1088/0264-9381/16/12/314.

van der Waerden, B. L. "Babylonian Astronomy. II. The Thirty-Six Stars." *Journal of Near Eastern Studies* 8, no. 1 (1949): 6–26. http://www.jstor.org /stable/542436.

Van Doren, Benjamin M., Kyle G. Horton, Adriaan M. Dokter, Holger Klinck, Susan B. Elbin, and Andrew Farnsworth. "High-Intensity Urban Light Installation Dramatically Alters Nocturnal Bird Migration." *Proceedings of the National Academy of Sciences of the United States of America* 114, no. 42 (2017): 11175–11180. https://doi.org/10.1073/pnas.1708574114.

Van Helden, Albert. "Longitude and the Satellites of Jupiter." In *The Quest for Longitude: The Proceedings of the Longitude Symposium, Harvard University, Cambridge, Massachusetts, November 4–6, 1993*, ed. William J. H. Andrewes, 86–100. Cambridge, MA: The Collection of Historic and Scientific Instruments, Harvard University, 1996.

Vanderwal, R., and R. Fullagar. "Engraved Diprotodon Tooth from the Spring Creek Locality, Victoria." *Archaeology in Oceania* 24, no. 1 (1989): 13–16.

Venkatesan, Aparna, James Lowenthal, Parvathy Prem, and Monica Vidaurri. "The Impact of Satellite Constellations on Space as an Ancestral Global Commons." *Nature Astronomy* 4, no. 11 (2020): 1043–1048. https://doi.org/10.1038/s41550-020-01238-3.

Villanueva, G. L., M. Cordiner, P. G. J. Irwin, I. de Pater, B. Butler, M. Gurwell, S. N. Milam, et al. "No Evidence of Phosphine in the Atmosphere of Venus from Independent Analyses." *Nature Astronomy* 5 (2021): 631–635. https://doi.org/10.1038/s41550-021-01422-z.

Walker, D. P. *Spiritual and Demonic Magic: From Ficino to Campanella*. University Park: Pennsylvania State University Press, 2000.

Weibel, Deana L. "The Overview Effect and the Ultraview Effect: How Extreme Experiences in/of Outer Space Influence Religious Beliefs in Astronauts." *Religions* 11, no. 8 (2020): 418. https://doi.org/10.3390/rel11080418.

Westfall, Richard. *Never at Rest: A Biography of Isaac Newton*. Cambridge: Cambridge University Press, 1980.

Whewell, William. "On the Connexion of the Physical Sciences. By Mrs Somerville." *Quarterly Review* 51 (1834): 54–67.

Whiteside, D. T. "Isaac Newton: Birth of a Mathematician." *Notes and Records of the Royal Society of London* 19, no. 1 (1964): 53–62. http://www.jstor.org/stable/3519861.

Whiteside, D. T., M. A. Hoskin, and A. Prag, eds. *The Mathematical Papers of Isaac Newton*. Cambridge: Cambridge University Press, 1967.

Whitman, Walt. "When I Heard the Learn'd Astronomer." Poetry Foundation. https://www.poetryfoundation.org/poems/45479/when-i-heard-the-learnd-astronomer.

Whittaker, Ian. " Is SpaceX Being Environmentally Responsible?" *Smithsonian Magazine*. February 7, 2018. https://www.smithsonianmag.com/science-nature/spacex-environmentally-responsible-180968098.

Wigner, Eugene P. "The Unreasonable Effectiveness of Mathematics in the Natural Sciences." In *Mathematics and Science*, edited by Ronald E. Mickens, 291–306. Singapore: World Scientific, 1990

Wikander, Ola. "The Burning Sun and the Killing Resheph: Proto-astrological Symbolism and Ugaritic Epic." In *Sky and Symbol: Proceedings of the Ninth Annual Sophia Centre Conference*, edited by Liz Greene and Nicolas Campion, 73–83. Bristol: Sophia Centre Press, 2013.

Wilkes, M. V. "Herschel, Peacock, Babbage and the Development of the Cambridge Curriculum." *Notes and Records of the Royal Society of London* 44, no. 2 (1990): 205–219. http://www.jstor.org/stable/531607.

Williams, Glyndwr. "Tupaia, Polynesian Warrior, Navigator, High Priest—and Artist." In *The Global Eighteenth Century*, edited by F. A. Nussbaum, 38–51. Baltimore: Johns Hopkins University Press, 2003.

Williams, J. E. D. *From Sails to Satellites: The Origin and Development of Navigational Science*. Oxford: Oxford University Press, 1993.

Wilx, Andy, and Anita Ganeri. *Star Stories: Constellation Tales from Around the World*. London: Templar Books, 2018.

Wisan, Winifred Lovell. "Galileo and God's Creation." *Isis* 77, no. 3 (1986): 473–486. http://www.jstor.org/stable/231609.

Woodhouse, Jayne. "Eugenics and the Feeble-Minded: The Parliamentary Debates of 1912–14." *History of Education* (Tavistock) 11, no. 2 (1982): 127–137. https://doi.org/10.1080/0046760820110205.

Woolley, Richard. "Captain Cook and the Transit of Venus of 1769." *Notes and Records of the Royal Society of London* 24, no. 1 (1969): 19–32. http://www.jstor.org/stable/530738.

Worden, Alfred M. *Hello Earth; Greetings from Endeavour*. Los Angeles, CA: Nash Pub., 1974.

World Meteorological Organization. *International Cloud Atlas*. Geneva: World Meteorological Organization, 1956.

Wragg Sykes, Rebecca. *Kindred: Neanderthal Life, Love, Death and Art*. London: Bloomsbury Sigma, 2020.

Wright, Thomas. *An Original Theory or New Hypothesis of the Universe. Founded upon the Laws of Nature*. Cambridge: Cambridge University Press, 2014. Originally published in 1750.

Yaden, David B., Jonathan Iwry, Kelley J. Slack, Johannes C. Eichstaedt, Yukun Zhao, George E. Vaillant, and Andrew B. Newberg. "The Overview Effect: Awe and Self-Transcendent Experience in Space Flight." *Psychology of Consciousness: Theory, Research, and Practice* 3 (2016): 1–11. https://doi.org/10.1037/cns0000086.

Yeomans, D. K., J. Rahe, and R. S. Freitag. "The History of Comet Halley." *Journal of the Royal Astronomical Society of Canada* 80 (1986): 62–86. https://ui.adsabs.harvard.edu/abs/1986JRASC..80...62Y.

INDEX

Abbot, Charles, 53
AI. *See* artificial intelligence
Airy, George, 109
Airy Transit Circle, 109
Alighieri, Dante, 8, 235
Allen, Paul, 241
Almagest (Ptolemy), 98, 108, 153
Amtorians, on Venus (fictional), 54
Anaxagoras, 218
Andagarinja women's ritual, 85
Anderson, Chris, 208
Andía y Varela, Ignacio, 122–123
Andromeda Galaxy, 46, 174, 176
Antikythera mechanism
  as computer, 105
  discovery of, 104
  working of, 106–107
Apollo (Greek god), 47–48
Apollo (NASA program), 238–239, 241, 246

Aratus, 84
Archimedes, 105–107, 137, 179
arcminute, 96, 129
Aristarchus of Samos, 152
Arrhenius, Svante, 54
artificial intelligence (AI), 17, 88, 184, 208, 227
artificial satellites, 17
  cost of launching, 241
  impact on astronomy of, 242
  mega-constellations of, 243
  ruining the sky, 236–237
Asimov, Isaac, 26, 28, 55
asteroid
  Ceres as, 186
  dinosaurs wiped out by, 14, 58–60
  Gaussian distribution and, 186–187
  mining of, 244
astral determinism, 226
astrarium, 103–104, 107–108

astrology
  Brahe and, 155
  conjunctions in, 157, 226
  constellations in, 222–223
  etymology of, 222
  heavenly events recorded in, 8, 19
  Jupiter's qualities in, 228–229
  Jupiter-Saturn conjunctions in, 32,
    155–157
  Laplace and legacy of, 228
  Mercury's qualities in, 228
  in Newton's life, 227
  predictions in, 224–225
  Reagan, N., consulting, 225
  in Scientific Revolution, 226
  Urban VIII condemnation of,
    225
  Western, 223
astronomical unit (AU), 127, 137
astronomy, 17–18, 202, 208
Atlas, 3, 84
atmosphere, of Venus, 55
atomic clock, 108, 110
atom's structure, 191
AU. *See* astronomical unit
Aubrey, John, 102
Augustine (Saint), 163, 215, 226
Aurelian, 214
Austin, Alfred, 84

Babbage, Charles, 129, 193–197,
  202
Babylonians
  number system of, 95
  solar eclipse and, 27–28
  uš introduced by, 96
Baedeker, Karl, 230
Banks, Joseph, 127, 130–135
Barlow, James, 54–55
Barrow, Isaac, 164
Bayeux tapestry, 34
Beagle HMS, 139, 195, 204

Beck, Peter, 239
Bell, Derrick, 18
Bellarmine, Robert, 163
bell-shaped curve, 187–188, 198,
  205
Bessel, Friedrich, 173–174, 200,
  202, 205
Bezos, Jeff, 241, 245, 252
Big Bang, 4–7
  energy distribution after, 205
  Laplace's demon observations
    and, 205
  theoretical physicists on, 15
  time after, 15–16
  universe after, 15–16
Blue Origin, 252
Board of Longitude, 126, 137–138
Bode, Johann, 186
body mass index, 188
Boltzmann, Ludwig, 198, 206
Bonaparte, Napoléon, 177–178,
  184
Bradbury, Ray, 54
Brahe, Tyge (Tycho), 8
  astrology and, 155
  on comets, 32
  evidence-based arguments
    from, 164
  Kepler's alliance with, 159,
    179
  Mars data from, 184
  mathematicus and, 226
  as naked-eye observer, 155
  precision pursuit of, 184
  on supernova SN 1572, 156
broom stars, comets as, 236
Brown, Frederic, 239
Bruno, Giordano, 163, 173
Bryant, William Cullen, 118
the Bull. *See* Taurus
Burrough, Edgar Rice, 54
Byron (Lord), 168, 195

Caesar, Julius
  calendar reform by, 215
  comet as soul of, 33
  metaphor using, 119
  Polaris metaphor and, 121
Cailleach na Mointeach ("the old
    woman of the moors"), 48
calendars
  Caesar reforming, 86–87, 215
  intercalary days and, 86–87
  Inuit and Mayan, 87
  Lebombo bone and, 75
  lunar, 86–87
  lunar phases for, 31, 76–77, 86
  Roman arrangement of, 87
  spiral-shaped in Antikyhtera
    mechanism, 107
  Sumerian, 96
Caligo
  clouds on, 57–58, 60
  description of, 58
  photosynthesis on, 59
  tales of, 20–21, 62–64, 91–92,
    112–114, 144–146, 180–181,
    209–210, 233–234, 257–259
  vitamin D problem on, 59–60
  as world without stars, 18, 57
Callanish (or Calanais), 46–47
Campanella, Tommaso, 171, 224–225
Campbell-Stokes sunshine
    recorder, 51
cannocchiale (telescope), 150–151, 156,
    159–162
Cannon, Annie Jump, 175
Caracalla, 224
carbon dioxide, 55
Carter, Jimmy, 262
celestial navigation, 142–143
celestial spheres, 177
Cepheid variables, 175
ceremonial place, 82
Cesi, Federico, 162

Chaplin, Charlie, 101
Charles V (Emperor), 107
ChatGPT, 208
Chaucer, Geoffrey, 214, 220
Christmas, 32, 170, 216
chronometer, 138
Cicero, Marcus Tullius
  awed by geared planetarium,
    105–106
  Milky Way dispute and, 171
  Moon's orbit and, 218–219
circadian rhythms, 56
Clean Air Act, 53
climate change, 19, 232, 247, 251–252
clocks, 93
  atomic, 108, 110
  Jacopo de' Dondi designing face
    of, 99
  marine chronometer and,
    137–138
  radio galaxies and setting,
    110–111
  sundial used for, 100
  Swiss watch industry and,
    199–200
  Uranium-238 as, 262
clockwork universe, 104, 176–177,
    179, 201
clouds
  on Caligo, 57–58, 60
  cover, 58–59
  on exoplanets, 55–56
  glow-in-the-dark, 56
  Jupiter's pattern of, 58
  Sun and, 51, 58–59
  on Venus, 54
Cloud-Watcher (fictional), 20–21,
    42, 62
Cohen, Thierry, 240, 248
collective unconscious, 217, 222
colonization, of Mars, 252
Columbus, Christopher, 124

comets
  Brahe on, 32
  as broom stars, 236
  Caesar's soul as, 33
  earliest accounts of, 32–33, 60
  gravity's power on, 166–168
  Great Comet of 1680–1681, 168, 170
  Halley's, 32–34, 170, 224
  kites mimicking, 164
  of May 44 BC, 33
  NEOWISE, 236
  Newton observing, 166–168, 236
  of 1682, 167
Constantine (Emperor), 215
constellations
  in astrology, 222–223
  in three-dimensional space, 37
  navigation using, 17, 80–81, 117
  Ophiuchus, 31–32
  Orion, 2–3, 40, 80, 84–85, 248–249
  Taurus, 3, 83–85
  time deforming, 37
  Ursa Major, 40
  Ursa Minor, 117, 120
Cook, James, 179
  as *Endeavour* commander,
    116, 127
  K-1 chronometer tested by,
    138–139
  lunar tables used by, 129–130
  mission of, 128
  New Zealand return of, 136
  secret instructions to, 131
  Tahiti arrival of, 116
  opinion of Tupaia by, 130–135
  Venus transit data from, 128, 137,
    170, 173
Copernican model, 154–155
Copernican Revolution, 149
Copernicus, Nicolaus
  evidence-based arguments by, 164
  forbidden texts of, 163–164

  heliocentric universe of, 154, 203
  Novara working with, 226
  On the Revolutions of the Celestial
    Spheres by, 154–155
  scientific method used by, 19
  solar system theory of, 157, 216
coronaphiles, 24
coronavirus pandemic, 18
cosmology, 5, 7
creation, center of, 153, 172
cultural liaison, 133

da Vinci, Leonardo, 103
dark matter, 5, 15, 39, 148
Darwin, Charles
  attending Babbage's party, 196
  Beagle's mission with, 195–196
  FitzRoy's voyage with, 139–140
  Laplace's demon influencing,
    202
  scientific work regrets of, 249
  species transmutation from, 197
  on time and chronology, 203–204
Darwin, George, 205
De revolutionibus orbium coelestium (On
    the Revolutions of the Celestial
    Spheres), 154–155
decans, 97–98
DeMoivre, Abraham, 167
Descartes, René, 172, 227
The Dialogue on the Two Chief World
    Systems (Galileo), 163
DiCaprio, Leonardo, 244
Dickens, Charles, 188
Diodorus of Sicily, 47–48
Dirac, Paul, 207
direct sum, 217
Disraeli, Benjamin, 183–184
Domitian, 224
de' Dondi, Giovanni, 102–103
de' Dondi, Jacopo, 99, 126, 137,
    202

doomsday preppers, 254
dreaming tracks, 79–80
Duino Mithraeum, 213
dung beetles, 72–73
Dunkin, Edwin, 36, 41

eagle hawk songline, 79–80
Earth
    asteroid wiping out life on, 14,
        58–60
    climate changing on, 14
    God's creation of, 228
    planetary orbits and, 158
    Sun used to measure size of, 47
Earth, age of
    Bible estimation of, 203,
        228
    Halley's method for, 204
    Kelvin on, 204
    salinity estimating, 204
    today's estimation of, 205
eclipse, lunar, 26–27, 98, 127, 155,
    175–176
eclipse, solar, 8
    astrologers' predictions and,
        224–225
    Babylonians and, 27–28
    at crucifixion of Jesus, 28–29
    description of, 27, 34–35
    Great American, 24
    Halley's predictions of, 27
    Moon in total, 26, 31
    in *Nightfall*, 26
    scientific explanation of, 31
    substitute king and, 29
    Thales of Miletus prediction of,
        28, 117
    types of, 27
    viewing of, 25–26
Eddington, Arthur, 206–207
Egypt, 30, 96–97
Eiffel, Gustave, 237

Einstein, Albert, 140
    Eddington's observations and, 207
    gravitational waves prediction
        from, 206
    machine age dangers from, 250
    relativity thoughts of, 5
    theory of general relativity from, 17,
        142, 183, 206–207
Eiseley, Loren, 86, 252, 255–256
Eliade, Mircea, 7, 16
Eliasson, Olafur, 212, 232
Emerson, Ralph Waldo, 23
    on beauty of night sky, 243
    on heavenly bodies presence,
        36, 41
    on presence of sublime, 244
    on stars, 222
Empyrean, 8, 156, 173
"Emu in the Sky" (Gawarrgay), 79
Endeavour HMS, 128, 130–136
    Cook as commander of, 116, 127
    Venus transit mission, 127
Enheduanna, 77
Enkidu, 94
Enterprise starship (fictional), 245
Equation of Time, 108
Eratosthenes, 47
d'Errico, Francesco, 73
errors
    avoidance of in lunar tables, 129
    astronomers' observations with,
        179
    in Babbage's collection of tables,
        192
    measurement, 186–187
    in *Nautical Almanac*, 192
    in navigation, 124
    random, 185
    social physics and, 189
    theory of, 185–186, 188
Euahlayi people, 79–80
eugenics, 187–188, 198–199

Euler, Leonhard, 185
evidence-based arguments, 164
Ewe tribe, 7–8
exoplanets, 55–56

Faroe Islands, 51–52
Finsen, Niels Ryberg, 51–52
First Nation Aboriginal people
  British invasion decimation of, 78
  dreaming tracks of, 79–80
  Euahlayi people and, 80
  Pleiades and, 85, 87
  star lore of, 78–79
FitzRoy, Robert, 139–140, 204
Flamsteed, John, 166–167
Fourier, Jean-Baptiste, 188, 191
Fred (reconstruction of Neanderthal
  man), 66
Frederick II of Denmark, 155

galaxies, in universe, 15
Galileo, Galilei, 5, 143, 206–208
  astral determinism by, 226
  blasphemy charge avoided by,
    150–151
  cannocchiale built by, 150–151, 156,
    159–162
  *The Dialogue on the Two Chief World
    Systems* by, 163
  evidence-based arguments by, 164
  God beliefs of, 150–151
  on God's celestial spheres, 177
  Jupiter's moons discovery by, 127,
    159–162
  Kepler's support for, 162
  Milky Way's stars observed by,
    171–172
  Saturn's rings observed by, 161
  telescope used by, 39–40,
    148–151, 179
  Venus observed by, 161
Galton, Francis, 198

Garibaldi, Giuseppe, 230, 236
Gascoigne, William, 109
Gauss, Carl Friedrich, 19, 185
  Laplace impressed by, 186–187
  normal distribution from, 186–187,
    197–199, 205
Gaussian distribution. *See* Normal
  distribution
Gawarrgay ("Emu in the Sky"), 79
general relativity, theory of, 17, 142,
  183, 206–207
Genovese, Chris, 40
geometry as explanation for solar
  system, 157–158
Gilbert Islands, 122–123
Gilgamesh, 2–3, 93–95, 98, 213, 218
Giotto, 32
Global Positioning System (GPS),
  17, 142
gnomon. *See* sundial
God, 105–106
  as center of creation, 153, 172
  clockwork universe role of, 176–177
  Earth's creation by, 228
  Ewe tribe on, 7–8
  Galileo on celestial spheres
    from, 177
  Galileo's beliefs in, 150–151
  geometry and, 157–158
  Israel fight of, 162
  Jesus and, 28–29
  laws of nature and, 197
  Newton and role of, 176–178
  planet movements by, 177
  role in universe of, 176–178
Godwinson, Harold (king of
  England), 33, 224
gold, symbol for, 214
golden record, 261–262
Gould, Stephen J., 60
GPS. *See* Global Positioning System
Great American eclipse, 24

Great Bear constellation. *See* Ursa
  Major
Great Comet of 1680–1681, 168, 170
great year, 120–121
Green, Charles, 127–128, 130
Greenwich Mean Time, 140–141
Greenwich Observatory, the Royal, 129
  Airy Transit Circle telescope at,
    109, 140
  as reference for chronometer, 138
  longitude found with, 126
  time lady of, 141
Gregory, James, 172
Gryspeerdt, Edward, 51
Guevara, Che, 230

H-4 chronometer (watch), 138
Halley, Edmond, 167–171
  Earth's age estimation method
    by, 204
  lofty limits of universe from, 176
  Newton's equations used by, 179
  solar eclipse predictions of, 27
  stars's change in position discovered
    by, 173
  on transit of Venus, 127–128
Halley's comet, 32–34, 170, 224
Hare, Brian, 73
Harney, Bill Yidumduma, 79, 81
Harrison, John, 138, 179
  Longitude Prize and, 137, 139
  marine chronometer created by,
    110, 137
  money received by, 139
Hawking, Stephen, 252
heavenly events, 8, 19
Heizer, Michael, 237
heliocentric universe, 154, 203
Herodotus, 28
hero's fate, 219
Herschel, John, 109, 173, 192–193,
    197, 203

Herschel, William, 109, 178, 186
Hesperians, on Venus (fictional), 54
Hipparchus, 98, 119, 127, 152
Hirsch, Adolph, 200
Hodgson, Francis, 240
Holocaust, 199
Homer, 3, 9, 17, 118
Hooke, Robert, 101–102, 167
l'Hôpital, Marquis de, 168
Horace, 117
Hornsby, Thomas, 137
Horologi, Giovanni de gli, 103, 137
Howard, Luke, 49–50
Hubble, Edwin, 175–176
Hubble Space Telescope, 40, 56, 148,
    242–243
human computers, 129, 192
Humanity Star (satellite), 239
Hutton, James, 202–203
Huygens, Christiaan, 161, 172

imaginary intelligence, 189–190
intercalary days, 86–87, 96
International Meridian
    Conference, 140
International Space Station, 236, 245
  dangers to, 243
  nationalism and, 240
  resupply of, 253
internet, 206, 240–241
Inuit
  calendar, 87
  navigation by stars of, 116
  star lore, 89
Irwin, Geoffrey, 121
Isaac's dials, 164
isolation experiments, 56–57

James I (king), 160
James Webb Space Telescope (JWST),
    148, 208
Jesus, 28–29

John of the Clock. *See* Horologi, Giovanni de gli

Joly, John, 204

Jupiter
astrological qualities of, 228–229
cloud pattern over, 58
Earth's orbit and, 158
Galileo discovering moons of, 127, 159–162
longitude found using moons of, 160
Saturn's conjunction with, 32, 155–157
Star of Bethlehem and, 32
Sun's distance to, 127
telescope viewing, 160
universal gravitation and, 168, 170, 179
universe's order and, 151

JWST. *See* James Webb Space Telescope

K-1 chronometer (watch), 138–139

Kapoor, Anish, 56

Kelvin (Lord), 204

Kendall, Larcum, 138

Kennis, Adrie, 66

Kennis, Alfons, 66

Kepler, Johannes, 8
astrological conjunctions studied by, 157, 226
Brahe's alliance with, 159, 179
cosmos models and, 184
evidence-based arguments by, 164
Galileo support from, 162
New Astronomy by, 159
Ophiuchus constellation observation of, 31–32
planetary elliptical orbits from, 166–167
planetary motion laws from, 159
as Platonic mystic, 158, 177

solar system distances from, 127, 158
solar system idea of, 159
supernova explosion described by, 31–32

Kessler, Donald, 244

Kinnebrook, David, 200

Kirk, James T. (fictional), 245

Kneller, Godfrey, 169

Knight, Chris, 74, 76–77

Kondos, Demetrios, 104–105

kosmos, 7

Krakatoa volcano, 50

Kreidberg, Laura, 56

Kubrick, Stanley, 49

Lagashians, 26, 28, 55

Laguarda Trías, Rolando, 125

Laplace, Pierre-Simon de, 19
admiration of Euler from, 185
astrological legacy according to, 228
Bonaparte's exchange with, 177–178
celestial mechanics from, 189
clockwork universe and, 179
Gauss impressing, 186–187
least-squares method explanation from, 185, 187
La méchanique céleste by, 178
micromanagement by, 184–185
as Newton of France, 178
normal distribution from, 187–188, 197–199, 205
on planetary orbits, 178
Quetelet influenced by, 188
Theory of Probabilities by, 194
universe's present state from, 190

Laplace's demon
AI and, 208
atom's structure from, 191
Big Bang observations and, 205
capabilities of, 190–191
Darwin, C., influenced by, 202

as imaginary intelligence, 189–190
influence of, 190
mechanical embodiment of, 194–195
modern science blueprint from, 191
subatomic structures and, 206
Lardner, Dionysius, 192
Lascaux caves, 77–78, 83
latitude, determining, 122, 124
Laurel, Josie, 56
Lawrence, Jennifer, 244
laws of nature, 197
least-squares method, 185, 187
Leavitt, Henrietta Swan, 175–176
Lebombo bone, 75
Legendre, Adrien-Marie, 185
Leibniz, Gottfried, 165, 177
Leopardi, Giacomo, 5
light-year, invention of, 174, 202
Lippershey, Hans, 150
Little Bear constellation. *See* Ursa Minor
Locke, John, 191
Lockyer, Norman, 47
London, Jack, 52
"The Long Rain" (Bradbury), 54
longitude
Columbus's determination of, 124
demarcation line for, 124
determining, 126, 129, 137
Green measuring, 130
howling dog and, 126
International Meridian Conference on, 140
Jupiter's moons used for, 160
K-1 determination of, 138
Magellan's estimates of, 125
measuring with Meyer's tables, 129
prime meridian and, 109
Royal Greenwich Observatory used for, 126
at sea, 125

Longitude Prize
Board of Longitude and, 138
crackpot proposals for, 125
Harrison and, 137, 139
high-tech mechanism vying for, 137
never awarded, 139
'taking lunars' in contention for, 129–130
Lorentz, Hendrik, 140
Louis the Pious, 29
Louis XIV (king of France), 74, 216
Lovering, Joseph, 185
lunar calendar, 86–87
lunar tables, 129–130, 139
Luther, Martin, 216
Lyell, Charles, 196, 203–204

machine age dangers, 250
Magellan, Ferdinand, 124–125
Magi, 31–32
magnetic north, 120
major standstill, Moon in, 47–48
marine chronometer
first sea clock as, 137–138
GPS replacing, 142
H-4 prototype as, 138
Harrison creating, 110, 137
on HMS *Beagle*, 139
Marius, Simon, 174
Mars
astrology and, 228
Brahe's data on, 184
colonization of, 252
Earth's orbit and, 158
moons of, 221–222
orbital motion of, 158
planetary motion and, 159
Sun's distance to, 127
universe's order and, 151
Marshack, Alexander, 74–75, 77
Marx, Karl, 201–202

Maskelyne, Nevil, 174
  H-4 prototype confiscation and, 138
  Kinnebrook dismissed by, 200
  lunar distance tables of, 139
  Moon's position used by, 129–130
  Nautical Almanac by, 140, 192
  Piazzi asteroid discovery and, 186
mathematicus, 226
Maxwell, James Clerk, 198, 206
May 44 BC comet, 33
Mayan calendar, 87
McCaughrean, Mark, 26
McPhee, John, 250
measurement errors, 186–187
La méchanique céleste (Laplace), 178
Medici, Cosimo II de', 160
megamachine, 254, 263
Mercury
  astrological qualities of, 228
  Earth's orbit and, 158
  Sun's distance to, 127
  universe's order and, 151
  Venus and, 154
meridian zero, 140
meteors, 3, 249
metonic cycle, 106
Meyer, Tobias, 129
The Midnight Sky (Dunkin), 36
Milky Way
  Cicero and dispute about, 171
  city skyglow obscuring, 38
  dung beetles guided by, 72–73
  Galileo observing stars of, 171–172
  number of stars in, 262–263
  over London, 36
  over New York, 38
  over Rio, 240
  Tupaia's weather predictions and,
    133–134
Milton, John, 36, 220
Mitchell, Ed, 246–247
Mithen, Steven, 88

Mithra
  ancient cult of, 213
  trinity of, 215
  worship of, 214
modern science, 191
Montalbini, Maurizio, 57
Monte Verità, 39–40, 174
Moon
  at major standstill, 47–48
  Cicero and orbit of, 218–219
  cycles of, 7, 27, 70–72, 76, 106
  cycles of life and, 217–218
  Enheduanna and, 77
  as feminine principle, 219
  fishing under light of, 41
  human exploration of, 240–241, 246,
    252–253
  hunting using light of, 71–72, 74
  in lunar calendar, 86–87
  Maskelyne navigating using,
    129–130
  as Nanna, 96
  Newton using, 126–127, 129
  in prehistoric societies, 76
  telescope viewing, 161
  in total solar eclipse, 26, 31
  universe's order and, 151
  Virgin Mary's association with, 219
  women's cycle and, 74
moons
  Galileo discovering Jupiter's, 127,
    159–162
  of Jupiter, 160–161
  Mars with multiple, 221–222
Morrison, James, 123
Mumford, Lewis, 6, 18, 254
Musk, Elon, 239, 241, 252–253
naked-eye observer, 155
Nanna, Moon as, 96
NASA program, Apollo, 238–239,
    241, 246

Nautical Almanac (Maskelyne), 129,
140, 192–193
Navajo Native Americans, 8,
120–121, 183
navigation
celestial, 142–143
by Columbus, 123–124
constellations used for, 17,
80–81, 117
of dung beetles, 72–73
errors in, 124
Inuit using stars for, 116
Maskelyne using lunars, 129–130
meridian zero used in, 140
Phoenician, 117–119
Polynesian, 116, 121–123
of starlings, 73
Tupaia's prowess in, 133–134
Neanderthal man, 66
Neanderthals, 9, 66–70
nebula
Andromeda, 174
Genovese viewing, 40
Orion, 2, 80
NEOWISE comet, 236
Neptune, 158
New Astronomy (Kepler), 159
Newton, Isaac, 5–6, 34, 119–120
as absent-minded, 165
astrology in life of, 227
childhood of, 164–165
comets observed by, 166–168, 236
God's role in universe and, 176–178
Great Comet trajectory computed
by, 170
Halley using equations of, 179
Laplace's fame in France like, 178
Moon used by, 126–127, 129
portrait of, 169
Principia by, 170–171
Sirius's distance to Sun from,
172–173

sundials of, 164
telescope and physics of, 179
universal gravitation from, 168,
170, 179
Nightfall (Asimov), 26
Nordgren, Tyler, 24
normal distribution, 186–187,
197–199, 205
Norris, Ray, 85
North star. *See* Polaris
Northern Lights, 46
Northridge earthquake, 38
Notarbartolo, Albert, 238
Novara, Domenico Maria, 226

Odysseus, 9–10
"the old woman of the moors"
(Cailleach na Mointeach), 48
On the Revolutions of the Celestial Spheres
(Copernicus), 154–155
Ophiuchus, 31–32
orbital weapons, 244
Oresme, Nicole, 104, 176
Orion, 2–3, 40, 80, 84–85, 248–249
Osiander, Andreas, 154–155
Osiris, 3, 72, 97
Outer Space Treaty, 244
overview effect, 246–247, 255
Ovid, 65, 86, 216

Palm, Theobald Adrian, 59
parabolic antennas, 242
parallax
distant stars shifting in, 152–153
solar, 128
discovery of stellar, 205
importance for astronomy of, 173
Parkinson, Sydney, 136
passenger pigeons, 251–252
Paul V (pope), 162
pea soup fog, London, 52
personal equation, 200

Petrarca, Francesco, 102
Phoenician navigation, 117–119
photographic plate, 174
photosynthesis, on Caligo, 59
Piazzi, Giuseppe, 185–186
Pickering, Edward, 174–175
Pickersgill, Richard, 134
Pirates of Venus (Burroughs), 54
planets
  clouds on exoplanets and, 55–56
  Copernican view of orbits of, 154
  Earth and orbits of, 158
  etymology of, 7
  God and movements of, 177
  Kepler on elliptical orbits of,
    166–167
  Kepler's laws of motion, 159
  Laplace on orbits of, 178
  Mars and motion of, 159
  Ptolemaic system for, 152, 154,
    161, 229
  Sun surrounded by family of, 12
  universe's order and, 151
  zodiac's relationship with,
    222–223
Platonic solids, 158
Pleiades, 47, 70, 77, 83
  Aboriginal peoples and, 85, 87
  in Andagarinja women's ritual, 85
  Aratus on, 84
  in cultures around world, 80–81
  First Nation people and, 78
  in legends, 3
  in Pitjantjatjara's fertility
    rituals, 79
  only six visible, 84
  Subaru's connection with, 84
Pleione, 3, 84
Poincaré, Henri, 11, 60
  on astronomy, 17–18, 202, 208
  glow-in-the-dark clouds and, 56
  space-time and, 140

Polaris
  Caesar not using metaphor of, 121
  first use of name, 120–121
  movement of, 121
  determination of latitude from, 124
pollution, light, 17, 19, 243, 248
Polynesian navigators, 121
  in Gilbert Islands, 122–123
  methods of, 123
  renaissance of, 143
  star lore of, 116
Pope, Alexander, 169
powder of sympathy, 125
Powers, Richard, 251
precession
  ignored in astrology, 223
  of equinoxes, 81, 223
  perspective shift in, 119
  Ptolemy on, 119, 173, 223
  slow circle rotation in, 120
prime meridian, 109, 140
Principia (Newton), 170–171
Ptolemaic system, 152, 154, 161,
  229
Ptolemy, Claudius, 154
  Almagest by, 98, 108, 153
  planetary system by, 152
  interpretation of precession from,
    119, 173, 223
  universe size estimated by,
    151–153

Quetelet, Adolphe, 188–189,
  197–198, 202

radio galaxies, 110–111
random errors, 185
Reagan, Nancy, 225
Reagan, Ronald, 225
Regiomontanus, 103, 226
religious symbolism, 216
Rilke, Rainer Maria, 1, 236

Rittenhouse, David, 109
Rome, 87, 230
Roosevelt, Franklin D., 251
Rudolph II, 155, 159

SAD. *See* seasonal affective disorder
Sagan, Carl, 11–12, 55, 252, 261
Sahagún, Fray Bernardino de, 28
Salk, Jonas, 255
Sargon the Great, 77
saros cycle, 27
satellites
    artificial, 17
    as artwork, 237–238
    Hubble Telescope affected by,
        242–243
    Humanity Star as, 239
    internet access from, 241–243
    as space junk, 239, 242–244
Saturn
    Earth's orbit and, 158
    Galileo observing two 'stars'
        near, 161
    Jupiter's conjunction with, 32,
        155–157
    Star of Bethlehem and, 32
    Sun's distance to, 127
    universal gravitation and, 168,
        170, 179
    universe's order and, 151
Savonarola, Michele, 103
Scientific Revolution, 226
scientists, 101, 167
    origin of the name, 195
seasonal affective disorder (SAD),
    46, 49
Seneca, 169, 171
sever sisters. *See* Pleiades
sewing needle, invention of, 69
Shakespeare, William, 119, 121,
    219, 255
Shapshu goddess, 232

Shatner, William, 245–246
Shepard, Alan, 238
sidereal day, 108–110
Sirius, 248
    decans and, 98
    distance to Sun from, 172–173
    Egyptian's worshiping, 97
1682 comet, 167
sky
    beauty of night, 243
    etymology of, 7–8
    polluting of, 242–243
Smithson, Robert, 237
social media, 36
social physics, 198
solar panels, 242
solar system
    Copernicus's theory of, 157, 216
    Kepler on distances in, 127, 158
    Kepler's idea of, 159
    solar parallax in, 128
    Tycho Brahe's theory of, 159
Solla Price, Derek de, 101, 103 106
Songer, Nan, 110
space
    art, 237–240
    colonialism, 253–254
    cost of going to, 241
    exploration, 11–12
    as global commons, 240–244
    junk, 239, 242–244
    Poincaré on time and, 140
    race, 240–242
    tourism, 245–247
spaceflight, 240–241
SpaceX, 241–242, 244
species transmutation, 197
spider
    black widow, 110
    ranches, 109
    threads in telescope, 109, 141
star lore, 78–79, 89, 116

Star of Bethlehem, 31–32
The Sidereal Messenger, 162
stars
  Alnilam, 2
  Alnitak, 2
  Arcturus, 230, 236
  Betelgeuse, 2, 40
  Caligo as world without, 18
  celestial navigation and, 142–143
  comets as broom, 236
  Emerson on, 222
  Galileo observing, 161, 171–172
  Halley discovering changes in
    position of, 173
  Newton worrying about their
    gravity, 173
  Inuit navigating by, 116
  Milky Way's number of, 262–263
  Mintaka, 2
  observed before telescope, 84
  parallax in, 152–153
  Polaris, 81, 116, 120–121, 124
  pollution hiding groups of, 17
  precession of, 81, 119, 120, 173, 223
  Rigel, 2
  Vega, 147
stellar parallax, 205
Stonehenge, 28, 47, 83
Storm Dennis, 46
subatomic structures, 206
suborbital tourist flights, 244–245
Sumerian calendar, 96
Summer Triangle, 147
Sun. *See also* eclipse, solar
  cloud cover and, 51, 58–59
  size of Earth measured using, 47
  Egypt's obsession with, 30
  hero's fate represented by, 219
  isolation from, 56–57
  Jesus's association with, 28–29
  as masculine god, 30, 217, 219
  Mithra cult and, 213

  planetary family around, 12
  religious symbolism tied to, 216
  Shapshu goddess of, 232
  Sirius's distance to, 172–173
  soot-filled air and, 59–60
  telescope viewing, 4, 161
  ultraviolet light from, 52
  universe's order and, 151
  Vespasian's troops assisted by, 214
sundial
  clocks set using, 100
  earliest known, 96–97
  Hooke and, 101–102
  Newton collecting, 164
  sunwise direction of, 103
  types of, 98
supernova explosion, 31–32
Swiss watch industry, 199–200

Tacitus, 224, 227
Tahiti, 116, 130–131
Taiata, 132, 136
taking lunars, 129
Tate Modern gallery, 212, 231
Taurus, 3, 83–85
Taylor, Bayard, 83
telescope, 236. *See also* Hubble Space
    Telescope; James Webb Space
    Telescope
  cannocchiale as, 150–151, 156,
    159–162
  Cesi's invention of the word for,
    162
  cosmology benefiting from, 5
  Galileo using, 39–40, 148–149, 179
  Genovese using, 40
  invention of, 150
  Jupiter viewed in, 160
  Moon viewed in, 161
  Newtonian physics and, 179
  spider threads in, 109, 141
  stars observed before, 84

Sun viewed in, 4, 161
transit, 109
transit circle, 140
temperatore, 100
Tennyson, Alfred, 83
Terra Australis Incognita, 131
Thales of Miletus, 28, 117
theory of general relativity, 17, 142, 183, 206–207
Theory of Probabilities (Laplace), 194
Thompson, Nainoa, 143
Thompson, William, 77
Thoreau, Henry David, 40–41
time lady, of Greenwich, 141
time measurement, 96
time zones, from prime meridian, 140
tourist flights, suborbital, 244–245
transhumanism, 253
transit telescope, 109
Treaty of Tordesillas, 124
Tredgold, Alfred, 199
Trotta, Benjamin, 23, 108–111
clock reading by, 93
spelling test of, 231
telescope used by, 236
Trotta, Emma , 23, 148, 217, 237
Tupaia
Cook utilizing, 130–135
as cultural liaison, 133
death of, 136
Map, 115–116
navigational prowess of, 133–134
Tahiti exile of, 130–131
weather predictions by, 133–134
Twain, Mark, 183

ultraviolet light, 52
Ulysses, 118
uniformalism, 203
universal gravitation, 168, 170, 179
universal joints, 102
universe

age of, 5, 12–13, 205, 261
after Big Bang, 15–16
clockwork, 104, 176–177, 179, 201
cosmology of, 7
estimated size of, 151–153
galaxies in, 15
God's role in, 176–178
Halley on lofty limits of, 176
heliocentric, 154, 203
planets and order of, 151
present state of, 190
Ptolemy's estimated size of, 151–153
size of visible, 15, 174–176
Uranus, 158
Urban VIII (pope), 163, 224–225
Ursa Major constellation, 40
caribou shape of, 81
used by Ulysses, 118
Ursa Minor constellation, 117, 120
uš ("oosh"), 96

Venus, 32
Amtorians and Hesperians on (fictional), 54
astrology and, 228
atmosphere of, 55
clouds on, 54
Cook's transit data on, 137, 170, 173
Earth's orbit and, 158
figurines, 75, 77–78
Galileo observing, 161
Halley on transiting of, 127–128
of Laussel, 75–76, 83
Mercury and, 154
over Rome, 230
pressures and heat on, 55
Sun's distance to, 127
transit of, 128, 137, 170, 173
universe's order and, 151
Vespasian (Emperor), 33, 214
Visconti, Duke Gian Galeazzo, 103
vitamin D, lack of, 59–60

Viviani, Vincenzo, 163–164
Voyager 1 spacecraft, 11–12, 261–262

Warhol, Andy, 121, 239
watches, 138–139, 199–200
Watt, James, 107
Webb, James, 148
Western astrology, 223
Whitman, Walt, 201
Wigner, Eugene, 207

William the Conqueror, 33–34, 224
Worden, Al, 246
Wragg Sykes, Rebecca, 67
Wren, Christopher, 167
Wright, Thomas, 147, 173

Yeats, William, 217

Zeus, 3, 7, 84–85
zodiac, 99–100, 107, 157, 222–223

BASIC BOOKS

Basic Books UK is a dynamic imprint from John Murray Press that seeks to inform, challenge and inspire its readers. It brings together authoritative and original voices from around the world to make a culturally rich and broad range of ideas accessible to everyone.

RECENT AND FORTHCOMING TITLES BY BASIC BOOKS UK

The Nowhere Office by Julia Hobsbawm

Free Speech by Jacob Mchangama

Hidden Games by Moshe Hoffman and Erez Yoeli

The Ceiling Outside by Noga Arikha

Before We Were Trans by Kit Heyam

Slouching Towards Utopia by J. Bradford DeLong

African Europeans by Olivette Otele

How to Be Good by Massimo Pigliucci

The Mongol Storm by Nicholas Morton

For Profit by William Magnuson

Escape from Model Land by Erica Thompson

Truth and Repair by Judith L. Herman

Queens of a Fallen World by Kate Cooper

Elixir by Theresa Levitt

Power and Progress by Daron Acemoglu and Simon Johnson

1923 by Mark Jones

Credible by Amanda Goodall

The Master Builder by Alfonso Martinez Arias

A Theory of Everyone by Michael Muthukrishna

Justinian by Peter Sarris

The Women Who Made Modern Economics by Rachel Reeves

Starborn by Roberto Trotta